KV-052-137

R26221

TEXTILE TERMS *and* DEFINITIONS

Tenth Edition

Compiled by

The Textile Institute Textile Terms and Definitions Committee

Edited by

Professor J E McIntyre DSc CText FTI

P N Daniels BA MIInfSc

The Textile Institute

The Textile Institute
International Headquarters
10 Blackfriars Street
Manchester M3 5DR, UK

1st Edition		September	1954
2nd Edition	Revised and Enlarged	September	1955
3rd Edition	Revised and Enlarged	February	1957
4th Edition	Revised and Enlarged	January	1960
5th Edition	Revised and Enlarged	August	1963
5th Edition	2nd impression	August	1967
5th Edition	3rd impression	August	1968
6th Edition	Revised and Enlarged	August	1970
6th Edition	2nd impression	April	1972
7th Edition	Revised and Enlarged	May	1975
7th Edition	2nd impression	November	1978
7th Edition	3rd impression	November	1984
8th Edition	Revised and Enlarged	June	1986
8th Edition	Reprinted with revisions	September	1988
9th Edition	Revised and Enlarged	September	1991
10th Edtion	Revised and Enlarged	September	1995

ISBN 1 870812 77 8

Printed and bound in Great Britain by
Biddles Limited, UK

Contents

Preface to the Tenth Edition

For over forty years *Textile Terms and Definitions* has occupied an important place in the literature of the world textile industry. It has evolved, edition by edition, in order to keep pace with advances in science and technology, management and design, relevant to the manufacture, use and marketing of textiles. Many new terms have been introduced. Some outdated terms have, with regret, been removed, in order to keep the size and cost within reasonable limits. Readers with an interest in such now-historic terms should retain copies of previous editions. One day there may be a publication that brings them all together, but not just yet.

This tenth edition represents a considerable expansion in scope and therefore also in size. With the integration of members of the Clothing and Footwear Institute into The Textile Institute, it seemed appropriate to incorporate terms from the field of clothing, which had not been specifically covered in previous editions. A new panel, under the expert chairmanship of Dr D. J. Tyler, has made a systematic survey of this field, concentrating particularly upon technical terminology rather than upon general clothing and fashion terms. A panel has also been established to review terms and definitions from the field of footwear, with the objective of providing, in due course, a basis for their publication.

Twelve other specialist panels have considered existing and proposed definitions in the fields of finishing, floorcoverings, knitting, lace, manufactured fibres, narrow fabrics, natural fibres, nonwoven textiles, spinning, technical textiles, testing and quality, and weaving. They have recommended revisions, additions and deletions in their respective areas to the Committee, which is indebted to the Chairmen and members of these panels for their dedication and industry.

The aim of the publication remains that of reflecting actual usage rather than trying to modify it. We have attempted to indicate cases where it seems that one term is displacing others in common usage. Examples are to be found in the titles of two of the panels where *manufactured* displaces *man-made* (of fibres) and *technical* replaces *industrial* (of textiles).

There can be no doubt that usage quite often differs from country to country - just as it has long done in different parts of the same country. *Vive la différence*. But we must aim also to provide a basis for understanding each other. As in all activities of The Textile Institute, this project involves the participation of Textile Institute members from several countries. For this edition the Committee has also had considerable assistance from the Committee on Textile Terminology of the American Society for Testing and Materials.

The Committee, and particularly its Chairman, owe a particular debt of gratitude to Paul Daniels, the Institute's Information and Publishing Manager, for his advice and guidance and, above all, the care and attention he has given to the ordering, editing and production of this volume.

J. E. McIntyre
Chairman
Textile Terms and Definitions Committee

Copyright

We would like to acknowledge with thanks the agreement of the following people and organizations for allowing us to reproduce material in which they hold copyright.

American Society for Testing and Materials
Barber-Colman Company, Rockford, U.S.A.
British Nonwovens Manufacturers' Association, Brentford, England
British Standards Institution, London, England
British Textile Technology Group, Manchester, England
H. Carr and B. Latham, *The Technology of Clothing Manufacture*
Cobble Blackburn Limited, Blackburn, England
Peter Collingwood, *The Techniques of Tablet Weaving*
Department of Textiles, University of Manchester Institute of Science and Technology, Manchester, England
International Organization for Standardization, Geneva, Switzerland
Longman Group Limited, Harlow, England
Pitman Publishing Limited, London, England
Platt Saco Lowell (U.K.) Limited, Accrington, England
Shirley Developments Limited, Stockport, England
Society of Dyers and Colourists, Bradford, England
SP Tyres UK Limited, Birmingham, England
Dr H. Stalder, Maschinenfabrik Rieter AG, Winterthur, Switzerland

Notes

Terms which occur in the text in **bold type** are defined elsewhere in the book.

Fabric Structures
The following conventions and diagrams are used throughout this book to describe or represent fabric structures.

Woven Fabrics
For loom-state (unless otherwise stated) woven fabrics parameters are quoted in the following order and in the units given:

(i) Numbers of **ends** per cm x Numbers of **picks** per cm.
(ii) Yarn **linear density (tex)**.
(iii) **Twist level** (turns per metre) and **twist direction** (S or Z).
(iv) Yarn **crimp** (%).
(v) Fabric thickness.
(vi) Area density (g/m^2).
(vii) **Cover factor** (K) warp+weft.
(viii) Fabric width.

In all weave diagrams, marked squares indicate where the warp is lifted over the weft.

Knitted Fabrics, weft-knitted
In diagrams representing machine produced weft-knitted fabrics, having simple structures, the notation system outlined in BS 5441:1988 *Methods of test for knitted fabrics* is used. Dots represent the sites of needles; one double row of dots represent the needles of a rib machine if alternated (see Rib gaiting diagram below), or an interlock or purl machine if directly aligned (see Interlock gaiting diagram below). One double row of dots represent one cycle or knitted course. It is possible and sometimes desirable to build up more than one cycle on one set of dots. The order of a sequence of knitted courses is indicated by a number at the side of each double row of dots, starting at the bottom with the first course knitted.

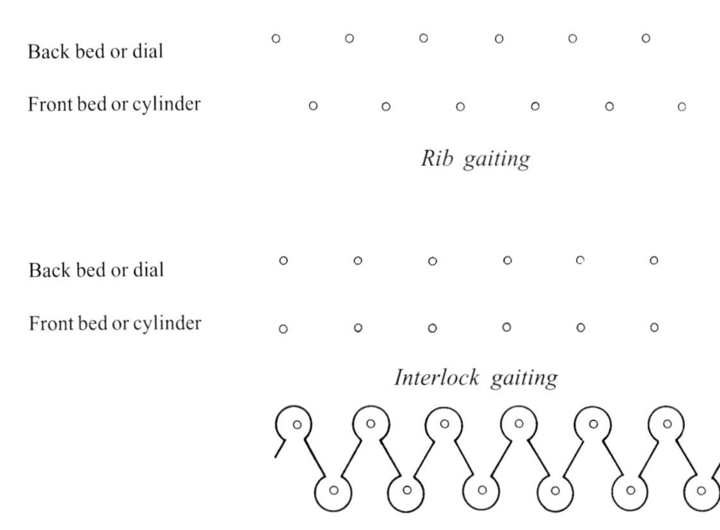

Rib gaiting

Interlock gaiting

Knitted open loops

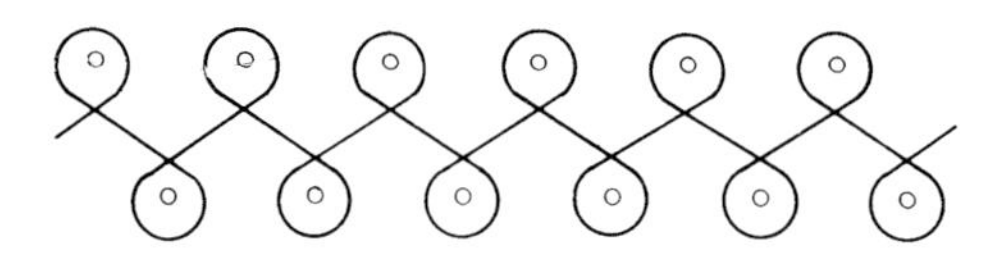

Knitted closed loops

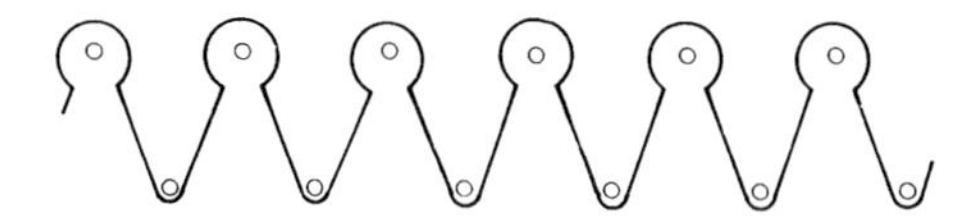

Course containing knitted loops (back bed) and tuck loops (front bed)

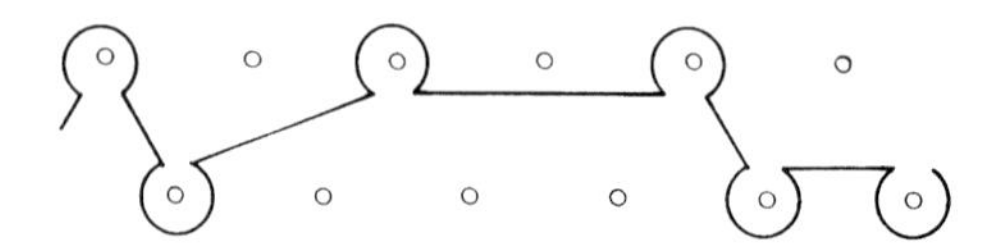

Course containing knitted loops and float loops

Knitted Fabrics, warp-knitted

In diagrams representing warp-knitted fabrics the dots represent the position of the needle heads on successive courses. Vertical rows of dots represent wales and horizontal rows of dots represent courses. The lowest row of dots used is the first course. The line drawn to represent the fabric structure shows the path of the yarn as it progresses from course to course through the fabric. Each guide bar is shown separately.

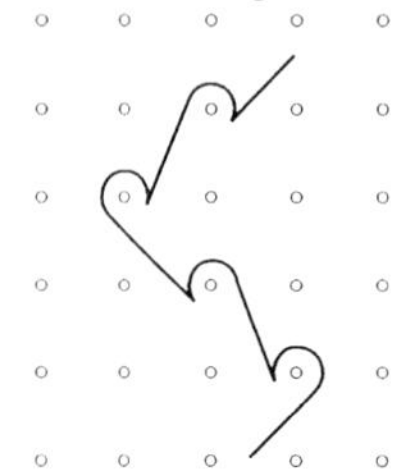

Open laps

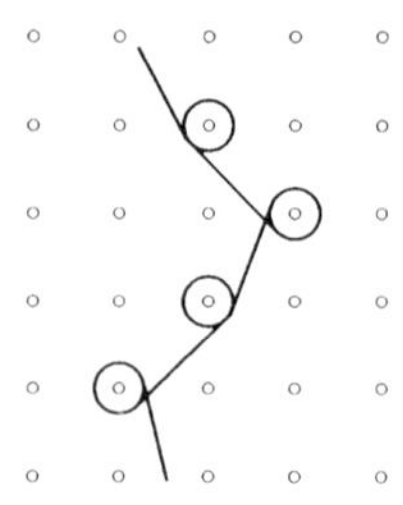

Closed laps

Abbreviations and Symbols

ASTM American Society for Testing and Materials

BS British Standard

E the number of knitting needles per inch on a circular machine. (See also *Note 1* under **gauge** 1.)

EC European Community

ER the number of knitting needles per two inches on a Raschel knitting machine. (See also *Note 2* under **gauge** 1.)

FTC Federal Trade Commission (U.S.A.)

ISO International Organization for Standardization

K cover factor (woven fabrics)

N (followed) by suffix) Symbol for specific length or count unit (see page 397)

R (sometimes with a subscript number) Denotes an organic mono- or bi-radical in chemical formulae

R (before linear density or count values) resultant. Indicates over-all value of combined components (see page 398)

SI Système International d'Unités

T (followed by suffix) Symbol for linear density unit (see page 396)

Textile Terms and Definitions

abaca
See **manila**.

abraded yarn
A continuous-filament yarn that has been subjected to abrading action, generally to provide it with the hairiness characteristic of a staple-fibre yarn.
Note: Unintentional abrading of yarn is a defect.

abrasion mark
See **chafe mark**.

abrasion test
A test used to simulate and measure the wear performance of textile yarns, fabric or floorcovering in use.
Note: There is little agreement between results obtained on different machines when used to test the same product, and the results obtained do not necessarily simulate effects produced during wear. One such type of machine is the 'Martindale'.

absolute humidity
The mass of water vapour present in unit volume of moist air. A typical unit is gram per cubic metre ($g.m^{-3}$).

absolute loom efficiency
See **loom efficiency (absolute)**.

accelerant
A substance, often a swelling agent, which, added to a dyebath or printing paste, accelerates the diffusion of a dye into a substrate. Accelerants may also be used to increase the rate of reaction in chemical finishing.

accessory
1. (Clothing) A subsidiary item of dress, such as a scarf, belt, gloves, umbrella.
2. (Sewing machine) A supplementary part or component of a sewing machine used to diversify its function.

accordion fabric
A weft-knitted plain-based fabric, showing a figured design in two or more colours, that is produced by knitting and missing, and in which tuck loops are introduced to eliminate long lengths of floating thread at the back.

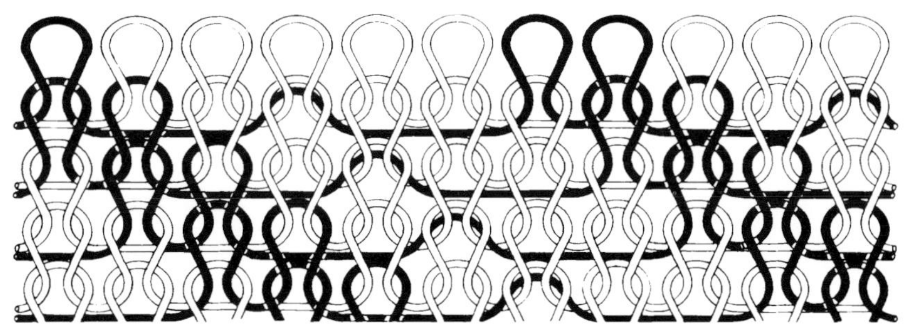

Accordion fabric

accordion pleats
See under **pleats**.

accuracy (testing)
The degree of agreement between the true value of the property being tested and the average of many observations made using the test method.
Note: The word 'bias' is also used in this sense.

acetate (fibre) (generic name)
A manufactured fibre of cellulose ethanoate (acetate) wherein less than 92%, but at least 74% of the hydroxyl groups of the original cellulose are ethanoylated (acetylated). (See also Classification Table p.401.)
Note: In the U.S.A. acetate is defined as a manufactured fibre in which the fibre-forming substance is cellulose acetate. Where not less than 92% of the hydroxyl groups are acetylated, the term **triacetate** may be used as a generic description of the fibre.

acetic acid value
In the characterisation of cellulose acetates, the percentage by mass of combined ethanoyl (acetyl) radical expressed as ethanoic (acetic) acid.

acetylation
The process of introducing an ethanoyl (acetyl) radical into an organic molecule.
Note 1: The term acetylation is used to describe the process of combining cellulose with ethanoic (acetic) acid.
Note 2: A partial acetylation is sometimes applied to cotton in the form of yarn or fibre to change its properties.

acid ageing
ageing in which a volatile acid is present in the vapour.

acid dye
An **anionic dye** characterized by **substantivity** for **protein**, **polyamide** or other fibres containing basic groups. Acid dyes are often applied from an acidic or neutral dyebath.

acrylic (fibre) (generic name)
A manufactured fibre composed of synthetic linear macromolecules having in the chain at least 85% (by mass) of recurring cyanoethene (acrylonitrile) groups. (See also Classification Table p.401.)

add-on
The weight of solids left on a given weight of fabric after impregnation and drying. The percentage add-on is given by

$$(w_2 - w_1) \times 100 \div w_1,$$

where w_1 is the weight of material before impregnation, and w_2 is the weight of material after impregnation and drying. The use of the terms **pick-up** and **wet pick-up** to denote the weight of solids taken up by a fabric is to be deprecated.
Note: Besides impregnation, fabric can also be sprayed, lick-roller coated, or foamed and coated resulting in the deposition of a solute. The add-on is then calculated as above.

added mass; virtual mass (technical textiles)
In air-supported and tension-membrane structures, the mass of air which moves with the fabric structure, and which is often much greater than the mass of fabric itself.

addition polymer
See **polymer, addition** under **polymer**.

addition polymerization
See **polymerization, addition** under **polymerization**.

additive finish
A **finish** which increases the final mass per unit area of the dry textile material and imparts desired properties. (See also **weighting**.)

adhesion test
A test to assess the force required per unit width to separate layers of coated or laminated fabrics.

adhesive-bonded nonwoven fabric
Textile material composed of a **web** or **batt** of fibres, bonded by the application of adhesive material in liquid form. Methods of application include **saturation bonding**, **spray bonding**, **print bonding**, and **foam bonding**.

adjustable top feed
See under **feed mechanisms (sewing)**.

advertising tape
See **bolduc**.

aeroelastic (technical textiles)
Descriptive of the phenomena or analyses combining aerodynamics and elastic behaviour, in relation to air-supported and tension membrane structures.

aeroplane fabric; airplane fabric
Any fabric used as the outer covering of a heavier-than-air aircraft, but now usually a simple, single, closely specified fabric of tightly woven construction, which may include rip-stop threads to enhance tear resistance.
Note: Such fabrics were originally made of cotton or linen, and doped in place to shrink the fabric on to an air frame to reduce air permeability. They may be doped in order to provide environmental protection. Modern fabrics are typically polyester- or polyamide-based and may be coated with a thin polymer film or films, these fabrics being used on light and micro-light aircraft, gliders and hang-gliders.

affinity; dye affinity
The quantitative expression of **substantivity**. It is the difference between the chemical potential of the dye in its standard state in the fibre and the corresponding chemical potential in the dyebath.
Note: Affinity is usually expressed in units of joules (or calories) per mole. Use of this term in a qualitative sense, synonymous with substantivity, is deprecated.

afgalaine; afghalaine
A plain-weave, all-wool dress fabric, containing (i) woollen warp with woollen weft (usually woven: 17x17; 78x78tex woollen warp and weft; 285 g/m^2; K=15.0+15.0; 1.52m wide), or (ii) worsted warp with woollen weft (usually woven: 16x13; 74tex worsted x 88tex woollen; 250 g/m^2; K=13.8+12.2; 1.52m wide).
Note: In both types, the warp ends are usually S and Z-twist alternately, with S-twist woollen weft and usually woven 1.52m wide with 17 ends and picks per cm and 8tex woollen warp and weft.

A-frame
A movable batching unit, in which a horizontal roller is supported by two A-shaped frames. The unit is used to wind fabric in beam form for either storage or wet processing.

after-welt; anti-ladder band; anti-run-back course; garter band; shadow welt (knitting)
A band on a stocking, following the **welt**, in which there is a variation of quality, stitch, and/or yarn.

afterflame
Continuous flaming of a material after removal of the ignition source.

afterflame time; duration of flaming
The length of time for which a material continues to flame, under specified test conditions, after the ignition source has been removed.

afterglow
Glow in a material after the removal of an external ignition source or after the cessation (natural or induced) of flaming of the material.

afterglow time; duration of afterglow
The time for which a material continues to glow, under specified test conditions, after cessation of flaming or after removal of the external ignition source.

agave
Genus of spiny-leaved plants yielding various types of leaf fibres, e.g., **cantala**, **henequen**, **sisal**.

ageing
1. Originally a process in which printed fabric was exposed to a hot moist atmosphere. At the present time, the term is almost exclusively applied to the treatment of printed fabric in moist steam in the absence of air. Ageing is also used in the development of certain **colorants**, e.g., aniline black.
2. In the manufacture of **viscose**, the oxidative depolymerization of alkali cellulose in order to produce a controlled decrease in the chain length of the cellulose. The term is sometimes, although incorrectly, used to describe the ripening of viscose (see **ripening** 2).
3. The oxidation by exposure to air of drying-oil sizes and finishes, e.g., in the production of oiled silk and oilskins and in linseed oil sizing.
4. The deterioration of rubber and plastic coatings and proofings and of some lubricants on textiles, caused by gradual oxidation on storage and/or exposure to light.
5. Progressive change with time in the structure and properties of polymers, including wool fibres.

ageing (testing)
Storage of a material under defined conditions, to determine by subsequent tests the effect of these conditions on the properties of the material. The conditions may be chosen to accelerate any natural changes that may occur.

ager
A chamber used for ageing (see **ageing** 1).

agra gauze; agré gauze
A plain-weave open-sett silk fabric having a gauze-like appearance that receives a stiff finish and is used for trimmings.

agrotextile
Any textile material used in agriculture, horticulture or fisheries.

air house (technical textiles)
A small air-supported structure.

air laying
A method of forming a **web** or **batt** of staple fibres in which the fibres are dispersed into an air stream and condensed from the air stream on to a permeable cage or conveyor to form the web or batt.

air permeability test
A measure of the rate of passage of air through unit area of fabric at a specified pressure difference.

air-gap wet spinning
See **wet spinning (manufactured fibre production)**.

air-jet spinning
See **jet spinning**.

air-jet weaving machine
See under **weaving machine**.

air-supported structure (technical textiles)
A structure, for example a fabric roof, supported by internal air pressure.
Note: People and goods may enter and leave the enclosure through air locks. The roof may span considerably more than 200 m. An inflation pressure differential of the order of 1% atmospheric pressure is used.

air-jet textured yarn; air-textured yarn
See **textured yarn**, *Note 1* (viii).

airloop fabric, warp-knitted
A warp-knitted fabric with reverse locknit lapping movements, the yarns from the back guide bar being overfed to give a short pile on the surface of the fabric.

airplane fabric
See **aeroplane fabric**.

Albert cloth
A reversible centre- or self-stitched double-cloth overcoating, woven with a different design on each side, in stripes or checks.

alfa fibre
Fibre from grass leaves of *Stipa tenacissima.*

alginate (fibre) (generic name)
A manufactured fibre composed of metallic salts of alginic acid. (See also Classification Table p.401.)

alhambra quilt
A jacquard figured fabric with a plain ground weave that requires two warps. The figuring warp

alhambra quilt (continued)
is usually two-ply and coloured, the ground warp singles and undyed. The weft is often made on the condenser system, soft spun, and of coarse count.

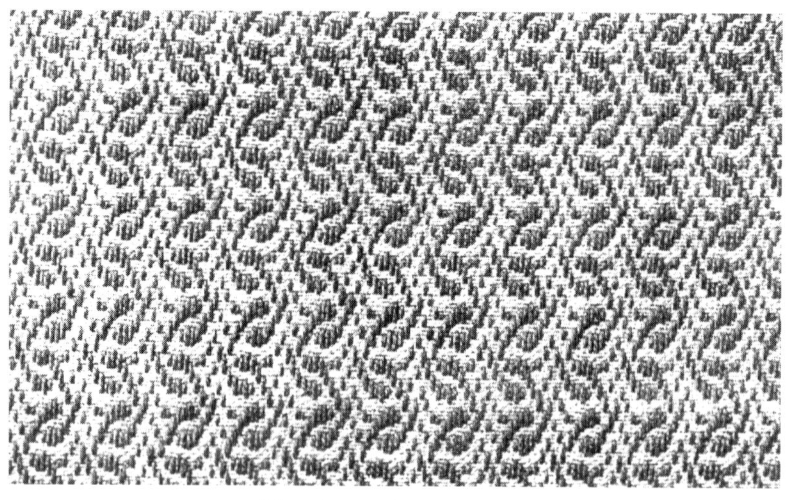

Alhambra quilt (actual size)

alkali solubility
1. The solubility of wool in sodium hydroxide solution which provides an index of the change in its chemical properties brought about by certain agencies.
2. A measure of the amount of non-cellulosic constituents (readily available as food-stuffs for micro-organisms) which is left in commercially boiled flax.

alkali-cellulose
The product of the interaction of strong sodium hydroxide (caustic soda) with purified cellulose.
Note: In the manufacture of **viscose** rayon, the cellulose may be cotton linters or wood-pulp. After pressing, alkali-cellulose usually contains approximately 30% of cellulose and 15% of sodium hydroxide, the remainder being water. During the steeping of the cellulose in sodium hydroxide (18-20% w/w) to form the alkali-cellulose, soluble impurities, including soluble cellulose, are removed.

allovers (lace)
Lace in which the repeats merge into a whole without marked divisions in the pattern.
Note: Allovers may be made the full width of the machine and cut to selling width after finishing.

alpaca fabric
A fabric made from **alpaca fibre**.
Note: The term has been used to describe fabrics made from black cotton warp and alpaca weft, subsequently piece-dyed. This usage is deprecated.

alpaca fibre (hair)
Fibre from the fleece of the alpaca (*Lama pacos*).

alpha metric (α_m)
See **twist factor**.

alumina (fibre)
A fibre of aluminium oxide. (See also **ceramic fibre** and Classification Table, p.401.)

amazon fabric
A light-weight dress fabric with a full, soft handle made from worsted (usually merino) yarns as warp and soft-spun worsted or woollen yarns as weft. The weaves employed are 5-end satin or 2x1 twill, a typical construction for the satin structure being 35 x 17; R22/2tex x 48tex.

American cloth
A light-weight, plain-weave fabric, usually of cotton, coated on one side with a mixture of linseed oil and other materials so as to render it glossy and impermeable to air or water.

American cord
See **rat-tail cord**.

anaphe
See **silk, wild**.

angel lace, warp-knitted
A patterned warp-knitted fabric made with separating threads that are usually of secondary cellulose acetate subsequently dissolved out to leave narrow strips for trimming. It is generally produced on a tricot machine using atlas lapping movements to produce a scalloped edge.

angle of lay
See **lay, angle of**.

angle of lead; winding-on angle
In ring spinning or ring twisting, the angle formed at the traveller between a package radius and the tangent to the package surface.

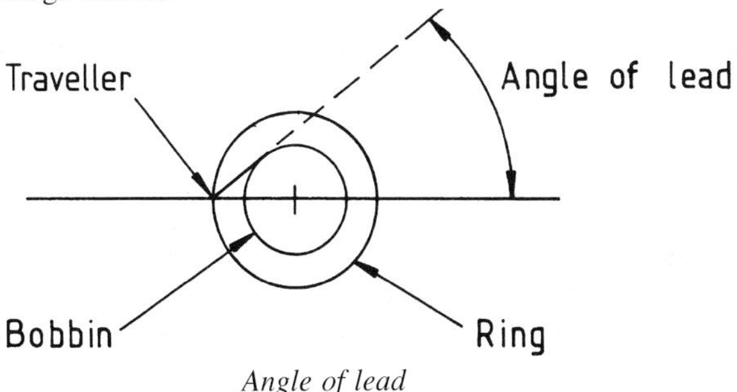

Angle of lead

angle of wind
The angle contained between a wrap of yarn on the surface of a package and the diametrical plane of the package.
Note: Other angles are: yarn-crossing angle, yarn-reversal angle (see diagram).

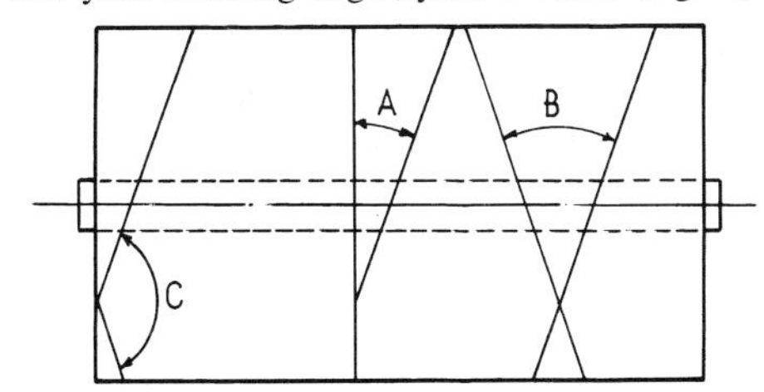

Angles made by yarn wraps on the surface of a package (cheese)
A-Angle of wind B-Angle of crossing C-Angle of reversal

angola
1. (Yarn) A yarn spun on the woollen system from a mixture of wool and cotton or other fibre.
2. (Fabric) A plain or twill fabric made from a cotton warp and an angola-yarn weft.

angora fabric
A fabric woven from **angora yarn**.
Note: The use of this term to refer to fabric made of cotton warp and mohair weft is deprecated.

angora fibre (hair)
Fibre from the angora rabbit (*Oryctolagus cuniculus*). (See also **rabbit fibre (hair)**.)
Note: The hair of the angora goat is known as **mohair**.

angora yarn
An extremely soft yarn made from the hair of an angora rabbit. The yarn is nominally of 100% angora fibre, although a small proportion of other fibres, e.g., up to 5%, is allowed in certain countries dependent on their fibre content labelling laws.

anidex (fibre) (U.S.A.)
A manufactured fibre made from a synthetic linear polymer that consists of at least 50% by mass of one or more esters of a monohydric alcohol and propenoic acid (acrylic acid). (See also Classification Table p.401.)

anionic dye
A dye that dissociates in aqueous solution to give a negatively charged coloured ion.

anisotropic
Having different values for a property in different directions.

anti-ladder band (knitting)
See **after-welt**.

anti-redeposition agent
A substance which, when added to a wash liquor, prevents the redeposition of soil particles on a clean or washed substrate.

anti-run-back course (knitting)
See **after-welt**.

antichlor
A chemical used to inactivate residual chlorine in materials.

anticockle treatment
A mild **setting** treatment imparted to wool knitwear usually by heating the textile material in the presence of an aqueous solution of a reducing agent. The object of this treatment is to prevent distortion and **cockling** of the knitted structure during subsequent wet processing, e.g., coloration at a higher temperature.

antifoam
A substance that prevents foam generation, e.g., in jet dyeing machines, or causes its collapse, e.g., in print pastes.

antisoil finish
A substance applied to the surface of, or incorporated into, a substrate to prevent soiling.

antistatic agent
A substance applied to a substrate to prevent the accumulation of an electric charge.

antung
A Chinese word meaning a silk fabric, machine woven from machine-reeled, **net** Tussah silk in a plain weave and usually of medium or heavy weight.

apparel
Personal outfit, garments, clothing or attire, including headwear and footwear.
Note: This definition includes all apparel even if made of non-fibrous materials. Some dictionaries imply the inclusion of other, non-clothing habiliments and attached or carried accessories such as jewellery, handbags or walking sticks within the definition of apparel.

apparent wall thickness
See **wall thickness**.

appearance retention test
A test designed to assess the likely change in appearance in a floorcovering, or other textile product, in use.
Note: For carpets various types of test apparatus may be used, or specimens may be placed on the floor in areas of high density of pedestrian traffic.

appliqué
A cut-out design or shape attached to the face of a fabric for ornamentation, and frequently of a different type and/or colour of material.

apron
A device used to control the movement of fibres in a **drafting** system. It is more common to utilize two aprons (double apron drafting).

aqueous extract
Liquid obtained from the immersion of a textile specimen in water under prescribed conditions.

aramid (fibre) (generic name)
A manufactured fibre composed of synthetic linear macromolecules having in the chain recurring amide groups, at least 85% of which are joined directly to two aromatic rings, and in which imide groups may be substituted for up to 50% of the amide groups. (See also Classification Table, p.401.)
Note: In the U.S.A. the imide substitution is not included in the definition.

arctic fox fibre (hair)
Fibre from the arctic fox (*Vulpus lagopus, Canis isatis*).

Argyle gimp
A woven figured narrow fabric having three series of wefts and a warp that is usually continuous-filament yarn. Two series consist of three gimp cords laid flat; the ground or third series consists of two gimp cords and forms a plain weave. The two series of three gimp cords form a double-wave raised pattern by passing through the warp every sixth pick alternately and returning over the top of the warp. The over-all width is about 16mm.

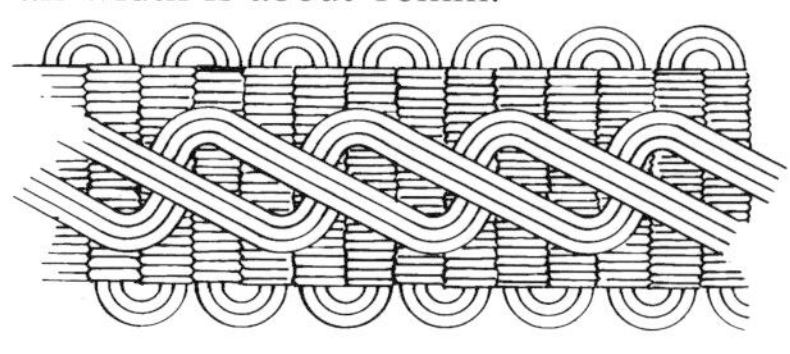

Argyle gimp

arm (sewing machine)
The part of the **sewing machine head** which is above the **bed** normally comprising a vertical arm and a horizontal arm (see diagram). It is a casing which houses and provides mounting points for the moving parts inside.

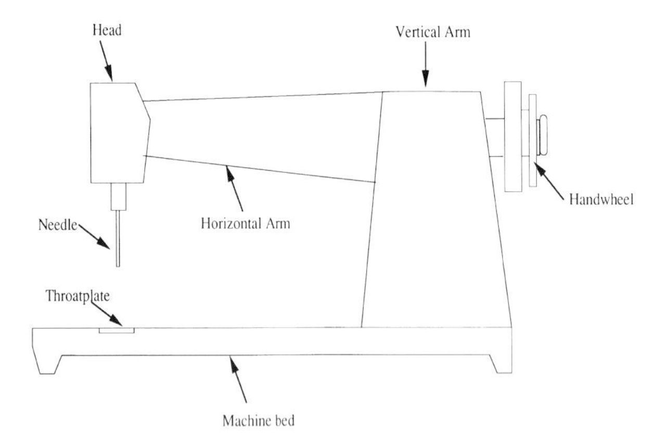

Sewing machine

armuré
A French term for a small pattern in pebbled or embossed effect, hence: (i) armuré weave: a weave designed to produce this effect, for example, a weave of a broken or wavy rib character. In some cases, a definite figure rather than a textural surface is produced. If the ribs are broad, they may have the long floats on the back stitched, and (ii) armuré fabric: a fabric in an armuré weave.

Armuré (actual size)

artificial sports surface
See **sports surface, artificial**.

artificial weathering test
See **weathering, artificial, test**.

artist's canvas
A fabric made of cotton, linen, jute, or hemp, prepared with size and suitably primed specifically for artists' painting grounds.

asbestos
A generic name used to describe a family of naturally occurring fibrous hydrated silicates.

assembly beaming; dry taping
The winding of warp yarn from several back or section beams (each containing part of the total number of ends) on to a weaver's beam.

assembly winding
The winding of two or more yarns as one on to a single package usually in preparation for a subsequent twisting process.

astrakhan
The skin of still-born or very young lambs (originally from Astrakhan in Russia) the curly wool of which resembles fur.

astrakhan fabric
A curled-pile fabric made to imitate the fleece of a still-born or very young astrakhan lamb.

astrakhan fabric, warp-knitted
A warp-knitted fabric in which a thick curled yarn is attached to the ground fabric by the threads of two guide bars while three other guide bars knit the ground fabric.

astrakhan fabric, weft-knitted
A weft-knitted plain based fabric with curled yarn inlaid on a tuck-miss basis. (See also **laid-in fabric, weft-knitted**.)

astrakhan yarn
A type of **curled yarn** used in astrakhan fabric.

astrakhan, woven
Pre-treated pile yarn is lifted over wires, inserted at regular intervals, and woven into a plain weave ground.

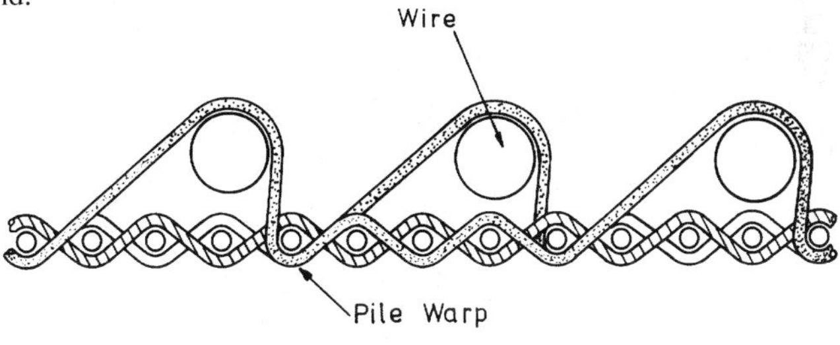

Astrakhan fabric: section through weft

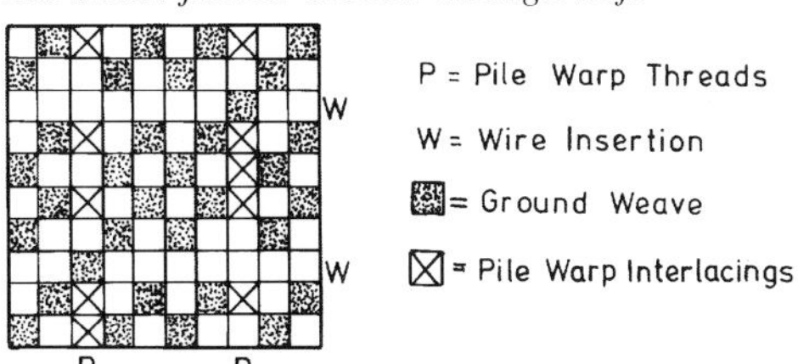

Astrakhan weave

atactic polymer
See **polymer, atactic** under **polymer**.

atlas fabric, single bar; vandyke fabric; shadow stripe fabric
A warp-knitted fabric characterized by having one set of threads traversing in a diagonal manner, one wale per course for a number of courses, returning in the same manner to the original wale. Open or closed laps may be used. A typical example is shown.

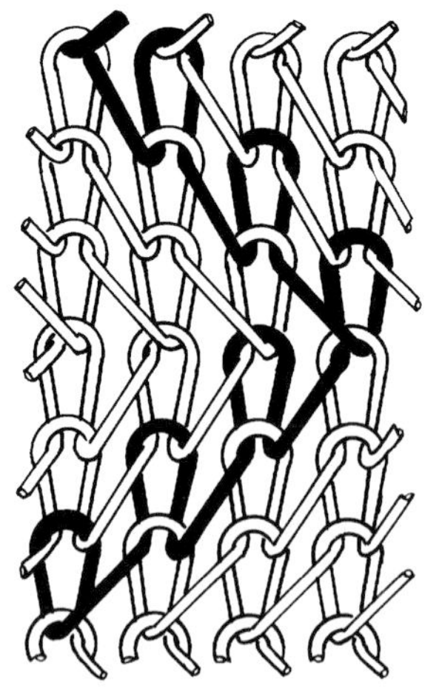

Single bar atlas fabric

atlas fabric, two bar
A warp-knitted fabric having two sets of threads making identical single atlas movements in opposition.

atmosphere for testing
See **standard atmosphere for testing (textiles)**.

auto-leveller
A device that is fitted to carding and drawing machines to automatically reduce the variation of the linear density of the output material. The result is achieved by monitoring the linear density and, if necessary, changing the draft to compensate for any deviation from a pre-set value.

autoclave
A vessel in which textile material may be treated with steam under pressure. Facilities are often provided for the creation of a vacuum either before the introduction of steam (when the purpose is to achieve better steam penetration) or subsequently (when the purpose is to assist in drying).

automatic loom
See **weaving machine**.

automatic weaving machine
See **weaving machine**.

Axminster carpet
A machine-woven carpet, with cut pile, in which successive weft-wise rows of pile are inserted during weaving in a pre-arranged colour sequence. There are four main types of Axminster weave:

> **chenille Axminster**
> A carpet that has a pile of chenille weft. (See also **chenille yarn** under **fancy yarn**.)
>
> **gripper Axminster**
> A carpet in which tufts of yarns are inserted at the point of weaving by means of grippers.

The colours are selected by jacquard-operated carriers that present the appropriate ends of yarns to the grippers before the tufts are severed from the yarns. Jacquards are normally of eight, ten or twelve frames which allows designs of eight, ten or twelve colours without **planting.**

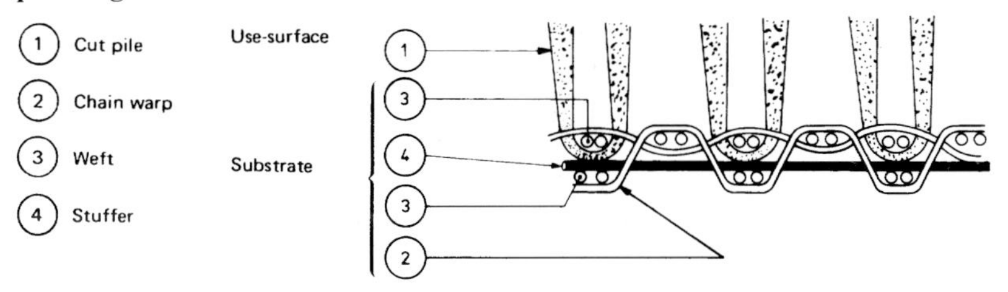

Gripper Axminster carpet (warpways section)
The above diagram is derived from figure 9 of BS 5557

spool Axminster
A carpet in which the yarn for each weft-wise row is wound on a separate spool according to the design. This allows an unlimited number of colours to be used in the design. The tufts are severed from the yarns presented at the point of weaving after insertion in the backing structure.

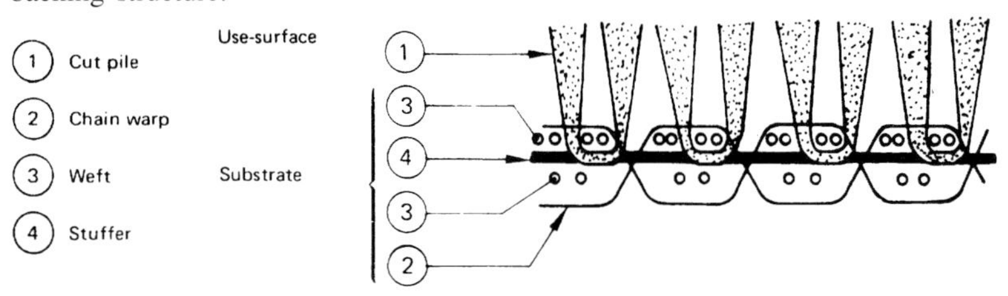

Spool Axminster carpet (warpways section)
The above diagram is derived from figure 8 of BS 5557

spool-gripper Axminster; gripper-spool Axminster
A carpet in which the yarns for each weft-wise row are wound on a spool as in spool Axminster weaving, allowing for unlimited use of colour. The tufts severed from the yarns are inserted at the point of weaving by grippers as in gripper Axminster weaving.

auxiliary; auxiliary product
A chemical or formulated chemical product which enables a processing operation in preparation, dyeing, printing or finishing to be carried out more effectively or which is essential if a given effect is to be obtained.

azlon (fibre) (U.S.A.)
A manufactured fibre in which the fibre-forming substance is composed of any regenerated naturally occurring protein.
Note: The ISO **generic name** is **protein**. (See also Classification Table p.401.)

azoic dyeing
The production of an insoluble azo compound in a substrate by interaction of a diazotized amine (azoic diazo component) and a coupling component (azoic coupling component).

BA wool; Buenos Aires wool
Wool originating from Argentina.

baby flannel
A light-weight **flannel** used for children's garments.

back (fabric)
The reverse of a fabric as opposed to the **face**.

back (weft-knitted fabric)
See **plain fabric, weft-knitted**.

back beam
A beam from which the warp is fed during sizing.

back crossing heald
See **leno weaving**.

back grey
1. Fabric used on a roller printing machine between the blanket and the fabric to be printed, sometimes known as the face fabric, in order to: (i) absorb any printing paste that percolates through the printed fabric, or (ii) to impart resilience to the face fabric allowing sharper prints.
2. Fabric used in screen printing to support light weight, open weave fabrics to be printed in order to maintain stability (e.g., prevent curling) of the face fabric. The back grey and face fabric are gummed together prior to the combined fabric itself being gummed to the printing table conveyor belt.
Note: Synthetic-polymer fibre fabrics can be gummed to the back grey to prevent slipping. (See also **bump grey**.)

back loop (weft knitting)
See under **knitted loop (weft knitting)**.

back rack
See **cratch**.

back rail; back rest; whip roller
The bar, rail, or roller at the back of a loom over which the warp threads from the beam pass.

back rise
The distance, measured along the surface of a bifurcated garment, from the **crotch** to the centre back of the waistline. (See also **front rise**.)

back standard
See **leno weaving**.

back warp (or weft); backing warp (or weft)
Additional warp (or weft) on the back of a fabric, bound or stitched to the ground structure so that it does not interfere with the appearance of the face, giving greater weight, thickness, warmth, etc.

back-filling
The application of a filling material (see **filling** 1) to the back of a fabric.

back-wind (pirn winding)
The final stage of pirn-winding in which the full pirn continues to revolve whilst the yarn-traverse guide-eye carries the yarn from the nose of the pirn to the base, whereupon the rotation of the pirn is stopped. Back-wind prevents lively or slippery yarns from falling off the nose chase and thus reduces transport and handling problems and hence waste.

back-winding (spinning)
The process of rewinding, e.g., from **hank** or **cone** on to a more suitable **package** for the next process.

backing
A strip of material placed on the inside of a part of a garment to act as a reinforcement.

backtanning
An after-treatment to improve the wet fastness of dyed or printed silk or polyamide materials, using either natural or synthetic tanning agents.

backwashing
The washing of dyed or undyed wool sliver before or after **gilling** and/or combing.

bad cover (fabric)
1. A fabric appearance in which the spaces between the threads are more pronounced than is required. The degree of cover can be affected by loom adjustments, sett, or count of reed, or by the construction of the yarns used.
2. The appearance of a finished fabric in which the surface is not covered, or the underlying structure not concealed, by the finishing materials used, to the degree required.

bagging; tacking
The sewing together of the two selvedges of a fabric to form a tube in order to prevent selvedge curling, to encourage **ballooning**, and thereby reduce rope marking (see **rope marks**) in the fabric face during wet processing.

baize; baze
A woven woollen felt used for covering tables and screens.

baking; thermofixation
The use of dry heat to achieve the fixation of **colorants** or chemical finishes on textile materials.

balance
The relation of one section of a garment to another, particularly that of back and front lengths, with the aim of achieving a garment in harmony with individual posture.

balance marks
Visual guides, generally in the form of notches, used to maintain the correct **balance** of a garment during assembly, and usually positioned in side seams, shoulder seams, armholes and elsewhere as necessary.

balance wheel (sewing machine)
See **hand wheel (sewing machine)**.

balanced twist yarn
lied or cabled yarn which, solely by virtue of the twist combination chosen, has no unbalanced twisting couple.

balanced weave
A weave in which the average float is the same in the warp and weft directions, and in which the warp and weft floats are equally distributed between the two sides of the fabric.
Note 1: Balanced weaves include plain, 2/2 matt, 2/2 twill, **mock leno**, and many crêpe weaves.
Note 2: Unbalanced weaves include warp-faced and weft-faced twill weaves, satins, and sateens.

bale breaker
A machine used for **opening** cotton or other short-staple fibres removed directly from a compressed bale. Layers of compressed fibres are taken from the bale and fed into a machine where the tearing action of two coarse spiked surfaces (rollers or lattices), moving in opposite directions, produces a more open mass of tufts.

bale dyeing
Dyeing of loose stock (usually synthetic-polymer fibre) in the form of an unpacked bale.

ball fall
A measure of the viscosity of a liquid, expressed as the time in seconds required for a standard sphere to fall through a column of liquid of standard length under standard conditions.

ball top
A cross-wound self-supporting package of combed sliver produced on the worsted system.

ball warp
Parallel threads in the form of a leased twistless rope wound into a large ball by hand or by machine.

ball warping
See **warp**.

ball-warp sizing
The application of size to warp yarn in ball-warp rope form.
Note: Subsequent squeezing and drying are essential features of the process.

ballet toe (knitting)
A type of **reverse toe** in which the toe yarn on the upper side covers the ends of the toes only, and the toe is usually extended and more pointed.

ballistic tearing strength
The force required to prolong a tear over a given length of fabric determined from the energy absorbed in tearing at high speeds. It is usually conducted at shock loads using a falling pendulum.

balloon (yarn)
The curved shape that a yarn length, whether stationary or in forward motion, forms in space when it is made to circulate (i.e. whirl) at a fixed or varying radius around an axis.
Note 1: The forward motion of the yarn length is commonly called the yarn delivery speed.
Note 2: Balloons commonly occur when a yarn length circulates a package, mainly out of contact with the package, during over-end **twisting**, winding and unwinding operations (e.g., uptwisting, downtwisting, ring spinning, yarn doubling or folding, see diagram). Balloons can also occur when yarn lengths whirl free of lateral constraints (e.g., air-jet spinning).
Note 3: Balloons may be single or multiple and their size, basic shape and angular velocity are determined by the geometry of the system and the rate of twisting. These factors control the balance of forces producing the balloon. The forces are centripetal, coriolis, air-drag and yarn tension. Other factors that may influence balloon size, shape and angular velocity are package size, ring size, traveller weight, linear density, hairiness of the yarn and the presence of balloon control rings or balloon separators.

balloon fabric
Any fabric which forms a functional part of the lift-creating and, where different, the outer envelope of a lighter-than-air aircraft. It is usually a simple, single fabric, of tightly woven

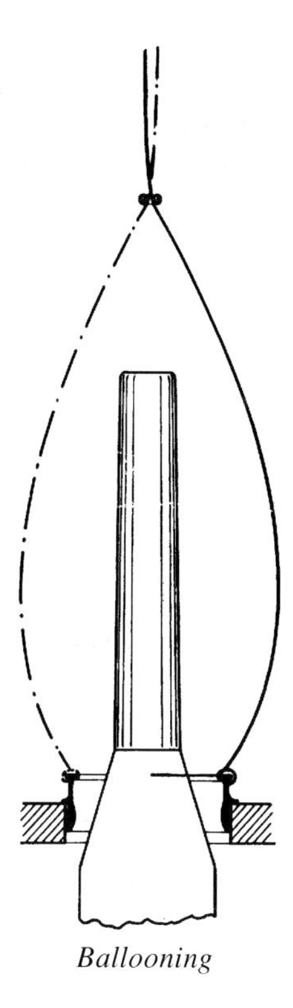

Ballooning

construction, and may include rip-stop threads to enhance tear resistance, although some plied fabrics are used.
Note: Such fabrics were originally made of cotton, linen or silk and doped or coated with rubber to reduce gas permeability, but modern fabrics are typically polyamide or polyester coated with a polymer or polymers, and/or laminated with a thin natural or synthetic-polymer film or films. This reduces permeability to the lifting gas employed, confers protection against ultra-violet radiation, and provides weather resistance.

balloon separators
Devices, typically plates, designed to keep gyrating balloons of yarn separate during ring spinning, winding, or on withdrawal from packages during doubling or twisting of yarns.
Note 1: Separators are primarily intended to prevent collisions between adjacent gyrating yarn balloons and between yarn balloons and machine parts.
Note 2: Separators may restrict and confine intentionally the dimensions of yarn balloons, and also affect yarn tensions during processing.

balloon-control rings
One or more rings positioned between the **ballooning eye** and the ring on a **ring-spinning** machine or **ring-twisting** machine, in order to contain the **ballooning yarn** and maintain a suitable yarn tension. (See also diagram under **balloon (yarn)**.)

ballooning (finishing)
The entrapment of air, either deliberately or accidentally, by a fabric during wet processing.

ballooning eye; twizzle; lappet
A yarn guide that forms the apex of the yarn **balloon**.

ballooning yarn
A yarn which exhibits a **balloon** during processing.

banded laces
Laces, produced on a warp-knitting or twist-lace machine, that are narrow bands or strips and used for trimming garments. The bands are usually held together for finishing purposes and subsequently separated. There are three main classes:
(i) edgings: Narrow bands of lace with one straight edge for sewing purposes and the other made ornate by scalloping and using pearls. They are used for sewing on to edges of a garment.
(ii) insertions: Banded laces with two straight sewing edges, used for inserting between two panels of fabric.
(iii) galloons: Banded laces scalloped or fancy at both edges, used to sew on the top of a fabric for decorative purposes.

bannister harness
See **split harness**.

Bannockburn
A firmly woven Cheviot tweed in 2/2 twill weave (straight twill or herringbone) having single and two-ply yarns alternatively in both warp and weft. Originally the two-ply yarns were made by plying a single yarn identical to the one used in the fabric with a white yarn for use in the warp and with a dark yarn for use in the weft, but modern Bannockburn tweeds favour the count of the coloured single yarns approximating to the resultant count of the ply yarns, which are white-colour and/or light-dark woollen **grandrelle** or **marl** yarns, warp and weft.

bar (knitted fabric)
See **barré (fault)** 2.

bar (woven fabric)
An unwanted bar, running across the full width of a piece, which differs in appearance from the adjacent normal fabric. It may be shady or solid in appearance, and may or may not run parallel with the picks. This is a general term covering a number of specific faults as follows:

> **pick bar**
> A bar in which the pick spacing is different from that in the normal fabric. Types of pick bar are:
> (i) starting place: an isolated narrow bar running parallel with the picks, starting abruptly and gradually shading away to normal fabric. This is due to an abrupt change in the pick spacing followed by a gradual return to normal pick spacing. Such a bar may occur on restarting weaving after (a) pick finding, (b) unweaving or pulling-back, (c) prolonged loom stoppage. These bars may also be referred to as 'standing places' or **pulling-back places** if the precise cause is known.
> (ii) weaving bar: a bar which usually shades away to normal fabric at both its edges. It owes its appearance to a change in pick spacing, and may repeat at regular intervals throughout an appreciable length or even the whole length of a piece. Such a bar is the result of some mechanical fault on the loom, e.g., faulty gearing in the take-up motion, bent beam gudgeons, uneven or eccentric beam **ruffles**, uneven bearing surfaces at some point in the let-off motion, etc. Bars of this type associated with the take-up or let-off motion are also referred to as 'motion marks'.
>
> **shade bar**
> A bar that has developed a different shade from the adjacent fabric during, or subsequent to, dyeing and finishing owing to damage or contamination of otherwise normal fabric or of weft yarn prior to weaving.

tension bar
A bar composed of weft yarn that has been stretched more or less than the normal weft prior to, or during, weaving. This abnormal stretch may have been imposed, during winding, by faulty manipulation or by some mechanical fault in the machine; during weaving, by incorrect tensioning in the shuttle; or it may have been caused by excessive moistening of the yarn at some stage, leading to greater stretch than normal under the same applied tensions. Such a bar may appear as a cockled bar in those cases where stretch has been sufficient (see **cockle (defect)**).

weft bar
A bar that is solid in appearance, is clearly defined, runs parallel with the picks, and contains weft that is different in material, linear density, fibre, twist, lustre, colour, or shade from the adjacent normal weft.

Motion marks or regular weaving bars

bar warp machine
See under **lace machines**.

bar-filling (weft knitting)
The operation of placing rib borders or other knitted garment pieces, individually or in succession, one on top of another on to a point bar on a needle loop-to-point basis. It is a preparatory operation to facilitate the transfer of single knitted pieces from the point bar either to separate transfer bars or to the individual needle bars of a straight-bar plain machine.

barathea
A fabric of pebbled appearance, usually of twilled hopsack or broken-rib weave, made of silk, worsted, or manufactured fibres, and used for a variety of clothing purposes.

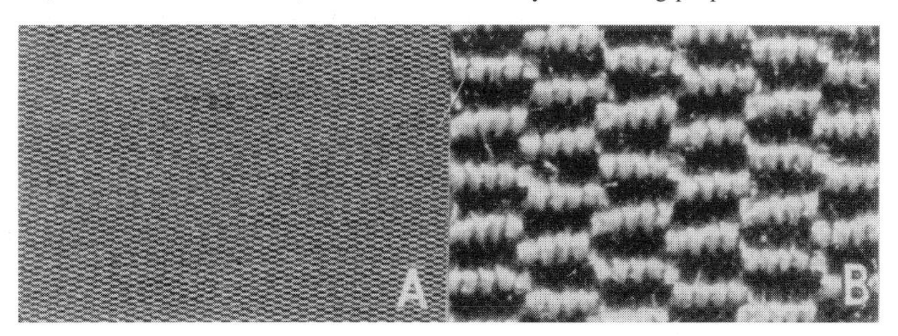

Actual size *Magnification x 8*

Barathea (silk type)

barathea (continued)

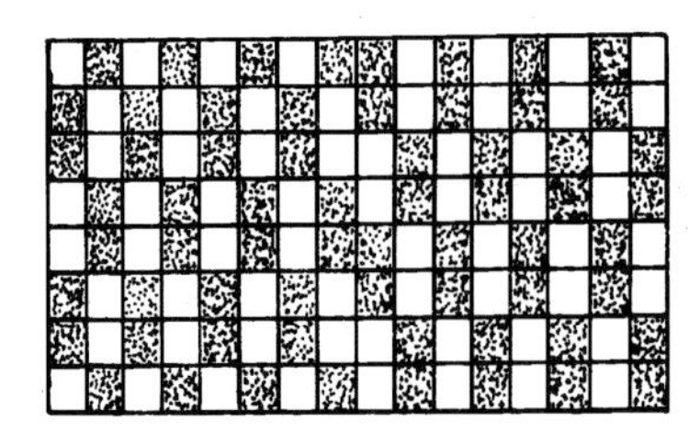

Barathea weave (silk type)

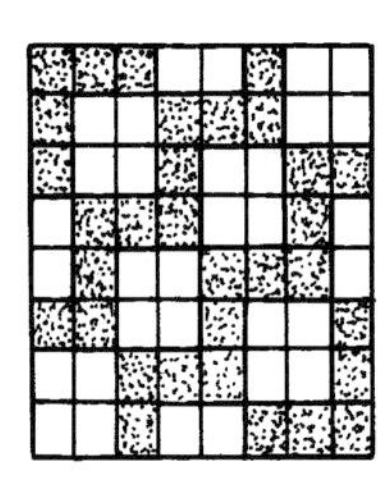

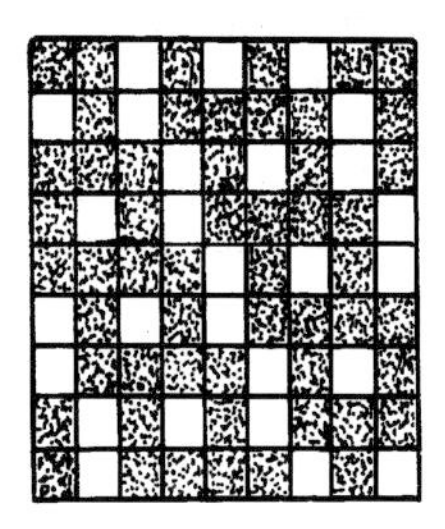

Barathea weaves (worsted type)

barbe
See under **fibre length**.

bare cloth
1. A fabric with **bad cover**.
2. A fabric, the surface of which is not raised, brushed, or napped to the required degree.

bark
See **beck**.

barmen lace
Narrow lace made on a type of braiding machine, on which individual carrier movements are controlled by a patterning mechanism.

Barmen machine
See under **lace machines**.

barr; barry (weaving)
Descriptive of a fabric containing bars (see **bar (woven fabric)**).

barras; hessen
A coarse linen fabric similar to sackcloth; originally produced in Holland.

barré (fault)
1. Unwanted stripes in woven fabrics, in the direction of the weft.
2. A fault in a weft-knitted fabric (usually knitted on a multifeeder machine) appearing as light or dark coursewise stripe(s) and arising from differences in lustre, dye affinity (or unlevel dyeing) in the yarn, yarn spacing or loop length, yarn linear density or defective plating. The fault when it occurs as a single defect in one course or one group of courses is termed a 'bar'.

barring-on (weft knitting)
1. The operation of placing the point bar in position and pushing the rib borders, or other knitted pieces, off the points on to the needles of a straight-bar machine.
2. The introduction of a point bar for the commencement of a welt.

bars (lace)
See **brides**.

basecloth (needling)
A textile fabric, normally woven, which may be included within a **needlefelt** to provide dimensional stability and strength and in some cases to facilitate the punching operation.

basic dye
A **cationic dye** characterized by its **substantivity** for standard acrylic, modacrylic and basic-dyeable polyester fibres.

basin waste; basinés; bassinas; bassinet; basinetto
Waste silk obtained from the inside of the cocoon, which remains after reeling.

basket
See **hopsack**.

bast fibre
Fibre obtained from the stems of various plants.

bastard reed
A reed in which the dent spacing at each side is slightly greater than in the centre.

batch polymerization
See **polymerization, batch** under **polymerization**.

batchwise process
Processing of materials as lots or batches in which the whole of each batch is subjected to one stage of the process at a time. (See also **continuous process**.)

batik dyeing; batik printing
A method of producing characteristic designs on fabric using a wax or gum resist.
Note: By the traditional technique, the wax is applied in the form of a design by a brush or using a tool from which the wax is poured. The fabric is then dyed, the dye usually penetrating through cracks in the wax to give a characteristic veined effect. After dyeing, the wax is partially or wholly removed, and the process is repeated a number of times using different dyes. The batik style is mass-produced today using a printing technique and wax as a resist, with indigo, mordant and azoic dyes.

batiste
A soft, fine, plain-woven fabric traditionally of flax but made in other fibres. A typical cotton construction is K=9.0-13.0+7.0-13.0, 100 g/m^2.

batt; batting; web
Single or multiple sheets of fibre used in the production of nonwoven fabric.

batten
1. A swinging frame that carries the cylinder of a jacquard machine.
2. The reciprocating part of a multi-space narrow fabric weaving machine which carries the shuttle-landings, shuttles and reeds.
3. See *Note* under **sley** 1.

battery
The part of an automatic weaving machine that holds full bobbins or pirns of weft in vertical or circular housings, or holds shuttles in a vertical housing.

baulk
That part of a **reed**, providing its lateral margins, that holds its wires or **dents** securely and at the desired **sett**.
Note 1: Pitch bound reeds are furnished with a baulk formed by twine secured by pitch. Most reeds have a cast metal baulk.
Note 2: Whereas a reed has two baulks a **comb** has one.

baulk finish
A finish in which woollen material is lightly milled in the grease, scoured, dyed, tentered to width, and lightly pressed.

Baumé, degrees
A scale used for measurement of the relative density (RD) of liquids by hydrometry. The following formula expresses the relationship between relative density and degrees Baumé (°Be) for liquids more dense than water:

$$^{\circ}\text{Be} = 145 - \frac{145}{\text{RD}}$$

bave
The silk fibre complete with its natural gum (sericin) as it is withdrawn from a cocoon formed by a silk worm. It comprises two **brins**.

BCF; BCF yarn; bulked continuous filament
A textured continuous-filament yarn, generally used either as a pile yarn in carpets or for upholstery fabrics. BCF yarn is usually made by hot-fluid jet texturing (see **textured yarn**, *Note 1* (iii)).

bead (narrow fabric)
A thickened selvedge of a **zip fastener tape** or **headband** which may be formed by weaving or knitting a cord or cords integrally with the fabric so that they lie substantially on each surface of the **tape** or, alternatively by weaving a tubular selvedge which incorporates **stuffer threads**.

bead wrapping fabric
See under **tyre textiles**.

beading
See **buttoning**.

beading (lace machines)
See **eyeletting (lace machines)**.

beam
A cylinder (usually of wood or metal) provided with end bearings and at each end of which may be mounted suitable flanges (see **warper's beam** and **weaver's beam**).

beam (lace machines)
1. (Furnishings) A subsidiary warp of parallel threads, wound in sheet form on to a beam to provide one set of threads in a net ground.
2. (Leavers) Parallel threads wound in sheet form on to a small beam tube to provide the threads for one steel bar. These threads may be used for patterning and/or netting.
3. (Warp) Parallel threads wound in sheet form on to a beam tube to provide the threads for pattern or warp bars (see **guide bars (lace machines)**). These threads are used for the structural ground or for patterning.

beam creel
A creel for mounting warp beams (back beams) from which sheets of ends may be withdrawn to feed the warp-sizing machine or dressing frame.
Note: Control is usually provided for the sheets of ends and the unrolling beams. Creels may be arranged to carry beams in a horizontal row or inclined towards the sizing machine or in tiers.

beam dyeing
Dyeing of textile materials wound on to a hollow perforated roller (beam) through the perforations of which dye-liquor is circulated.

beam ruffle
See **ruffle**.

beam warping
Winding a part of the total number of ends of a warp in full width on to a **back beam**.

beaming
The primary operation of warp-making in which ends withdrawn from a warping creel, evenly spaced in sheet form, are wound on to a beam to substantial length (usually a multiple of the loom warp length). (See also **direct warping**.)
Note 1: Several similar beams (termed a set of back beams) of the same length provide the total number of ends required in the warps to be made.
Note 2: The sheets from a set of back beams are usually run together as one sheet on to a succession of weaver's beams as an integral part of the sizing operation. Alternatively, the weaver's beams may be assembled from the back beams by **dry taping**.
Note 3: Beaming is suitable primarily for bulk production of grey warp. Fancy warps may be produced by arranging the colour sequence of the ends evenly throughout the set of back beams.

bear fibre (hair)
Fibre from the bear (Genus *Ursidae*).

bearded needle (machine knitting)
See under **needle (machine knitting)**.

beater; opening roller; combing roller
A rapidly rotating roller, which is covered with pins or card clothing, used to separate sliver into individual fibres. This type of unit is incorporated in the feed section of most open-end spinning machines.

beating
A term used in the wool industry for spare threads that are run from a warp during weaving. They can be used for replacing missing ends during the mending process (see **mending**).

beating (finishing)
A process which originally consisted of manually beating fabrics with sticks, but which can be mechanized. It is used in finishing certain raised wool fabrics.

beating-up
The third of the three basic motions involved in weaving, namely **shedding**, picking (see **picking** 1), and beating-up. It consists in forcing the pick of the weft yarn left in the warp shed up to the **fell (of the cloth)**.

beaver cloth
A heavy, firm-texture fabric, made from woollen yarns, which is milled, raised, and cut close on the face before receiving a **dress-face finish**. It is intended to simulate natural beaver skin.

beaver fibre (hair)
Fibre from the beaver (*Castor canadensis*).

beaverteen
A fabric having a very high weft sett used chiefly for heavy trouserings. In cotton, typical

beaverteen *(continued)*
constructions are: (i) 13x110; 18x20tex; K=5.5+49.2; (ii) 13x160; 18x16tex; K=5.5+64.0. The fabric is piece dyed and has a short, soft raised finish on the back. It is a heavier fabric than **imperial sateen**. (See also **fustian**.)

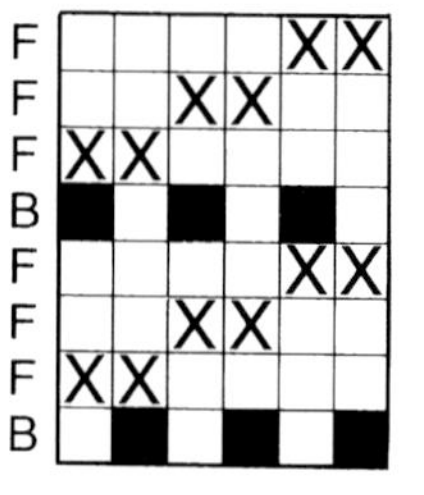

Beaverteen weave

beck; bark; kettle; trough
An open-topped vessel used to hold treating liquors, e.g. wash or dye liquor. (See also **bowl** 3 and **vat** 1.)

Becke lines
Lines which appear at the edge of a microscope image of a fibre caused by refraction at the fibre edge.

bed (sewing machine)
The part of the machine on which sewing takes place and in which the **throat plate** is located.

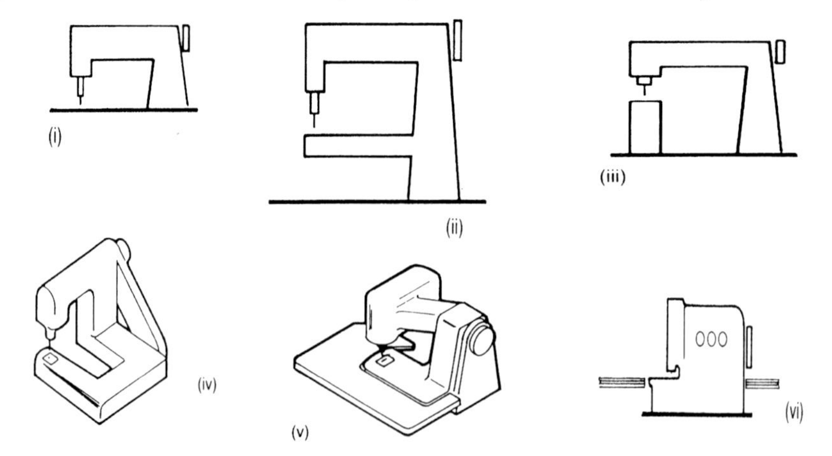

(i) Flat bed, (ii) Cylinder bed, (iii) Post bed, (iv) Feed-off-the-arm, (v) Blind stitch, (vi) Overedge
(Source: H. Carr and B. Latham. *The Technology of Clothing Manufacture,* 1994)

blind stitch bed
A bed which allows the needle to enter the material almost tangentially, so that stitches can be formed which are not readily visible on the face of the material.

cylinder bed
A cylindrical bed fixed horizontally above the surface of the table in which the machine is mounted to facilitate the handling and sewing of tubular articles.

feed-off-the-arm bed
A U-shaped bed.

flat bed
A bed which provides a flat surface in the same plane as the table in which it is mounted.

overedge bed
A bed which provides space for the material on only one side of the needle. This arrangement allows the formation of seams in which one or more threads are passed around the edge of the material.

post bed
A bed comprising a raised post which enables sewing to be carried out above the level of the table on which the machine is mounted.

Bedford cord
A fabric showing rounded cords in the warp direction with pronounced sunken lines between them, produced by the nature of the weave. The weave on the face of the cords is usually plain, but other weaves may be used. There are weft floats the width of the cords on the back. Wadding ends may be used to accentuate the prominence of the cords.

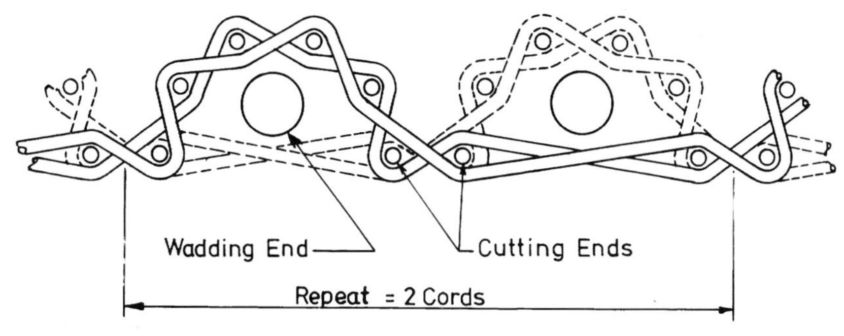

Wadded Bedford cord: section through warp

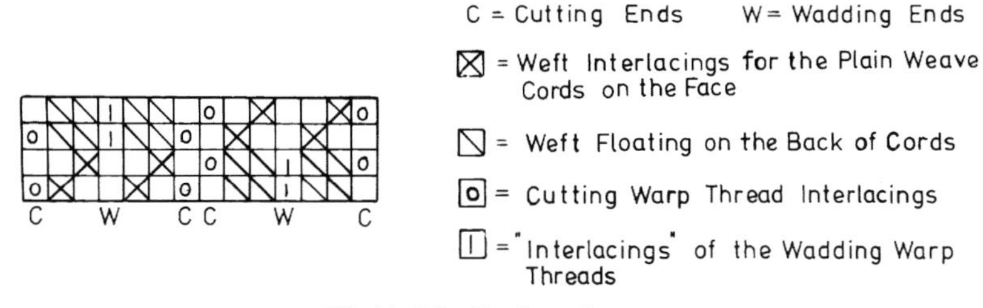

Wadded Bedford cord weave

beet
A bundle or sheaf of tied flax crop or straw.

beetle
To produce a firm, close, and lustrous fabric of cellulosic material, particularly linen or cotton, by subjecting the damp material, batched on a wooden or metal beam or roller, to repeated blows of wooden or metal hammers or fallers.

beggar's inkle
See **inkle, beggar's**.

bellies (wool)
The coarser quality of wool from the underside of a sheep.

belly bands
See **wrapper fibres**.

belt (tyres)
See under **tyre textiles**.

belt loop
A loop of material or thread which is sewn into the waist of a garment to accommodate a belt.

belting (industrial and mechanical)
A generic name covering all forms of belts, and rolls of material from which belts are made up, that are designed for the transmission of power or for the purpose of conveying or elevating.

> **endless woven belting**
> A woven narrow fabric, usually in plain or 2/2 twill weave, in which the warp consists of one continuous thread wound in a helix to the required length and woven without join or splice so that the first and last picks are adjacent.
>
> **solid woven belting**
> Belting consisting of more than one ply, the plies being interlocked in the weave, or bound together by binding threads in the course of weaving.
> *Note:* Solid woven belting is usually impregnated or otherwise treated to increase the coefficient of friction and the resistance to moisture and rotting, to improve linear stability, and to impart other desirable properties.

belwarp fabric
A fine worsted dress-cloth having a corkscrew twill weave with a clear finish.

bending length
The length of a rectangular strip of material that will bend under its own weight to a specified angle. (See also **flexural rigidity**.)
Note: Bending length is one of the factors that determine the manner in which a fabric drapes.

bengaline
A fabric with a more or less striking warp-rib appearance running across the fabric, produced from cotton, worsted, silk or other continuous-filament yarns, or in part from any of the materials named, a typical example having silk warp and worsted weft. The warp-rib or corded effect may be produced by: (i) suitable thickness and setting of warp and weft threads, (ii) suitable warp-rib weaves, and (iii) a combination of (i) and (ii).

berber
A term used originally to describe berber carpets, produced as squares, hand-woven by North African tribes with yarn spun by hand from wool from local sheep. They contain a proportion of naturally pigmented wool and may bear simple tribal motifs. In recent years the term has been commonly misused to describe manufactured carpets made from natural coloured wools, or dyed fibre, and having a homespun appearance.

Berlin
Wool rag in the finer machine-knitted grades.

bespoke
Descriptive of a tailored garment made to an individual customer's specification.
Note: In other sectors of the clothing industry, the term 'made-to-measure' is more usual.

bevel-woven material
A fabric in which warp bow is intentionally introduced, as in woven fabric disks (see diagram of warp bow under **bow (weaving)**).

bi-axial fabric, warp-knitted
A warp-knitted fabric in which additional, substantially straight threads generally run warpwise (along) and weftwise (across) the fabric. Such threads are inserted throughout the complete width and length of the fabric.

bias
An oblique direction to constructional length and width.
Note: The terms 'true bias' or 'true cross' refer to a bias of 45° to the selvedge. (See also **bias binding**.)

bias (testing)
See **accuracy (testing)**.

bias binding
A product made by cutting wide woven fabric at an angle of 45° to the selvedge. Most bias bindings have regularly spaced joins governed by the width of the original wide fabric, but if converted from tubular fabric, joins are avoided. Bias bindings do not fray and will stretch, and they are thus suitable for binding seams and conforming to curved contours. Bias bindings are usually folded into the centre, but other folded patterns as well as flat versions are produced.

bicomponent fibre
A manufactured fibre having two distinct polymer components. Both components are themselves usually fibre forming.
Note 1: Some of the more common bicomponent fibre cross-sections are described and illustrated below.
Note 2: In most bicomponent fibres, the two components adhere firmly to each other, though one may be removed later. However, in some types, the components are deliberately chosen to adhere poorly so that they separate in subsequent processing.
Note 3: A variety of complex conjugate and islands-in-the-sea bicomponent cross-sections have been used in the production of ultra-fine **microfibres** (see diagram below).
Note 4: Wool and some other animal fibres have a bicomponent structure with a side-by-side configuration of the **ortho-cortex** and **para-cortex**, which is associated with natural crimp.

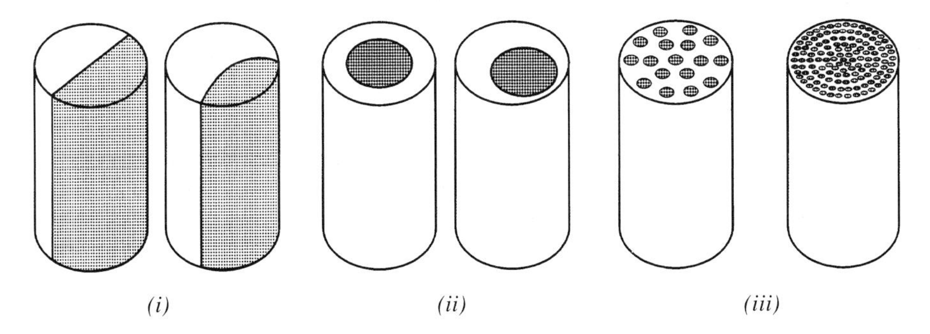

(i) Conjugate (bicomponent) fibres, (ii) Core-sheath bicomponent fibres, (iii) Islands-in-the-sea bicomponent fibre

bicomponent fibre (continued)

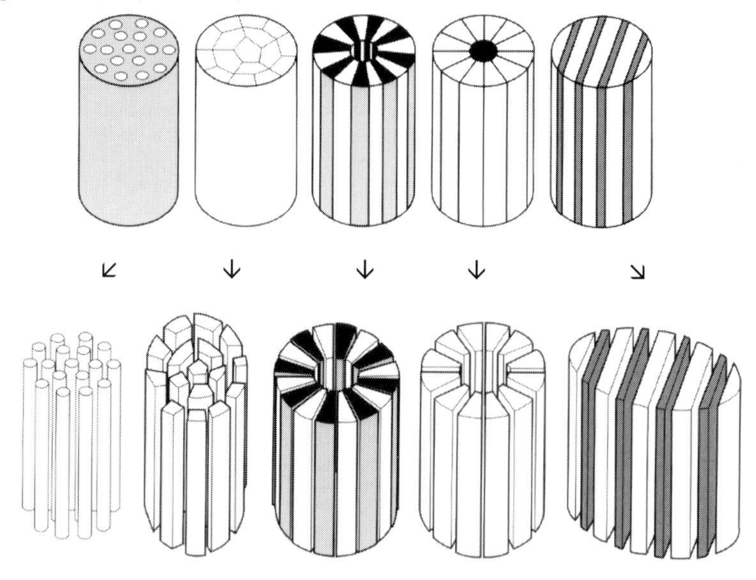

Examples of conjugate and islands-in-the-sea bicomponent fibres used in the production of ultra-fine microfibres

conjugate fibre; side-by-side bicomponent fibre
This configuration is mainly used to generate a helical crimp in the fibre through the differential shrinkage of the components (see diagram above). (See also **textured yarn**, *Note 1* (ix).)

core-sheath bicomponent fibre
This configuration may be used in the symmetrical or asymmetrical form to combine the properties of the two components, perhaps taking advantage of particular physical properties such as the strength or conductivity of the inner component and the aesthetic, textile, adhesive, or other properties of the outer component (see diagram above). The asymmetrical form may also be used for the generation of crimp.

islands-in-the-sea bicomponent fibre
This configuration has similar uses to the core-sheath type or may be used as an intermediate in the production of other specialized types of fibre such as **microfibres** or porous fibres where one component is dissolved away leaving the other (see diagram above).

matrix-fibril bicomponent fibre; fibril-matrix bicomponent fibre
In this type, fine **fibrils** of one component are embedded in a matrix of the other. The individual fibrils are randomly distributed over the fibre cross-section, are of varying but very restricted length and do not extend along the full length of the fibre. As with the core-sheath bicomponent fibre, this configuration is used to combine the properties of the two components. Matrix-fibril bicomponent fibres are sometimes known as 'biconstituent fibres' but the use of this term for this type of fibre is deprecated.

bicomponent yarn
A yarn having two different staple-fibre and/or continuous-filament components.
Note: Although normal usage excludes staple-fibre blends, it includes filament blends. In

addition the term is sometimes used to describe yarns made from **bicomponent fibres** for which the correct term is **bicomponent-fibre yarns**. The following may be regarded as bicomponent yarns:
(i) two-fold staple yarns, when the singles components are different;
(ii) two-fold continuous-filament yarns, when the singles components are different;
(iii) yarns in which a filament yarn is doubled or folded with a staple-fibre yarn;
(iv) **core-spun yarns**, wrapped yarns, or other core-sheath configurations;
(v) **filament blend yarns**.

bicomponent-fibre yarn
A yarn made from **bicomponent fibres**. (See also **bicomponent yarn**.)

biconical package
A conical package in which the traverse length is progressively reduced to produce tapered or rounded ends; such packages are referred to as 'tapered' or **pineapple cones**.

biconstituent fibre
See **matrix-fibril bicomponent fibre** under **bicomponent fibre**.

biconstituent yarn
See **filament blend yarn**

bight
1. The distance between the **seam line** and the adjacent edge of the material.
2. When sewing with a **swing needle machine**, the distance from the right hand line of needle penetrations to the left hand line of needle penetrations. (See also **seam allowance**.)

billiards fabric
A fine, heavily milled and dressed woollen fabric, used for covering billiard tables.

binder
1. Any thread, continuous filament or staple, added at the final stage of the construction of a yarn with the primary aim of giving cohesion to the assembly and for binding in special effects when these have been employed.
2. See **cut-off**.

binder; bonding agent (nonwoven)
An adhesive material used to hold together the fibres in a nonwoven structure. (See also **adhesive-bonded nonwoven fabric**.)

binding
A **narrow fabric**, woven, braided or knitted, used to protect, support or improve the appearance of a seam or edge.

glacé binding
A woven **narrow fabric** made from polished cotton warp and weft yarns. It is used principally in the men's tailoring trade, and usually in 3/1 twill weave, and 12.5mm wide. The stiffness imparted during weaving is sometimes enhanced by subsequent finishing.

simili binding; felling simili
A **binding** made from mercerized cotton yarns throughout, in a 3/1 broken-twill weave in imitation of satin, distinguished by well-pronounced selvedges raised on the face.

stay binding
A woven **narrow fabric** with cotton warp and weft, in 2/2 twill weave or derivatives, generally used for covering seams and strengthening garments.

binding point (weaving)
See **stitch (weaving)**.

binding thread; stitching thread
Additional warp or weft threads used in a fabric to join two or more layers of the fabric together or to fasten long floats of yarn to the body of the fabric.

biopolishing
Treatment of cellulosic fibre substrates, e.g., fabric or garments, with a cellulase enzyme under acid or neutral pH conditions to remove surface hairiness, decrease fabric weight and impart enhanced fabric properties. Biopolishing is an example of a subtractive finish.

birdseye
A fabric having a pattern of very small and uniform spots, the result of the combination of weave and colour.

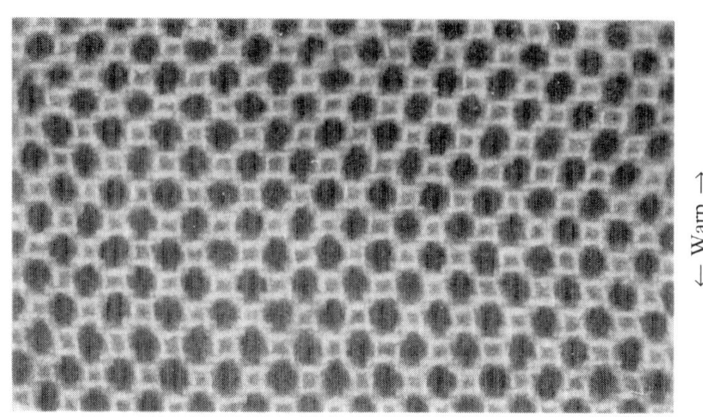

Birdseye (actual size)

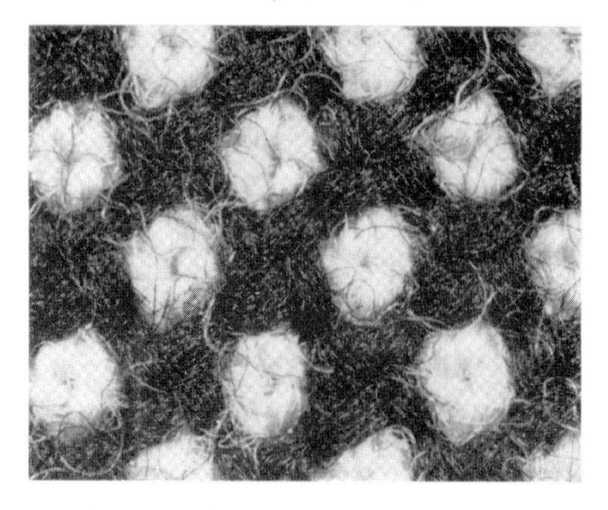

Birdseye (worsted-suiting type) (magnification x 5)

☐ Colour A on Face

▩ Colour B on Face

Birdseye: colour-and-weave effect

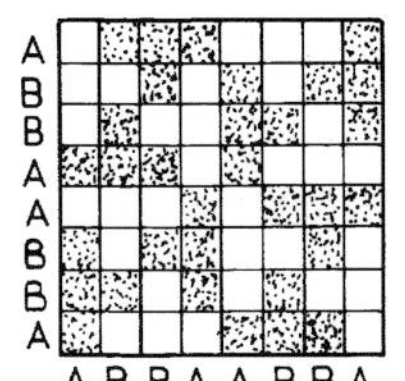

Birdseye: weave and colouring

birdseye backing (weft knitting)
The reverse side of a **rib jacquard** fabric characterized by courses in which knitted and float loops of one colour alternate with knitted and float loops of another, within and between successive courses.

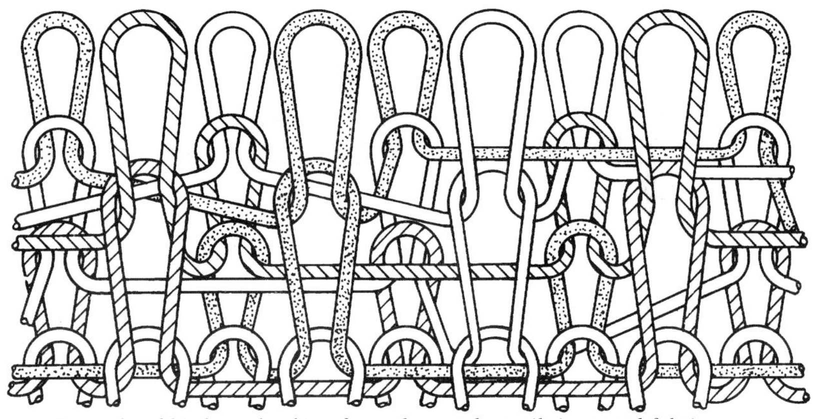

Example of birdseye backing for a three-colour rib jacquard fabric

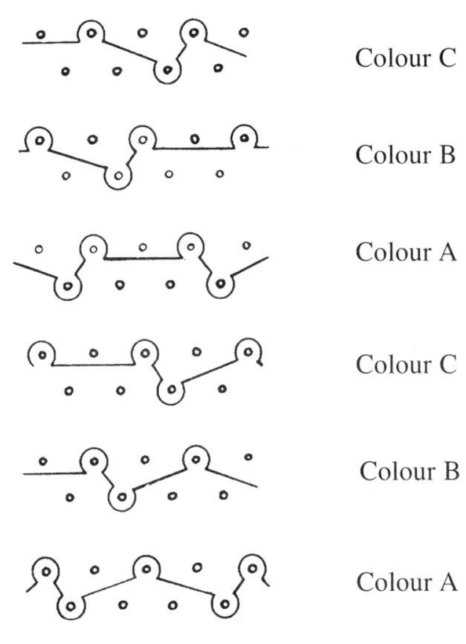

Birdseye backing

birefringence
The difference between the refractive index of a fibre measured parallel to its axis $n_{\parallel}$ and that measured perpendicularly to its axis $n_{\perp}$:

$$\Delta n = n_{\parallel} - n_{\perp}.$$

Note: Birefringence is frequently used as a measure of the degree of orientation of the macromolecules within a fibre.

biscuit; biscuit package
One of several narrow cylindrical cheeses of yarn wound on a single former side by side but not touching. Biscuit packages are used as the take-up in some synthetic-fibre extrusion systems.

bisu; husks
Silk waste remaining on cocoons at the bottom of the basin after reeling.

bivoltine silkworm
A variety of silkworm, which produces two generations per year.

blank needle (knitting)
See **dummy needle**.

blanket
A thick fabric that has good thermal insulation properties. It may be produced as a woven or a knitted fabric, or as a **needlefelt**. The woven structure can be a plain, twill or a multi-layer fabric. (See also **pattern blanket**.)

blanket cloth
A heavy overcoating with a soft raised finish.

blanket mark
A crimped, wavy, or pebbled mark embossed on a fabric by the blanket that covers the heated cylinder used in certain finishing processes.

blanket range
A sample of fabric woven in sections warp way and/or weft way to show as many designs and colourways as possible.

blanket seam
A printing fault characterized by a light-coloured, widthwise stitching mark on the print caused by the seam in the rubberized blanket. (See also **bump seam (fault)**.)

blaze
See **cocoon strippings**.

blazer cloth
Traditionally an all-wool woven fabric for apparel, in either solid colours or stripes, that may be milled and/or raised. Imitation blazer cloths introduce cotton in the weft. The term may be used loosely for other fabrics for blazers.

bleaching
A process for improving the whiteness of textile material, with or without the removal of natural colouring matter and/or extraneous substances.

bleaching agent
A chemical reagent capable of destroying partly or completely the colouring matter present in textile materials, and leaving them white or considerably lighter in colour.

bleeding
Loss of colorant from a coloured material in contact with a liquor, leading to an obvious coloration of the latter, or of adjacent areas of the same or other materials.

blending (spinning)
A process or processes concerned primarily with the mixing of various lots of fibres to produce a homogeneous mass.
Note: Blending is normally carried out to mix fibres, which may or may not be of similar physical or chemical properties, market values, or colours. Blending is also used to ensure consistency of end product.

blind stitch bed
See under **bed (sewing machine)**.

blinding
A marked and undesirable loss of lustre of fibres, especially acetate, caused by inappropriate wet processing.
Note: This may be caused by the formation, within or on the fibre, of voids or particles that scatter light.

blindstitch
A sewing stitch that is not visible on the face of a fabric or garment.

blister fabric, knitted
A three-dimensional relief effect fabric generally made on a rib basis. (See also **relief fabric, weft-knitted**.)

block copolymer
See **copolymer, block**.

block creeling
The simultaneous replenishment of a group of supply packages, e.g., the replenishment of roving packages on a ring-spinning machine.

blocking
The process of subjecting a **hood** or **body** to the combined action of pressure and steam to shape it into a hat or other shaped articles.

blocking
Unintentional adhesion between contacting surfaces of coated fabrics during storage or use.

bloom
A surface paleness observed when a coloured textile material is viewed **overhand**.

blotch
Any relatively large area of uniform colour in a printed design.

blowing (hat manufacture)
The process of separating loose fur from impurities such as skin by passing it through a current of air to produce material of clean and consistent quality.

blowing (steam)
A process in which steam is blown through a fabric, which is usually wound on a perforated roller.

blowing room; blowroom
The section in a cotton spinning mill where the preparatory processes of opening, cleaning and blending are carried out.

blown finish
A finish obtained by blowing dry steam through the fabric, usually a wool material, which is wrapped with an interleaving cotton fabric on a perforated roller.

bluette
A weft-faced 2/2 twill-weave fabric used for overalls, originally made from blue-dyed yarns, but now more frequently piece-dyed. Typical cotton construction: 17x43; 30x37tex; K=9.3+26.2; 66cm wide.

boarding
A process, involving heating under moist or dry conditions, carried out to confer a desired shape or size on hose or other knitted garments whilst on a former. When carried out before dyeing, the process is known as 'dye-boarding'.
Note: The article in the damp condition after scouring and/or bleaching and/or dyeing is dried on a specially shaped former, either by heating this former internally or by placing it between two steam-heated platens. If the drying is done in a hot chamber, the process is known as 'machine-finishing'. When the boarding is done by pressing between heated platens, it is usually known as 'trimming', 'pressing', or 'press-finishing'. The trimming, pressing or press-finishing operations carried out on dyed goods give the desired shape to the articles, but their prime purpose is to remove the moisture from the article without leaving it creased.

bobbin
A cylindrical or slightly tapered former with or without a flange or flanges, for holding slubbings, rovings or yarns. The term is usually qualified to indicate the purpose for which it is used, e.g., ring bobbin, twisting bobbin, spinning bobbin, condenser bobbin, weft bobbin or bottle bobbin. (See also **pirn**, **cop** and **tube**.)

bobbin (machine sewing)
1. A small tube, commonly with flanges, on which is wound the lower/under thread for a lockstitch sewing machine. The bobbin is held in a **bobbin case** during sewing. The bobbin is also referred to as a 'spool'.
2. A small cylindrical package of sewing thread.

bobbin, brass (lace machines)
Two machined brass discs, riveted at the hub to form a container for binding threads.

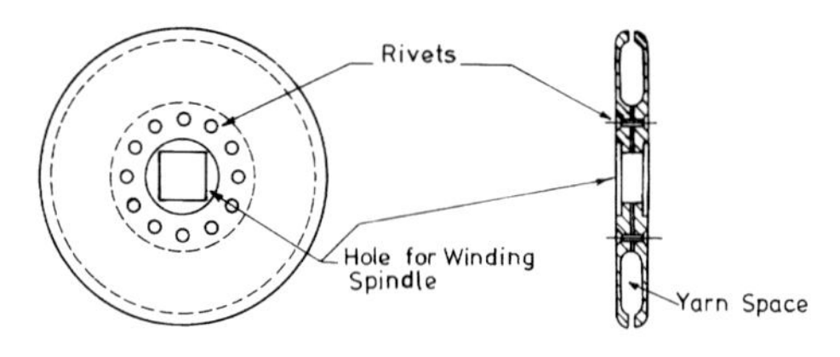

Brass bobbin

Note: The bobbin diameter has varied between 3.75cm and 7.5cm; the bobbin thickness has varied with gauge between 8.0mm and 3.2mm. The size and shape of the hole for the winding spindle vary with different types and sizes of bobbin.

bobbin, king (lace machines)
A cylindrical barrel with flanges at each end, one flange being notched on the outer flat face to engage with the ratchet on the Barmen lace machine, or on the circular braiding machine.

bobbin case (sewing machine)
A housing for the **bobbin**, which is inserted into a lockstitch machine prior to sewing.
Note: The lower/under thread leaves the bobbin case through a tensioner, which provides a low, adjustable tension during stitch formation.

bobbin case hook (sewing machine)
A device on the **bobbin case** which captures the upper/needle thread and carries it round the bobbin to form the stitch.

bobbin finings (lace)
A leavers-lace construction in which threads from the brass bobbin provide the filling-in of the **objects**. Both S and Z warps are necessary, and one thread of each twist is required for every bobbin thread. Although the warp threads traverse, the greater tension of the S-twist threads causes them to remain straight and to pull the bobbin threads sideways. The Z-twist warp threads interlace with the bobbin threads down the centre of the wale. In addition, thick threads from beams may be used, according to the pattern requirements, for outlining the objects.

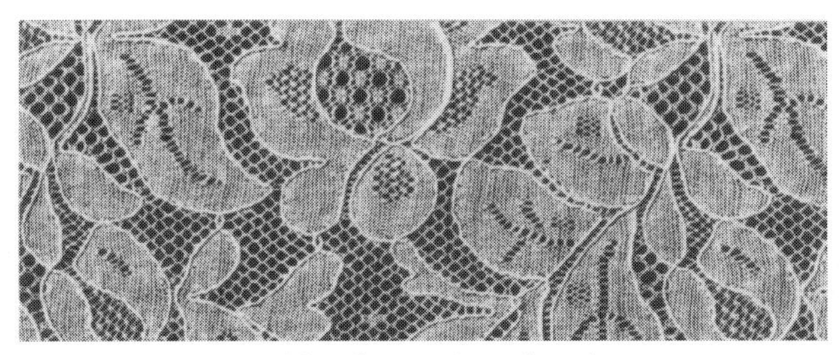

Bobbin finings (actual size)

bobbin lace; pillow lace
Hand-made lace produced in the twisting and crossing of threads that are fed from bobbins and worked into a pattern pricked on parchment or card pinned down to a pillow. As it is worked, the lace is secured in position by the insertion of pins into the pillow.

bobbin net; mosquito net
Originally, examples of **plain net**, but mosquito nets are now commonly made on warp-knitted machinery. (See also **sandfly net, warp-knitted**.)

bobbinet machine
See **plain net machine** under **lace machines**.

bodging-on (knitting)
A process of putting fabric on to points or needles, on a random basis.

bodice
The section of a woman's garment covering the trunk from neck to waist.

body
Synonymous with **hood**.

body carpet
Plain or unbordered patterned carpet in piece form, traditionally 0.69m (3/4yd) or 0.91m (1yd) wide, mainly used for making-up into larger areas by seaming or otherwise joining edge-to-edge.

body rise
See **depth of crotch** under **crotch**.

body tolerance
See **ease**.

boiling off
See **degumming**.

bolduc; advertising tape; printed string; tying tape; weftless tape
A weftless **narrow fabric** consisting of a number of warp threads held together with an adhesive.

boll
A seed-case and its contents on the cotton plant.

bolt (fabric)
See **piece**.

bolting cloth
A light-weight open fabric, characterized by its fine and uniform mesh, used for sifting flour or for screen-printing. Both warp and weft threads are accurately spaced and are woven in simple leno or other non-slip construction in order to maintain the mesh size.

Bolton sheeting
A sheeting fabric of 2/2 twill weave containing a condenser weft.

bombasine; bombazine
A lustrous dressweight twill fabric made with silk warp and fine worsted weft. It is generally piece-dyed black but also dyed to other colours. It was originally an all-silk fabric.

Bombyx mori
The cultivated silkworm which feeds on mulberry leaves.

bond strength test
A test to measure the force necessary to separate bonded layers of fabric or floorcovering. (See also **adhesion test**.)

bonded-fibre fabric
Synonymous with **adhesive-bonded nonwoven fabric**.

bonded-pile carpet
A textile floorcovering with a pile use-surface secured to a substrate by adhesion.

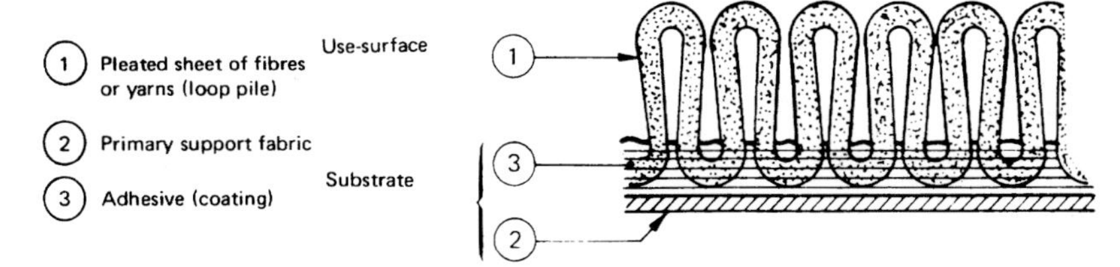

Bonded-pile carpet (loop pile) (longitudinal section)
The above diagram is derived from figure 13 of BS 5557

bonding
A process applied to bind and consolidate component plies in industrial synthetic sewing threads. (See also **bonded-fibre fabric**, **bonded-pile carpet** and **bonding agent**.)

bonding agent (nonwoven)
See **binder**.

book
A parcel of hanks of raw silk, usually having a total mass of 2kg.

book cloth
Fabrics of many qualities, used by bookbinders. They are generally of plain weave, usually coloured, heavily filled, and calendered or embossed between hot rollers.

border tie
See **jacquard tie (weaving)**.

boron (fibre)
An inorganic fibre produced by chemical vapour deposition (CVD) of boron on a fine heated tungsten wire or carbon filament. Boron fibres have potential uses in composite materials. (See also Classification Table, p.401.)

botany wool
A term applied to tops, yarns and fabrics made from **merino** wool. The term originated from Botany Bay in Australia.

bottle bobbin
A large-capacity wooden, metal, or composition bobbin, having a cylindrical barrel and a conical or flanged base, on to which yarn may be wound for withdrawal over the nose. The package when wound has a cylindrical body and conical nose.

bottom bars (lace machines)
Bars made from thin steel strips with circular holes punched in them to act as thread guides. They are narrower than **top bars** and work below the well of the machine. They are actuated by the bottom-bar jacquard and serve to modify the movement of threads controlled by stump bars in the well.

bottom douping
See **leno weaving**, *Note 4*.

bottoming
1. A thorough scouring, in preparation for bleaching, dyeing, or finishing.
2. Dyeing a substrate for subsequent **topping**.

bouclé fabric
A fabric with a clear-cut rough or granulated surface produced by means of fancy yarns and generally used for ladies coats, suits, and dresses.

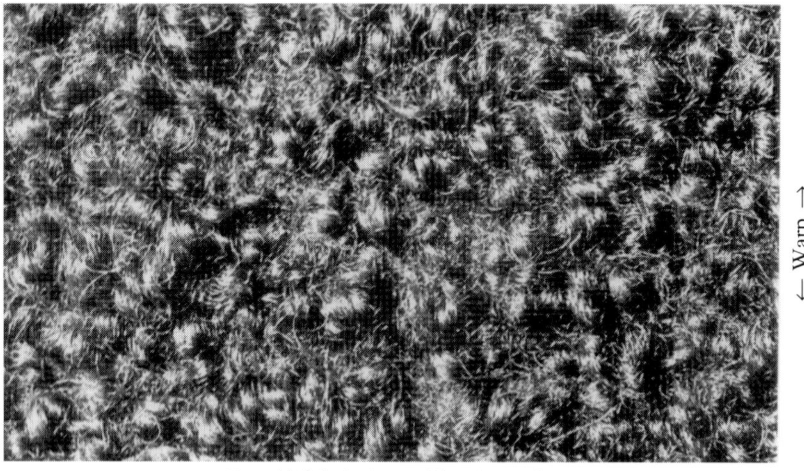

Bouclé fabric (magnification x 2)

bouclé yarn
See under **fancy yarn**.

bouclette fabric, warp-knitted
A three-bar construction made with full-set threading on each guide bar. The front and back bars make tricot lapping movements in opposition, while the middle bar lays-in over two needles in the same direction as the front bar. The middle bar is overfed so that its yarn is formed into an irregular pile.

bound seam
See under **seam type**.

boundary cable; edge cable (technical textiles)
Cable forming an edge to a **fabric field** in an air-supported or tension-membrane structure.

bourdon cord
A **cord** consisting of a central core helically covered by continuous-filament yarn used as a means of accentuating or outlining motifs in lace fabrics and in the manufacture of trimmings. (See also **gimp**.)

bourdonette
A **cord** produced by twisting several yarns together for use as a heavy thread in lace fabrics in order to simulate a **bourdon cord**.

bourette
See **noil (silk)**.

bourrelet (weft knitting)
A non-jacquard double-jersey fabric made on an interlock basis which is characterized by horizontal ridges on the effect side (see **plain fabric, weft-knitted**). The knitting sequence is generally a number of courses of interlock, followed by a number of courses knitted on one set of needles only. (See also **double jersey, weft-knitted**.)

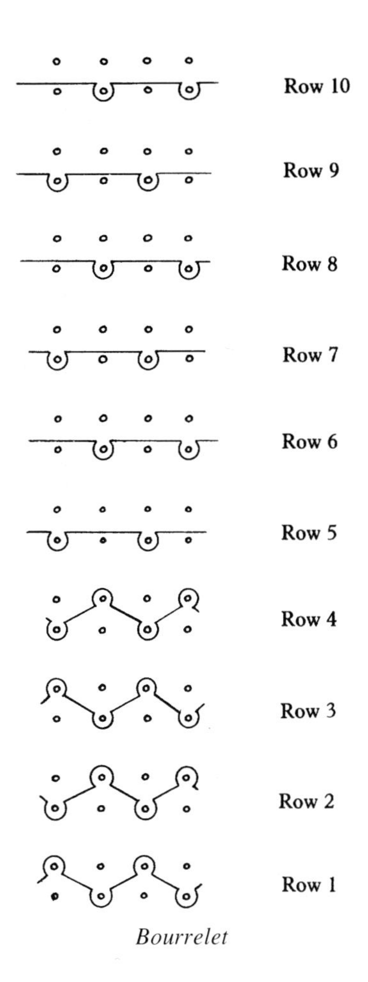

Bourrelet

bow (weaving)
Curvature of the warp or weft in a cloth. The cloth is said to be warp or weft-bowed according to which set of threads is curved.

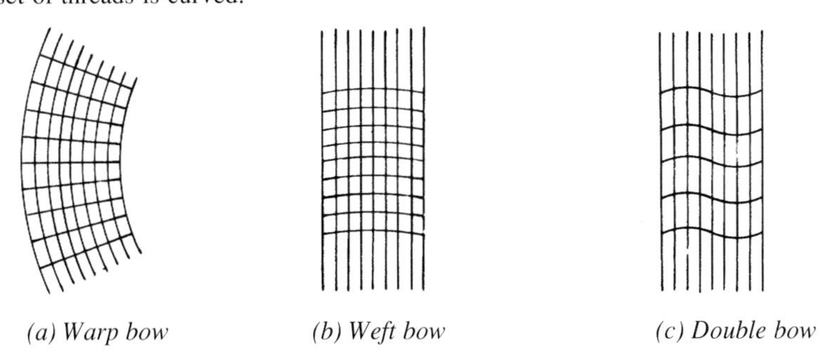

(a) Warp bow *(b) Weft bow* *(c) Double bow*

bowing (knitting)
A fault in a knitted fabric consisting of a curvature of the fabric courses, imposed during knitting or subsequent operations.

bowl

1. One of a pair of large rollers forming a **nip**.
2. (Throwing) A cylinder driving a take-up package by frictional contact.
3. An open vessel for such wet treatments as wool scouring, crabbing, etc.

Note: Equipment for uses 1 and 3 was originally fashioned from the boles (trunks) of trees but, while the term is derived from this word, the spelling is now always bowl.

box cloth

An all-wool, woollen-spun fabric with a fibrous surface and firm handle. The surface should be completely covered with fibres so that no threads show. It is woven in a variety of weaves, depending on the weight required. It is used for such purposes as leggings, coachman cloths and billiards cloth.

box motion

A mechanism available for weaving machines with shuttles that allows the use of more than one supply of weft. The mechanism holds two or more shuttles and either rotates (revolving box motion), or moves vertically (drop box motion), to place the required shuttle in the picking position.

box pleat

See under **pleats**.

box spinning; pot spinning; bucket spinning

Spinning with the use of a **Topham box**. (See also **cake**.)

brace web; suspender web

An elastic or rigid woven **narrow fabric** used for the straps of trousers supports known as braces or suspenders.

braid; plait

The product of **braiding**. Certain types of woven and knitted **narrow fabrics** are described as braids.

Braid

braided rope

See under **rope**.

braider

See **braiding machine**.

braiding; plaiting
The process of interlacing three or more threads in such a way that they cross one another in diagonal formation. Flat, tubular or solid constructions may be formed in this way.
Note: Tubular fabrics made by this process may be constructed with or without core, gut, filler, or stuffing threads, which when present are not interlaced in the fabric.

braiding machine; braider; Maypole braider
A machine provided with three or more **carriers** which are driven by means of horn gears along tracks which cross at intervals enabling the yarn drawn from the carriers to interlace to form a **braid**.

brass bobbin (lace machines)
See **bobbin, brass**.

brattice
A conveyor, formerly made of interlinked wooden slats, but nowadays often of metal, e.g., wire mesh, for the transportation of textile materials through machinery in tensionless state.

brattice cloth
A coarse cloth used for screens, ventilators, etc., especially in mines, in which case it is often coated.

breadth (lace)
1. A narrow lace, e.g., edging, insertion, galloon.
2. The pattern repeat carriage-way on a Leavers machine. It may be equal to, or a sub-division of, the **set-out**.

break (clothing)
The point where the **lapel** starts to roll over on a front-opening garment

break spinning
See **open-end spinning**.

breaker fabric
1. (Crossply tyres) See under **tyre textiles**.
2. (Conveyor belts) A layer of fabric between the main fabric core of the belt and the load-bearing surface of the rubber or PVC cover. It may be extended round the edges and may also be continued across the surface in contact with the pulleys.
Note 1: In special applications where thick covers are employed, two layers of breaker fabric may be used, one layer immediately above the carcase and the other layer disposed above it at a distance of about one-third of the thickness of the cover. This second layer is then termed a 'floating breaker' and extends the width of the belt.
Note 2: The fabric can be of an open leno weave or a tyre-cord fabric (see under **tyre textiles**). In the latter case, the cords may lie at right angles to the length of the belt.

breaking (bast fibres)
The deformation of the plant structure by flattening the stem, loosening the bond between the fibre bundles and the wood, and breaking the woody part into short pieces, to facilitate their removal from the fibre by **scutching**. Breaking by means of rollers is often referred to as 'rolling'.

breaking extension
See **extension at break**.

breaking force
See **tensile strength at break**.

breaking length
The theoretical length of a specimen (usually of yarn) whose mass would exert a force sufficient to break the specimen.

breaking machine; button breaker; stud breaker
A machine for continuously softening stiff fabrics. The fabric is drawn under tension over the edges of bars (knife-edge machine), or round rollers implanted with studs.

breaking point; ultimate rupture
The point on a force-extension curve which coincides with the maximum extension.

breaking strength; tensile strength
The maximum tensile force recorded in extending a test piece to breaking point. (See also **tensile strength at break**.)

breaking stress
The maximum stress developed in a specimen stretched to rupture. The force is usually related to the area of the unstrained specimen. If the actual stress, defined in terms of the area of the strained specimen, is used, then its maximum value is called the actual breaking stress.

breast (lace machines)
The straight edge of the **carriage** above the blade.

bribe
A synonym for a woollen **fent**.

brides (lace)
Connecting bars (or legs) used to join the **objects** in certain styles of lace where there is no net ground.

bright
Descriptive of textile materials, particularly manufactured fibres, the natural lustre of which has not been significantly reduced. (See also **delustrant**.)
Note: The term 'clear' is commonly used to denote the absence of delustrant in synthetic fibres, whereas bright may denote the presence of a very small amount of delustrant, insufficient to reduce the lustre of the fibre significantly.

bright (lace)
Openwork effects within a wale in lace furnishings, obtained without distorting the warp threads.

bright pick; bright yarn (defect)
A tight **pick** usually found in a fabric containing a continuous-filament weft.

Brighton honeycomb
See **honeycomb**.

brilliantine
A dress cloth of cotton warp and lustre worsted weft. It is generally of plain weave, but jacquard designs are sometimes used.

brin
A single filament of silk resulting from the degumming of the **bave** withdrawn from the cocoon.

brise bise
Lace curtaining designed to be hung horizontally across the lower portion of a window, close to the frame, provision usually being made for the insertion of a curtain rod or wire. It is traditionally manufactured in various sizes from about 38cm to 145cm wide, and is sold by length.
Note 1: It may be plain or patterned.
Note 2: Sectional panels of brise bise are designed to the required depth of the window; these can be cut to any number of panels according to the width of the window and, when cut, the edge is fast.

bristle
1. A short, stiff hair, e.g., on the back of a hog or in a man's short or unshaven beard.
2. Any rugged, stiff hair or filament, usually short and straight, and often pointed and/or prickly.
3. The essential filamentous components of brushes and brooms.
Note: For softer brushes, the term 'hairs' is usually preferred.

britch (wool)
Wool from around the tails and lower hindquarters of sheep, often heavily stained.

broad rib fabric, weft-knitted
See under **rib fabric, weft-knitted**.

broadcloth, cotton
A light-weight fabric of poplin type, used extensively on the North American continent for shirtings.

broadcloth, wool
A fabric made from fine woollen yarns in a twill weave, heavily milled (traditionally approximately 230cm in the loom for 140cm finished) and given **dress-face finish**. It is usually dyed in dark colours.

broadloom (carpets)
Descriptive of carpeting traditionally made 1.83m (2 yards) or more in width.

broadtail
See **karakul** 2.

broadtail cloth; caracul cloth; karakul cloth
A pile fabric woven to imitate a broadtail pelt (see **karakul** 2.).

brocade
A figured fabric, usually of single texture, in which the figure is developed by floating the warp threads, the weft threads, or both, and interlaced in a more or less irregular order. The ground is usually formed of a weave of simple character.
Note 1: Many furnishing brocades are made with a satin ground and a weft figure.
Note 2: More elaborate fabrics are also made with more than one warp and/or weft.

brocade *(continued)*

Brocade (actual size)

brocatelle
A heavily figured cloth used for furnishing purposes in which the pattern is brought into relief by the warp threads in a satin weave against a closely woven background texture. Two or more wefts are used and, in the better qualities, there is an extra binder warp.

Brocatelle (actual size)

broché
A brocade fabric that is figured by additional threads introduced by means of swivel shuttles. (See also **swivel weaving**.)

broken crow
See **crow twill**.

broken filaments
Individual filaments of a yarn inadvertently ruptured, usually through mechanical damage.
Note: A fibrous or hairy appearance due to broken filaments on the surface of a yarn package or fabric is often known as 'filamentation'.

broken pick
A pick that is inserted for only part of the fabric width.

broken twill
Any twill weave in which the **move number** is not constant, with the result that the continuity of the twill line is broken.

bronzing
A metallic sheen which can occur on a textile material or surface coating, usually when deeply coloured. (See also **oxidized oil staining**.)
Note: In textile coloration the effect is usually caused by excessive concentration of colorant at the surface of the substrate. This is normally undesirable, but is sometimes produced deliberately, for example, with some indigo dyeings.

brown lace
Lace in the condition in which it leaves the machine, before any bleaching, dyeing, or finishing treatment has been carried out. Known colloquially as 'in the brown'. (See also **grey**.)

bruised place
See **chafe mark**.

brushed fabric
See **raising**.

brushed fabric, warp-knitted
A fabric commonly produced from continuous-filament yarns in which the long underlaps of certain guide bars are raised (see **raising**) during finishing to form a pile consisting of broken filaments.

brushed yarn
A yarn which has been brushed to raise surface hairs. The brushing, which normally takes place by abrading the surface of the yarn with card clothing, is usually performed on bouclé, loop and other **fancy yarns**. Brushing may be used to achieve greater bulk and softer handle.

brushing
A finishing process in which the fabric is passed over one or more revolving brushes.

Brussels carpet
A loop-pile carpet, woven on a Wilton loom, over unbladed wires.

bucket spinning
See **box spinning**.

buckram
A stiff fabric, generally linen or cotton, made by impregnating a plain-weave open-sett fabric with fillers and stiffeners. Alternatively, fabric consisting of two stiffened fabrics bonded together, the fabrics being not necessarily of identical construction e.g., a low-sett coarse-count fabric might be bonded to a low-sett fine-count fabric. Its uses include linings, millinery, waistbands and bookbinding.
Note: In the 12th to 16th centuries the term had an entirely different meaning: it was applied to fine, delicate fabrics made of linen or cotton used for apparel.

buckskin fabric
A fabric similar in handle and appearance to, but heavier than, a **doeskin fabric**, made from fine merino wool, closely sett, heavily milled, dressed, and closely cut. Typical weaves are as shown.

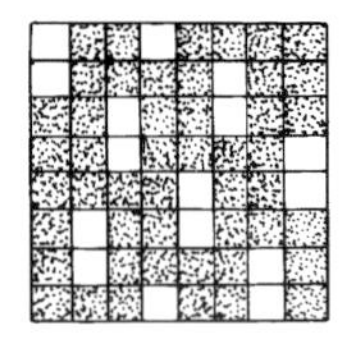
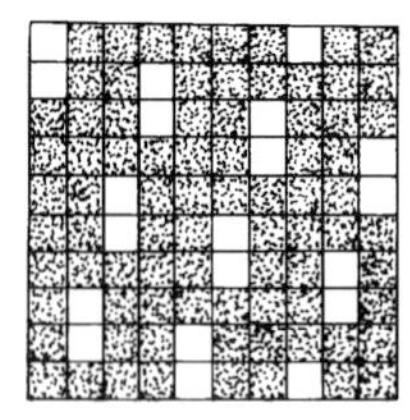
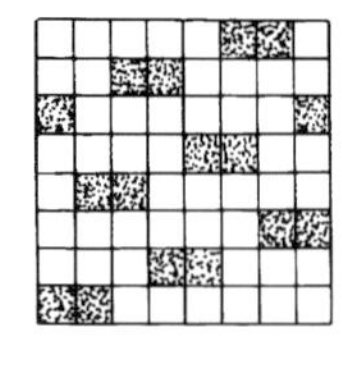
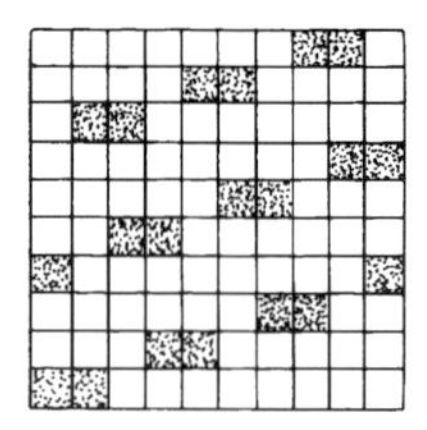

Buckskin: typical weaves.

Buenos Aires wool
See **BA wool**.

builder
A compound added to a detergent to increase detergency.

builder fabric
A square-woven heavy cotton **duck** made from very heavy ply yarns; it was formerly used in the carcase of rubber-tread tyres, but has now been largely replaced by tyre-cord fabric (see under **tyre textiles**).

bulge ratio
See **swell ratio**.

bulked continuous filament
See **BCF**.

bulked yarn
A yarn that has been treated mechanically, physically or chemically so as to have a noticeably greater voluminosity or bulk.
Note 1: (Staple yarns) The increased bulk may be obtained by the use of **bicomponent fibre** and/ or by blending together, during yarn spinning, fibres of high and low potential shrinkage, e.g., high-bulk acrylic yarns. During subsequent hot and/or wet processing, the greater contraction of the high-shrinkage fibres causes the yarn to contract longitudinally and the low-shrinkage fibres to buckle thus increasing the bulkiness of the yarn.
Note 2: (Continuous-filament yarns) The increased bulk may be obtained by one of the texturing processes; **textured yarn** is now the preferred term for yarns of this type.

bulky yarn
A yarn that has a greatly enhanced voluminosity in comparison with most yarns of a similar fibre type and linear density.

bullion cord; twine
A highly twisted assembly of yarns which may be spirally covered with continuous-filament yarns, used in the manufacture of **bullion fringe**.

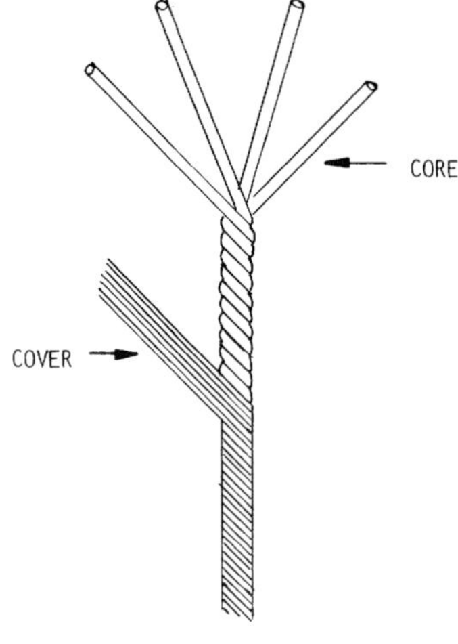

Bullion Cord

bullion fringe
A **fringe**, the weft of which consists of **bullion cord** or **twine** which is inserted double to form a loop, the sides of which twist together to form the tails or tegs which become the skirt.

bump cop
A solid yarn weft package used for very coarse weft yarns (150 tex and coarser) in shuttle loom weaving. It is forced into the hollow of the shuttle where it is held by grooves and ridges on the inside of the shuttle walls and base. The yarn is withdrawn from the centre of the package.

bump grey; bumps
Fabric bought to be used specifically as a **back grey**.

bump seam (fault)
A light-coloured width-wise printing fault caused by a stitching in the **back grey**.

bump top; bumped top
A package made by press-packing layers of coiled sliver. This may also be achieved by vacuum packing.

bumper (hat manufacture)
A machine used for acid milling in the production of **hoods** or **bodies**.

bumping (hat manufacture)
See **planking**.

bumps
See **bump grey**.

bunch (flax)
The aggregate of pieces (see **piece (flax)**), which are tied up with two or more ties preparatory to baling.

bunch, fabric
See **swatch**.

bunch, yarn
A length of yarn initially wound with a restricted traverse on to the base of a pirn. It provides a supply of a weft yarn, in an automatic shuttle weaving machine, from the time the need for weft replenishment is detected to the time when the replenishment is carried out.

Bundesmann test
See **water repellency**.

bunting
Fabric made for the manufacture of flags, originally a special type of wool fabric, of plain weave, but now tending to be a nylon-wool blend, produced from yarns spun on the worsted system.

burl
1. A wool trade term for an imperfection.
2. A small knot or lump (see **nep** and **slub**) in a fabric.
3. Small cellulosic or synthetic fibre impurities in a fabric (see **burl dying**).
4. To remove an imperfection.

burl dyeing
The coloration, at a low temperature, of (i) cellulosic impurities in dyed wool fabrics, or (ii) cotton warps in union fabrics in which the wool is dyed.

burlap
See **hessian**.

burling
A hand process associated with **mending**, which involves taking out from the back of the fabric all slubs and minor imperfections; drawing out thick threads of warp and weft; opening knots; and pushing irregularities such as loose ends and curls to the back of the fabric.

burn-out style; burnt-out style; devoré style
Production of a pattern on a fabric by printing with a substance that destroys one or more of the fibre types present.

burry wool
Wool contaminated with vegetable impurities adhering to the fleece.

bursting (knitting)
A hole in a fabric caused by the breakage of a yarn during knitting. (See also **cutting (knitting)**.)

bursting test
A test in which a fabric specimen, held in place by a coaxial ring clamp of given diameter, is distended to rupture either by the distension of an elastic diaphragm when subjected to fluid pressure or by the action of a ball moving through the ring clamp.

bursting distension
The distension of a specimen at the bursting strength. It is the height immediately prior to rupture of the centre of the upper surface of the specimen above the plane in which the clamped upper surface of the specimen lies.

bursting strength
The multi-directional resistance to rupture of a circular fabric specimen.

butt
To level the root ends of flax straw at any stage by vibrating it upright on a flat surface, either by hand or mechanically.

butt (weaving)
The thicker base end of a pirn used in the automatic bobbin change mechanism. It is the part of the pirn that is gripped by the jaws in the shuttle.

butt seam
See under **seam type**.

button
A knob or disc which can be attached to the garment as a means of fastening or ornamentation.

button breaker
See **breaking machine**.

button stand; buttonhole stand
The distance from the finished edge of a garment to the centre of the button or buttonhole.

buttonhole
The slit through which a button is passed to close and hold parts of a garment or other made-up article in the correct position.
Note 1: Uncut buttonholes, with or without buttons, may be used for ornamentation.
Note 2: The edges of buttonholes are usually hemmed or overlocked.

buttoning; beading
The formation of balls of fibre on warp yarns during weaving.
Note: Satisfactory shedding of the warp may be prevented by these buttons, which may also result in stitching (see **stitch (defect)**) or end breakages or both.

cable
To twist together two or more folded yarns. (See also **folded yarn** and **cabled yarn**.)

cable cord
A cord consisting of three **case cords** that have been over-twisted and are then twisted together in the reverse direction. The smaller sizes are often referred to as 'lacing cords' and the larger sizes, traditionally over 6mm diameter, as 'French crêpe cord'.

Cable cord

cable laid rope
See under **rope**.

cable stitch
Two or more groups of adjacent wales that pass under and over one another to give a cabled effect.

Cable stitch

cabled yarn
Two or more folded yarns twisted together in one or more operations.
Note 1: Combinations of folded yarn(s) and single yarn(s) may be described as cabled yarns, e.g., a single yarn twisted together with two folded yarns to give softness to the resulting yarn.
Note 2: In the tyre-yarn and tyre-cord sections of the textile industry, cabled yarns are termed cabled cords or cords. These terms include two-fold, continuous-filament, manufactured-fibre yarns, a traditional example being: 1650 denier (180 tex) rayon cord, singles twist 12 turns/inch (Z) and cable twist 12 turns/inch (S), (470 turns/m).

cake
The package, roughly cylindrical in shape, of continuous-filament yarn produced in the viscose spinning industry by means of a **Topham box**. (See also **mock cake**.)

cake sizing
The application of **size** to yarn wound in the form of a **cake**.

calender
1. A machine in which heavy rollers (bowls) rotate in contact under mechanical or hydraulic pressure. The bowls may be unheated, or one may be a thick-walled steel shell heated internally. All bowls may rotate at the same surface speed, or one highly polished and heated bowl may rotate at a higher surface speed than the rest. In certain specialized machines, e.g., those for knitted goods, two adjacent bowls may be heated, or, in the case of a laundry calender, one roller may work against a steam chest shaped to the curvature of the roller. (See also **friction calendering**.)
2. To pass fabric through a machine as in 1 above, normally to smooth and flatten it, to close the intersections between the yarns, or to confer surface glaze. Special calenders with an engraved heated bowl imprint a pattern in relief (see **emboss**) or modify the fabric surface to give high lustre (see **Schreiner**).
3. In coating fabric with rubber or plastics, to use such a machine with the bowls a definite

distance apart, so that the rubber or plastics mass is attenuated to a thin uniform sheet, which is then pressed into a firm adhesion with one side of the fabric passing through. Sometimes this operation is referred to as **calender spreading**.

calender bonding
A method of making a **thermally-bonded nonwoven fabric** in which calender rollers are used to apply heat and pressure to a fibre **web** or **batt**, thus causing bonding by the softening or melting of the heat-sensitive material. Embossed calender rollers are in common use for **point bonding**.

calender spreading
See **calender** 3.

calico
A generic term for plain cotton fabric heavier than muslin.

cam shedding
See **tappet shedding**.

cambric
A light-weight, closely woven plain fabric usually given a slight stiffening. Constructions of typical fabrics were: (i) handkerchief cambric (loomstate): 35x35; 10x7.4tex Egyptian cotton; 3x7.8%; 74 g/m^2; K=10.0+9.5; 96cm; (ii) linen cambric (bleached): 38x35; 11x11tex linen; 4.1x3.1%; 83 g/m^2; K=12.6+11.6; 94cm.

camel fibre (hair)
Fibre from the fleece of the camel (*Camelus bactrianus*) or dromedary. This comprises strong, coarse, outer hair and soft, fine undercoat, both of which are used in the manufacture of textile products.

camelot
An imitation **camlet** fabric produced in cotton and wool.

camlet
A fine, lustrous, plain-weave fabric made of silk, hair, or wool fibres and in a variety of qualities for suitings and furnishings.

Campbell twill; Mayo twill
A weave used extensively in the finer woollen and worsted trade.

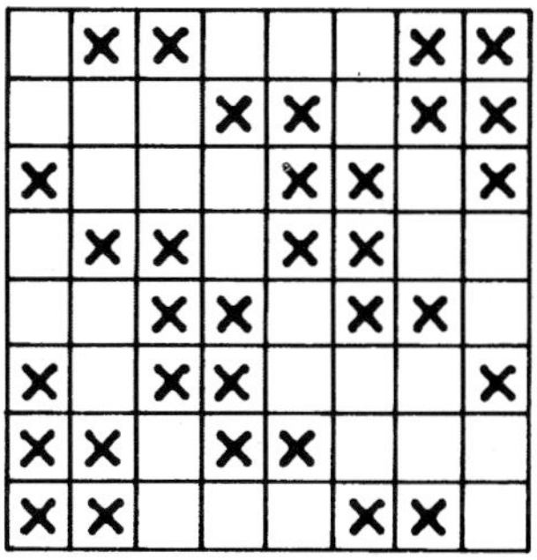

<table>
<tr><td></td><td>X</td><td>X</td><td></td><td></td><td></td><td>X</td><td>X</td></tr>
<tr><td></td><td></td><td></td><td>X</td><td>X</td><td></td><td>X</td><td>X</td></tr>
<tr><td>X</td><td></td><td></td><td></td><td>X</td><td>X</td><td></td><td>X</td></tr>
<tr><td></td><td>X</td><td>X</td><td></td><td>X</td><td>X</td><td></td><td></td></tr>
<tr><td></td><td></td><td>X</td><td>X</td><td></td><td>X</td><td>X</td><td></td></tr>
<tr><td>X</td><td></td><td>X</td><td>X</td><td></td><td></td><td></td><td>X</td></tr>
<tr><td>X</td><td>X</td><td></td><td>X</td><td>X</td><td></td><td></td><td></td></tr>
<tr><td>X</td><td>X</td><td></td><td></td><td></td><td>X</td><td>X</td><td></td></tr>
</table>

Campbell twill

can, drying
See **drying cylinder**.

canary stain
A yellow stain in wool that is not removed in scouring. This is probably caused by decomposition of cystine in the wool under certain conditions.

candlewick
See **wick**.

candlewick yarn
A coarse folded yarn usually of 100% cotton for use as pile in candlewick products.

cannage; tear drop; teariness
Local difference in light reflection caused by variations in curvature of warp crimp. The fault occurs in plain-weave fabrics made with a continuous-filament warp and may arise if the warp is too stiffly sized or if the warp tension during weaving is too low or varies.

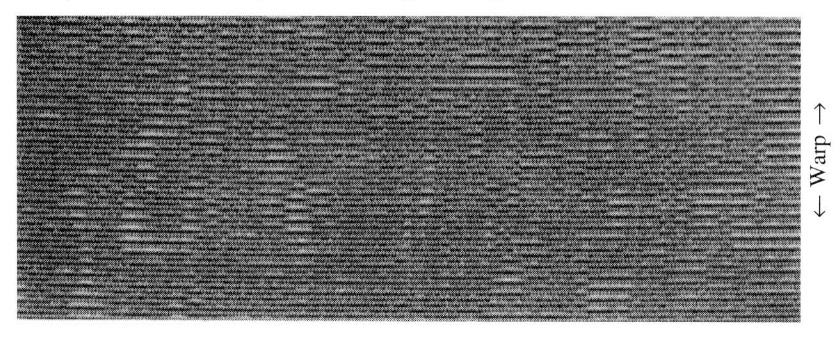

Cannage (actual size)

cannelle repp
A silk **repp** fabric made with two warps. A single warp forms the ground weave while a two-fold yarn floats over eight picks to create a rib effect.

cannetille
A warp-rib cotton fabric used for drapes and furnishings. It is woven with alternate ends under very low and very high tensions.

cantala
A fibre obtained from the leaf of the cantala plant *Agave cantala*.

canvas
A fabric usually made from cotton, flax, hemp, or jute in weights traditionally ranging from 200 to 2000 g/m^2.The weave is plain or double-end plain. In cotton canvas, the yarns may be singles but are frequently plied; in other canvases, the yarns are generally singles. The warp predominates, and a feature of the heavier canvases is the very close packing of the warp, which is highly crimped over a straight weft. The term canvas covers cloths with a great variety of uses, but the salient features of all are strength and firmness. (See also **duck**.)

cap spinning
A spinning system in which the spindle supports a stationary cap, the lower edge of which guides the yarn on to the revolving spinning package which is traversed. (See also **spinning**.)

Cape wool
South African wool (merino type).

capstan
1. A stationary guide, pin or post around which a running yarn, thread, string, rope etc. may be wrapped in order to change their tension.
2. A rotatable roller, usually driven and sometimes heated, around which a yarn etc. is wrapped more than once. There is usually a small guide or guide roller alongside (which may be set at a slight angle) to separate individual wraps. If driven, the capstan controls the rate of delivery of yarn. (See also **godet** and **duo**.)

caracul
See **karakul** 2.

caracul cloth
See **broadtail cloth**.

carbine needle (machine knitting)
See under **needle (machine knitting)**.

carbon (fibre) (generic name)
A manufactured fibre containing at least 90% of carbon obtained by controlled pyrolysis of appropriate fibres. (See also Classification Table, p.401.)

carbonising
Chemical degradation of cellulosic matter to a friable condition in order to facilitate its elimination from admixture with other fibres.
Note: The process involves treatment with acid, as by the use of hydrochloric acid gas (dry process) or sulphuric acid (wet process), followed by heating.

carbonised rag fibre
Animal fibre recovered by either the wet or the dry **carbonising** process.

carcase (lace machines)
The frame and heavy engineering parts of the machine. The term includes such moving parts as the catch bars, driving shafts, cam shafts, etc.

card
A machine used for **carding**.

card clothing
Material comprising a base structure and wires, pins, or spikes protruding from one face.

card clothing, base wire
A rectangular-section metal strip with castellations along one edge that is mounted on card cylinders when converting from flexible fillet clothing to metallic or rigid wire.

card clothing, flexible (fillet)
A toothed structure comprising a foundation of laminated fabric and rubber, with wire teeth, pins or spikes protruding from one face. It is used in carding and raising processes.

card clothing, flexible (top)
A toothed structure, like fillet, but designed to fit the iron flats of a revolving **flat card**.

card clothing, metallic (rigid)
An L-shaped steel strip with hardened teeth punched along the upper edge, applicable

card clothing *(continued)*

to all types of carding cylinders, fixed carding flats or segments, and the opening rollers of open-end spinning machines. Individual pins mounted in the roller, covering the whole surface, form a uniform carding surface.

card cutting

The process of punching holes in jacquard or dobby cards according to a specified design.

card cylinder

See **cylinder**.

card flats; carding plates

A series of narrow plates, with one of their surfaces covered in pins, which are used to perform the carding action jointly with the card **cylinder** of predominantly a short-staple card (see **carding**).

Note 1: The flats are fitted closely to the card cylinder and may be either fixed or slow moving over the top of the card; collectively they cover approximately a third of the card cylinder's peripheral surface.

Note 2: The pins of the flats face the saw-tooth wire clothing of the card cylinder and are angled in the opposing direction.

Note 3: Fixed flats or carding plates may also be used on nonwoven cards.

card loom weaving

The simplest form of weaving used as an introduction to the craft in schools. The warp yarns are held taut across a piece of cardboard, while the weft yarns are inserted by means of a bodkin.

card web

See **web** 1.

carded yarn

A yarn produced from fibres that have been carded but not combed.

carding

The disentanglement, cleaning and intermixing of fibres to produce a continuous **web** or **sliver** suitable for subsequent processing. This is achieved by passing the fibres between relatively moving surfaces covered with **card clothing**.

Note: The process follows **opening** and **blending**. Fibres may be fed to the card by a chute-feed, a weigh pan or in the form of a lap and are subjected to a **draft** of 50-150. The card produces a web which may be used in the manufacture of nonwoven fabrics, slubbings or, more commonly, slivers which may be subjected to further processing prior to spinning or which may be directly spun.

short-staple carding

The production of a web or sliver on a card, which has slow moving and/or stationary **flats**, and is usually used for processing cotton.

Note: When flats are slow moving over the top of a card, collectively, they are usually referred to as revolving flats and the card is termed a 'revolving flat card'.

woollen (condenser) carding

The production of **slubbings** from blended tufts of fibres, suitable for spinning. The tufts are carded to produce an even **web** of blended loose fibres which is divided into ribbons of uniform width and rubbed to form slubbings by the rubbing aprons of the card condenser. Roller and clear cards are used.

worsted carding
The production, from scoured loose wool, of a continuous sliver suitable for subsequent **gilling** and **combing**. Roller and clear cards are used

caroa; craua; croa; coroa
A fibre from the leaf of the plant *Neoglazovia variegata.*

carpet
A textile floorcovering having a textile use-surface formed from yarns or fibres projecting from a substrate.

carpet backing
See **substrate (carpet)**.

carpet square
Traditionally, a carpet in rectangular form (with or without a border) approximately at least 1.83m (2yds) at the shortest dimension and normally loosely laid.
Note: In North America carpet square is synonymous with **carpet tile**.

carpet tile
A textile floorcovering of predetermined shape and size intended to be used in a modular mode. To achieve dimensional stability, the tile incorporates a substrate comprising layers of, for example, glass fibre and bitumen, or PVC. (See also **carpet square**.)

carriage (lace machines)
A thin metal frame for carrying a **brass bobbin.**

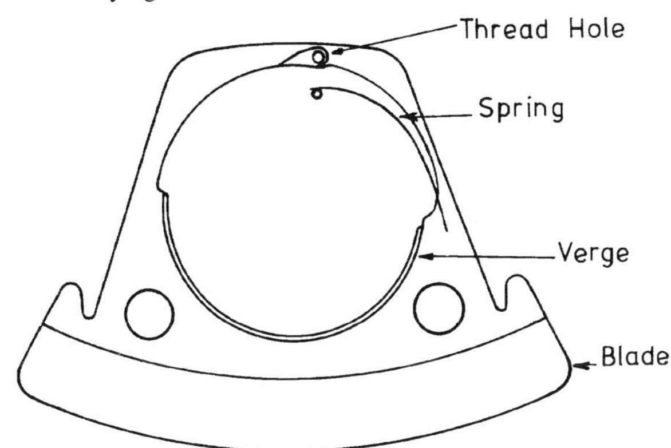

Carriage (lace machines)

Note: This carriage is typical of one used in lace machines. There are variations in size, shape, and thickness, but the main features noted above are always found.

carriage lace; coach lace (narrow fabric)
A woven **narrow fabric** on the face of which is an uncut pile design, generally incorporating one weft, but two or three wefts are sometimes used for further ornamental effects. It is used to give finish to the upholstery of railway carriages and of other vehicles.

carriage way (lace)
The direction across the width of the lace, at right angles to the dressing selvedges (see **pattern repeat**).

carrier

1. (Braiding) A moving holder for one yarn package (occasionally several), which moves in a track on a braiding machine.
2. (Coloration) A type of accelerant, particularly used in the dyeing or printing of hydrophobic fibres with disperse dyes.
3. (Fibre) A fibre component that is blended with the main constituent fibre to improve processing behaviour.
4. (Spinning) A positively driven, smooth metal roller set between the major drafting rollers on some worsted drawing boxes and spinning frames to control the fibres during drafting.

Note 1: This is used in conjunction with a **tumbler**.
Note 2: One, two, or three lines of carriers may be employed, depending on the fibre length of the material being processed.

5. (Yarn) A yarn introduced at some stage of processing to support the main component, generally as an aid to further manufacture.

carrot
The active reagent used for **carrotting**.

carrotting
The modification of the tips of fur fibre (rabbit fur) by chemical treatment to improve their felting capacity. The reagent originally used was mercury in nitric acid but mixtures of oxidizing and hydrolysing agents are now more common.

carved pile
See under **pile (carpet)**.

case cord
A soft and pliable **cord** consisting of two or more components twisted together, each component being made of a **core** which is helically covered at an acute angle by multiple ends of continuous-filament yarns.

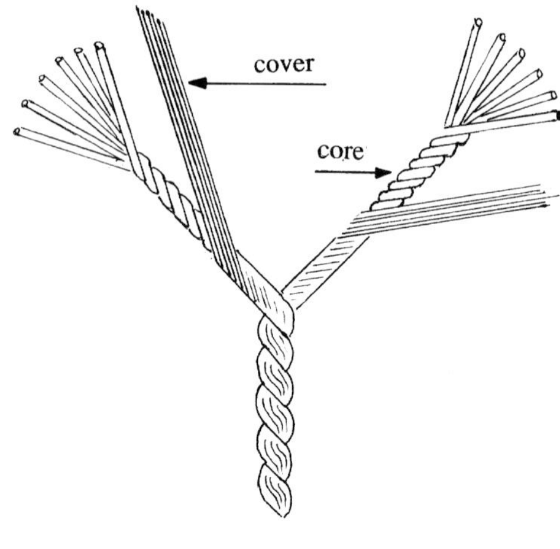

Case cord

casein
The principal protein in milk.
Note: It has served as the raw material for some regenerated protein fibres.

casement cloth
A light-weight to medium-weight weft-faced curtain fabric of cotton or manufactured fibre yarns, typically: 21x25; 18.5x37tex; 4x17%; 150 g/m^2; K=9.1.0+18.5.(See also **limbric**.)

cashgora fibre (hair)
Hair from the undercoat of the cashgora goat, originally produced by crossing a male Angora goat (*Capra hircus aegagrus*) with a female Cashmere goat (*Capra hircus laniger*).
Note: The fibre is of low to medium lustre, generally finer than 23 microns (µm) in diameter.

cashmere fibre (hair)
Originally hair from the downy undercoat of the Asiatic goat (*Capra hircus laniger*) with a mean diameter of 18.5 microns (µm) or less. Similar hair from animals bred selectively from the feral goat populations of Australia, New Zealand and Scotland, is regarded as cashmere provided the fibre diameter is similar.

casing
See under **tyre textiles**.

cassimer; cassimere
A closely woven 2/2 twill fabric with worsted warp and woollen weft, and having a closely cut, smooth face. This term is sometimes applied to any woollen fabric that does not have any other distinguishing name.

casting out
A term used in weaving to indicate that some hooks and/or **healds** of a jacquard machine, or some healds on certain heald shafts, are not used.

catalyst
A relatively small amount of a substance which facilitates a chemical reaction but is not consumed by it.

catch bar (lace machines)
A bar running the lace-making width of the Leavers and lace furnishing machines to which is attached a blade (see diagram). There are two such bars, one at the front and one at the back of the machine. Their complementary motions propel the carriages from the front combs through the well into the back combs and *vice versa*.

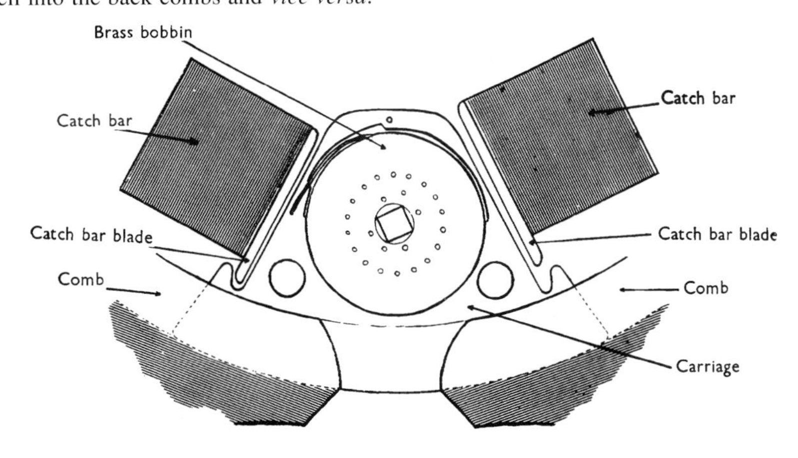

Catch bars

cationic dye
A dye that dissociates in aqueous solution to give a positively charged coloured ion. (See also **basic dye**.)

cauliflower ruche
See under **ruche**.

causticizing
Treatment of cellulosic fabrics with caustic soda solution of such concentration and under such conditions that a full **mercerization** effect is not obtained, but the colour yield in dyeing or printing is significantly enhanced. (See also **deweighting**.)

CAV
See **critical application value**.

cavalry twill
A firm warp-faced fabric in which the weave gives steep double twill lines separated by pronounced grooves formed by the weft. The name was originally applied to firm heavy-weight fabrics for making riding breeches for cavalry, but was later extended to cover fabrics used for raincoats and other clothing purposes.

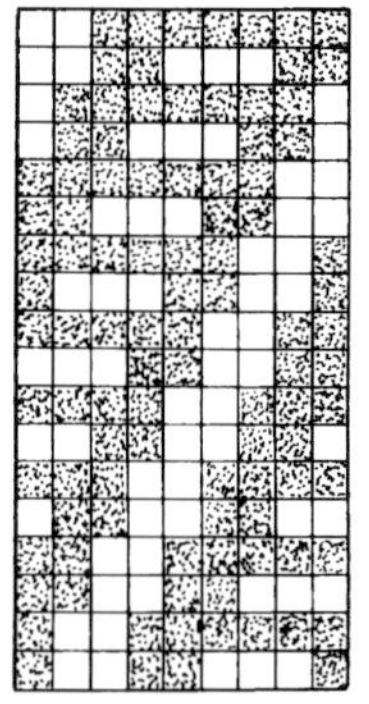
Cavalry twill weave

Cavalry twill (magnification x 5)

cavings (flax)
The reject from the bottom ridge of a **roughing-out** machine consisting mostly of rough bits of broken straw and some root ends.

cellular fabric
A fabric constructed so as to have a close and orderly distribution of hollows or holes. In woven fabric, this can be achieved by (i) **honeycomb**, (ii) leno (see **leno fabric**), or (iii) **perforated weave**.
Note: In certain sections of the trade, the term is used restrictively to describe leno cellular fabrics.

cellulose acetate
An ester formed from cellulose and ethanoic acid (acetic acid) (see **acetate (fibre)**).
Note: Purified cellulose is ethanoylated (acetylated) by ethanoic anhydride (acetic anhydride) in the presence of a catalyst (such as sulphuric acid or perchloric acid) in a solvent such as dichloromethane (methylene chloride) or ethanoic acid, which also acts as a diluent. The reaction proceeds until primary cellulose acetate containing about 60% by mass of combined ethanoic acid is formed. Secondary cellulose acetate is formed from this product by partial hydrolysis. It is obtained by adding water in excess of that required to react with the residual ethanoic anhydride, which thus allows hydrolysis to take place. When the hydrolysis is allowed to proceed until approximately 54% of combined ethanoic acid remains in the product, the cellulose acetate is soluble in propanone (acetone) and is sometimes known as acetone-soluble cellulose acetate.

cellulose diacetate
Strictly, an ester of cellulose and ethanoic acid (acetic acid) containing 48.8% by mass of combined ethanoic acid. This, however, is not a commercial textile product. The same term is sometimes used loosely to describe propanone-soluble (acetone-soluble) **cellulose acetate**. (See also **acetate fibre**.)

cellulose triacetate
Strictly, a cellulose acetate containing 62.5% by mass of combined ethanoic acid (acetic acid) but the term is generally used for primary cellulose acetate containing more than 60% by mass of combined ethanoic acid. (See also **triacetate fibre**.)

cellulose xanthate
An alkali-soluble salt formed by reaction between carbon disulphide and cellulose in the presence of a strong alkali.

centre (point) tie
See **jacquard tie (weaving)** under **jacquard (mechanism)**.

centre gimp (lace)
A Leavers-lace construction in which the filling threads, called **gimps**, lie between the front and back warp threads. The ground net is made by the interaction of two warp threads with each bobbin thread. The **objects** are filled by traversing gimp threads, according to the requirements of the pattern. Thick threads may be used for outlining the objects.

centre line (lace machines)
A datum line across the working width of the machine. The setting and adjustment of the machine are related to the centre line.
Note 1: The centre line of the machine lies just below the facing bar across the working width of the machine.
Note 2: The highest position reached by the points is adjusted to the centre line of the machine. (See also *Notes* under **circle (lace machines)**.)

centre loop (defect)
See **weft loop (defect)**.

centre selvedge
See **leno edge** under **selvedge, woven**.

centre weft fork
See **weft fork**.

centre-stitching
See **double cloth, woven**.

centrifugal hydroextraction
The removal of water by centrifugal force from wet textiles contained in a perforated rotor or 'basket'. (See also **expression (percent)**.)

centrifugal spinning
1. A method of spinning in which the yarn passes down a central guide tube and is then collected by centrifugal force on the inner surface of a rotating cylindrical container. (See also **box spinning**.)
2. (Manufactured fibre production) A method of fibre formation in which a molten or dissolved polymer is thrown centrifugally in fibre form from the edge of a rapidly rotating surface.

ceramic (fibre)
A refractory fibre composed of a metal oxide, metal carbide, metal nitride or their mixtures. Alumina and silica are the most commonly used. (See also Classification Table, p.401.)
Note 1: In this context, silicon and boron are regarded as metals.
Note 2: Ceramic fibres are used in electrical, thermal and sound insulation, in high temperature filtration and as reinforcement in some composites.

Cerifil
A system of spinning in which the yarn is wound on to the spinning tube via a 'winder'. The winder, which is rotated by the yarn, replaces the ring and traveller which are used in ring spinning and confines the yarn balloon.
Note: Cerifil® is a trade mark originally owned by the Cerit Research Organisation.

ceylon
A coloured woven fabric for blouses and shirtings made from a cotton warp and a cotton-wool weft.

ceylonette
An imitation **ceylon fabric** made entirely from cotton, with 22 ends x 18 picks/cm and 12 tex warp and weft.

chafe mark; bruised place; abrasion mark
A localized area where a fabric has been damaged by friction.

chafer fabric
See under **tyre textiles**.

chaff
A component of **trash** in cotton in the form of a heterogeneous assortment of vegetable fragments, most of them being small pieces of leaf, leaf bract (a small form of leaf growing beneath the boll), and stalk.

chain
1. (Carpet) Warp threads, usually woven in pairs, between the warpway lines of pile and alternating over and under the weft.
2. (Axminster: spool-loom overhead) A portion of the overhead mechanism of a **spool Axminster** or gripper-spool weaving machine. The number of links corresponds to the number of rows of pile in one complete repeat of the design, or multiples thereof; the correct row is presented (by means of the transfer arms, or gripper) to the weaving point as required.
3. See **warp**.

chain fork
See **lucet**.

chain mercerizing
See **mercerization**.

chain twill
An ordinary 3/3 twill.

chain warp (floorcoverings)
Warp threads, usually woven in pairs, alternating over and under the shots, which are then bound enclosing the stuffer yarns and the tufts or loops forming the pile.

chain warping
See **warp**.

chainette
A tubular cord produced on a circular knitting machine having no more than 20 needles.

chainless mercerizing
See **mercerization**.

chair web; upholstery web
A woven narrow fabric, for supporting the seat of chairs, couches etc. The common types are:
(i) English, British, or black and white, mainly or wholly of jute warps, dyed or stained black, with cotton wefts.
Note 1: Bleached hemp yarns are occasionally used in the selvedges for extra strength.
Note 2: Some specifications call for hemp or flax tow warps.
Note 3: Paper yarns are sometimes used in the warps to limit extensibility.
(ii) Plain or Indian, of undyed jute throughout.

challis
A lightweight, soft-handling, plain-weave dress fabric, generally of wool, using single worsted-spun yarns in warp and weft. It is often used as a base for printing.

chambray
A lightweight plain-weave cotton fabric having a coloured warp and white weft, producing a mottled appearance. It may also be made in striped, checked or figured patterns. The finer quality chambrays can have a silk weft, whereas synthetic yarns or cotton blends may be used in cheaper fabrics. Chambrays are used for women's and children's garments, pyjamas, shirts and sportswear.

channel (clothing)
A narrow passage formed between plies of material in a garment or by attachment of an additional ply of material, usually to house a **draw cord**.

char length
A term used synonymously for **damaged length**.
Note: In some national standards, char length is defined by a specific test method.

charged system
A method of dry cleaning in which an oil-soluble reagent such as petroleum sulphonate is added to the solvent so that a significant amount of water can be added to obtain a substantially clear dispersion of water in the solvent. In a high-charged system the concentration of added reagent, a so-called **detergent**, is 4 per cent while in a low-charged system the concentration ranges from 3/4 per cent to 2 per cent.

charmante satin
A double-wefted fabric, the face being a 1/2 twill and the back a weft sateen developed from thick, low-twist weft.

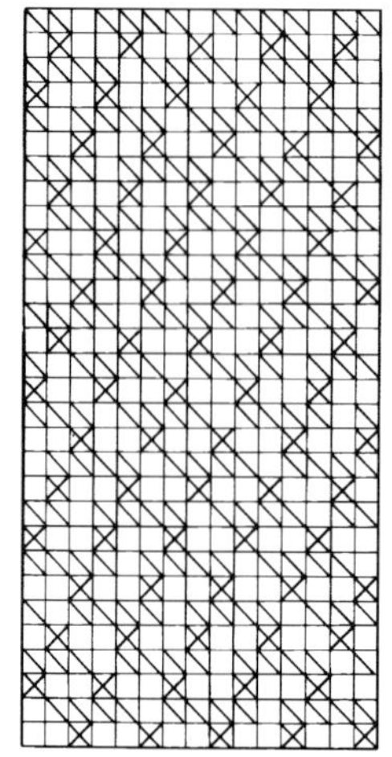

⊠ = Weft Interlacings for the 1/2 Z Twill Face
⧅ = Weft Interlacings for the 1/4 Sateen Back

Charmante satin weave

charmelaine
A 1x2 twill dress fabric made from botany wool warp and weft that has a twill nepp effect by virtue of a high-sett two-fold fine warp and a low-sett thick hard-twisted weft.

chase
The conical part of the body of yarn in cop, bobbin, or pirn form on which the thread is coiled during one traversing cycle.

chase length
1. The length of the conical portion of a package, measured along its surface.
2. The length of yarn wound on to a package in one complete traversing cycle.

check (fabric)
1. Two or more stripes of colour in the warp and weft direction resulting in a square or crossing line pattern.
2. A stripe of surface effects (i.e., caused by using different weaves and/or yarns) in the warp and weft direction resulting in a square or crossing line pattern.

check spring (sewing machine)
A spring-loaded guide on a lockstitch sewing machine which takes up slack in the upper/needle thread during the sewing cycle.

checking
See **shuttle checking**.

cheese
A cylindrical package of yarn, cross-wound on to a flangeless support. During winding the traverse length may be progressively reduced to produce tapered or rounded ends; such packages are then referred to as tapered, pineapple, or biconical cheeses.

cheese cloth; tobacco cloth
An open light-weight fabric of plain weave, usually made from carded cotton yarns.

cheese warp
Parallel threads in the form of a leased twistless rope, wound mechanically with a quick traverse on to a spool in the form of a large cheese.

cheese warping
See **warp**.

chemic; chemick
Calcium or sodium hypochlorite solution.

chemical fibre
See **fibre, chemical** under **fibre**.

chemicking
Bleaching non-protein fibre materials by means of a dilute hypochlorite solution.

chenille Axminster
See **Axminster carpet**.

chenille fabric
A fabric containing chenille yarn in the weft.

chenille yarn
See under **fancy yarn**.

Cheviot
1. A breed of sheep originating in the Cheviot Hills in Great Britain.
2. A tweed made from Cheviot wool or wools of similar quality.

chiffon
Originally a very light, sheer, open-mesh fabric made from silk yarns in plain weave; now made also from manufactured fibres. The term is loosely used adjectivally to describe the lightest types of particular fabrics, e.g., 'chiffon velvets', 'chiffon taffetas'.

china grass
See **ramie**.

chinchilla
1. A thick, spongy, weft-pile, double fabric, which has a face of long floats of soft twist yarns that are teazled to produce a long nap that is rubbed into curly nubs.
2. A woollen knitted fabric with a napped surface.

chine
Descriptive of a fabric or **ribbon**, usually plain weave, woven with a printed warp.

chintz
A glazed, printed, plain-weave fabric, originally and usually of cotton lighter than cretonne.
Note: The term 'fully glazed' applies only to a chintz that has been stiffened by starch or other substance and friction-calendered; the term 'semi-glazed', or 'half-glazed' applies to chintz that has been stiffened by friction-calendering alone.

chlidema square
A border square made entirely from **body carpet** 0.69m or 0.91m wide where additional designing procedures are used to produce a border-and-corner design.

chlorination
When used with reference to textile processing, a term indicating the reaction of a textile material with chlorine.
Note: The chlorine may be in the form of the gas, or its solution in water, or it may be obtained from a suitable compound. Wet chlorination implies that the goods are treated in aqueous solutions containing chlorine or a chlorine-yielding reagent. Dry chlorination implies treatment under non-aqueous conditions, e.g., by chlorine gas.

chlorofibre (fibre) (generic name)
A manufactured fibre composed of synthetic linear macromolecules having in the chain more than 50% (by mass) of chloroethene (vinyl chloride) or 1,1-dichloroethene (vinylidene chloride) groups (more than 65% in the case where the rest of the chain is made up of cyanoethene (acrylonitrile) groups, the modacrylic fibres being thus excluded). (See also Classification Table, p.401.)

chopped weft; cut weft
Weft that has been fractured by the reed while being beaten up during weaving.

chrome dye
A mordant dye capable of forming a chelate complex with a chromium atom.

chrome mordant process
A method of dyeing in which the fibres are mordanted with a solution of a chromium compound and subsequently dyed with a suitable chrome dye.

chromophore
That part of the molecular structure of an organic dye or pigment responsible for colour.

chrysalis
The form taken by silkworms in the passive stage of development between worm and moth. It is dark brown and fragments of it can often be detected in silk waste, especially noils.

chute feeding
Pneumatic distribution of fibre flocks from opening and cleaning lines to a set of cards.

CIE spectral tristimulus values
See **colour measurement**.

circle (lace machines)
An arc, whose axis is the **centre line** of the machine, that determines the setting of all parts whose placing or movements are co-ordinated with the swinging motion of the carriages.
Note 1: The specified circle of the machine determines the radius of the carriage and comb blades.
Note 2: The resultant movement of the carriages caused by the action of **catch bars, landing**

bars, **locker bars**, **driving bars**, and **fluted rollers** is concentric with the circle.
Note 3: The setting of the circle strips and the facing bar is determined by the circle.

circular knitting machine
A **knitting machine** in which the needles are set radially or in parallel in one or more circular beds. Used without further qualification, the term generally refers to a weft-knitting machine of this type.
Note: Machines with diameters of less than 165mm are often termed 'small-diameter machines'.

circular weaving machine
See **weaving machine**.

classing
See **wool classing** and **cotton classing**.

clean cut
A clean cut is created in a woven fabric where two adjacent ends or two adjacent picks interlace in an exactly opposite order to each other, e.g., in a herringbone twill.

clean dry mass
The mass of textile fibre free from moisture and other non-fibrous matter.

clear
See *Note* under **bright**.

clear finish
A type of finish on fabrics containing wool. The surface of the fabric is relatively free from protruding fibres and the weave and the colours of the constituent yarns are clear and distinct. Examples of clear finished fabrics include serges and many other worsted fabrics.

clearing
The removal from a dyed or printed textile of surplus colorant which, if allowed to remain, would mar the appearance or quality of the textile.

clearing (yarn)
The process of removing imperfections form a yarn. The fault is normally replaced by a knot or **splice**.

clearing cam (knitting)
A cam that displaces a latch needle in its track so that a loop may pass over the spoon of the latch on to the needle stem.

clicking top sliver; feather-edged slivers
Sliver that contains a large number of fibres projecting from the main body.
Note: During the unwinding of the sliver from the top in the creel, the projecting fibres 'click' to neighbouring slivers, which causes them to become disarranged and bent back to form neps in the subsequent products.

clip (wool)
1. The yield of wool from one shearing.
2. One season's yield of wool.

clip cone
See **two-for-one twisting**.

clip spots
Small spots of extra warp and/or weft between which the floating material is clipped or sheared off after weaving.

clips (lace)
Threads, used for repeated **motifs**, which are floated over the surface of the lace between motifs. They are clipped off in the finishing of the lace (see **float (lace)**).

clo
A unit of thermal resistance of garments, adopted in the U.S.A.
1 clo = 1.55 togs
Note: clo is based on human physiological factors and represents the thermal resistance of a complete clothing assembly under specified conditions. (See also **tog**.)

cloqué fabric, knitted
See **relief fabric, weft-knitted**.

cloqué, woven
A compound or double fabric with a figured blister effect brought about by the use of yarns of different character or twist, which respond in different ways to finishing treatments.

Cloqué (actual size)

closed lap (warp knitting)
A lapping movement in which the underlap takes place in the direction opposite to that of the preceding overlap. This results in the same thread crossing over itself at the base of the loop.

closed loop (knitting)
A loop at the base of which the thread crosses over itself.

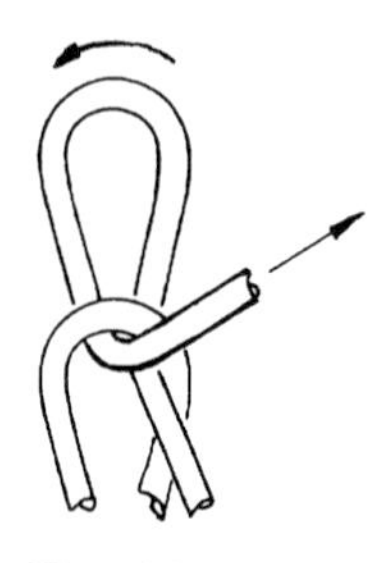

Closed loop

closed shedding
A method of **shedding** in which all warp threads are brought to the same level after the insertion of each pick of weft.
Note: There are two main types: bottom-closed and centre-closed shedding. The terms 'bottom' and 'centre' indicate the position of the warp threads when at rest.

cloth
A generic term embracing most textile fabrics.
Note: The term was originally applied to wool cloth suitable for clothing.

cloth (lace)
Solid effects in lace furnishings obtained by moving spool threads across two or more wales and back again without pillaring between the throws.

cloth roller
The roller on to which woven fabric is wound after passing from the **take-up roller**.

cloth take-up motion
See **take-up motion**.

clothing wool
Wools of short fibre, not suitable for combing, and used in the manufacture of woollens.

cloud yarn
See under **fancy yarn**.

cloudiness
1. A defect in webs and slivers consisting of areas of different densities.
2. In a dyed fabric, a defect consisting of random, faintly defined uneven dyeing.
3. In a bleached fabric, a defect consisting of opaque patches, usually visible only in transmitted light, due to residues remaining after bleaching.

co-mingled yarn
intermingled yarn in which two or more separate yarns are intermingled.

co-spun yarn
See **filament blend yarn**.

coach lace (narrow fabric)
See **carriage lace (narrow fabric)**.

coated fabric
1. A material composed of two or more layers, at least one of which is a textile fabric and at least one of which is a substantially continuous polymeric layer. The layers are bonded closely together by means of an added adhesive or by the adhesive properties of one or more of the component layers.
2. A textile fabric on which there has been formed *in situ*, on one or both surfaces, a layer, or layers, of adherent coating material.

cobble
To correct a defectively processed fabric.

cobbler
A defectively dyed or finished fabric returned for correction.

cobwebbing
A fault in winding that causes some threads to lie in a straight line across the end of a package. It is caused by the yarn slipping beyond the normal extremity of the traverse during winding, and may result in yarn breaks during unwinding. It is sometimes called 'stitching', 'crossing', or 'webbing'.

cockle; puckering (defect)
The wrinkled appearance of a fabric in which non-uniform relaxation or shrinkage has occurred. *Note:* This defect may result from variations in the tension of the ends or picks at the time of weaving, from variations in the degree of stretch imposed on the yarn during earlier processes, or from differences in the contraction of two or more yarns used accidentally in the fabric. The defect may be distributed over a large area of fabric or may be confined to isolated stripes, bars, or streaks.

cockle (yarn)
Isolated loops of fibres protruding from a yarn surface owing to the relaxation of adjacent fibres that have previously been stretched during processing.

cockle; hockle (cordage)
Deformation of the strands in a laid **rope** caused by the removal of kinks by pulling them out of the rope by force causing helical deformation of the strands themselves.

cockled bar
See **tension bar** under **bar (woven fabric)**.

cockled selvedge
See **selvedge, slack**.

cockled yarn (linen)
A weak, malformed yarn caused by a fault in the spinning-frame drafting rollers that allows the yarn to be formed from two or more ribbons of fibres delivered at different speeds to the twisting device of the spinning frame.

cockling (knitting)
An irregular surface effect caused by loop distortion.

cocoon (silk)
An egg-shaped casing of silk spun by the silkworm to protect itself as a chrysalis.

cocoon strippings; blaze; floss silk; keba
The first threads secreted by the silkworm when it finds a place to form its cocoon.

cohesion (fibre)
The resistance to separation of fibres in contact with one another in a fibre assembly, e.g., in a **sliver**.

cohesion (silk)
The degree of agglutination of cocoon filaments forming raw silk thread, determined by subjecting the raw silk threads to friction in a cohesion tester and counting the number of strokes required to spread and open out the constituent filaments.

cohesive force
The force required to overcome fibre cohesion (see **cohesion (fibre)**) in a fibre assembly.

cohesive set
See *Note 1* under **setting**.

coiler
A delivery device that deposits a sliver into a cylindrical can in the form of helical coils so as to permit easy withdrawal with the minimum of fibre disturbance.

coir
A reddish-brown-to-buff coarse fibre obtained from the fruit of the palm *Cocos nucifera*.
Note: There are three types of coir fibre: the longest and finest, which is usually obtained from the unripe fruit, is spun into yarn for making mats and ropes; a coarser fibre, known as 'bristle fibre', is used for filling brushes; and a shorter fibre is used for filling mattresses, and for upholstery etc. Bristle and mattress fibre, together with a small quantity of fibre suitable for yarn, is obtained mainly from Sri Lanka; yarn fibre only is obtained from continental India.

cold crack temperature
The temperature below which the coating on a coated fabric and/or the base fabric cracks when the folded fabric is cooled under specified test conditions.

cold drawing (synthetic filaments and films)
See **drawing, cold (synthetic filaments and films)**.

cold flatting
See **cramping**.

collapsed balloon spinning; suppressed balloon spinning
A system of **ring spinning** in which the rotating yarn balloon is greatly reduced in diameter by contact with the top of the spindle.
Note 1: The system is normally used for the economic spinning of semi-worsted and woollen carpet yarns by permitting the use of larger package sizes and/or higher spindle speeds.
Note 2: Special spindle top extensions are used to obtain the desired effect.

collar
The flat, upright or turned-over section of a garment which fits around the neck.

collar fall
The section of a turned-over **collar** from the crease to the outer or **leaf edge**.

collar stand
The upright portion of the **collar** which is attached to the neck edge of a garment.

colorant
A colouring matter, a **dye** or **pigment**.
Note: The Society of Dyers and Colourists recommends this noun as a generic term.

colorimetric system
See **colour measurement**.

colour; color
1. (Of an object) The particular visual sensation (as defined in 2) caused by the light emitted by, transmitted through, or reflected from the object.
2. (Sensation) That characteristic of the visual sensation which enables the eye to distinguish differences in its quality, such as may be caused by differences in the spectral distribution of the light rather than by differences in the spatial distribution or fluctuations with time.
Note: The colour of a non-selfluminous object is dependent on the spectral composition of the

colour; color *(continued)*
incident light, the spectral reflectance or transmittance of the object and the spectral response of the observer. Colour can be described approximately in terms of **hue**, **saturation** and lightness, or specified numerically by chromaticity co-ordinates, e.g., those defined by the *C.I.E. Standard Observer Data* (1931). Alternatively, colour can be specified by reference to visual standards, e.g., the *Munsell Colour Atlas*.

colour constancy
The ability of a coloured object to give the same general colour impression when viewed under different illuminants, the observer having been chromatically adapted in each case.
Note: The most common comparison is made between the impression under artificial light, e.g., tungsten filament, and that under daylight.

colour fastness
The property of resistance to a named agency, e.g., washing, light, rubbing, crocking, gas-fumes. (See also ISO 105.)
Note: On the standard scale, five grades are usually recognised, from 5, signifying no visible change, to 1, substantial change. For light fastness, eight grades are used, 8 representing the highest degree of fastness.

colour measurement

colorimetric system
Quantitative system of colour specification based on scales derived from either additive or subtractive colour mixture.

trichromatic system
System of colour specification based on the possibility of matching a colour stimulus by the additive mixture of three suitably chosen reference stimuli.

spectral tristimulus values
Tristimulus values, on any given trichromatic system, of the monochromatic components of an equi-energy spectrum. The set of spectral tristimulus values defines the 'colour-matching functions' or 'colour-matching curves'.

CIE spectral tristimulus values
Colour characterised numerically by tristimulus values of the spectral components of an equi-energy spectrum in the CIE (XYZ) system. (See also **colour**.)

colour quality
A specification of colour in terms of both **hue** and **saturation**, but not **luminance**.

colour staining
The unintended transfer of colour from a textile to another textile with which it is in contact, usually in wet or damp conditions. (See also **bleeding** and **staining**.)

colour value; tinctorial value
The **colour yield** of a **colorant**, compared with a standard of equal cost.
Note: It is usually determined by comparing the cost of coloration at equal visual strength. Comparisons are normally made between products of similar **hue** and properties.

colour yield; tinctorial yield
The depth of colour obtained when a standard weight of colorant is applied to a substrate under specified conditions.

colour-and-weave effect
An effect developed by a small-group colour patterning of warp and/or weft in a woven fabric. The various combinations of the warp and weft floats of the constituent colours produce a distinctive effect which renders the **weave effect** virtually indistinguishable and often the colour order of the threads is not apparent.

coloured cotton
See **cotton, coloured**.

comb
A **reed** with one **baulk** used in warp preparation to order and maintain the warp ends in a parallel form.

comb (lace machines)
A complete assembly of **comb leads**.

comb lead (lace machines)
A number of curved steel or brass blades cast together in a lead-alloy base to form the support and guide for the movement of the **carriage**.

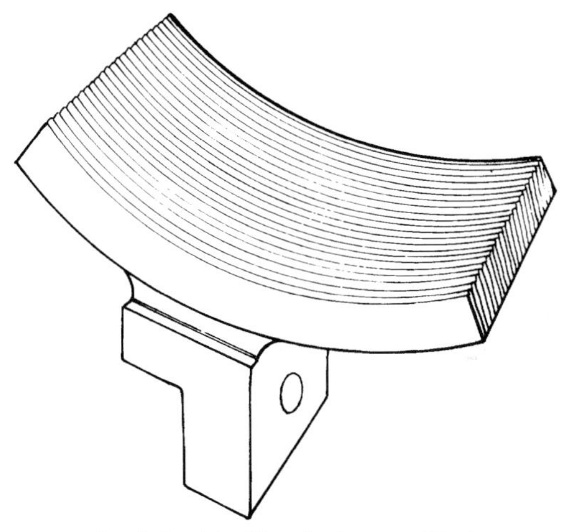

Comb lead (rolling locker machine)

combed yarn
Yarn produced from fibres that have been carded (or prepared) and combed.

comber board; cumber board
A board containing rows of holes (one for each harness cord) that determine the height, width, and spacing of the cords in jacquard weaving. The board may be built in sections.

comber web
See **web** 1.

combination (lace)
Openworks effects obtained in lace furnishings by nipping two or more pillars (or warps) together to make a hole wider than a normal wale.

combination yarn
A yarn in which there are dissimilar component yarns especially when these are of fibre and

combination yarn *(continued)*
filaments. (See also **composite yarn**.)

combined fabric; laminated fabric
A material composed of two or more layers, at least one of which is a textile fabric, bonded closely together by means of an added adhesive, or by the adhesive properties of one or more of the component layers. (See also **back grey**.)

combined heald and harness
An arrangement in which the ends in a **jacquard fabric** that are producing a basic weave for **ground** can be controlled by **healdshafts** that are operated by tappets or a **dobby**, while the ends that are figuring are controlled by the jacquard. The system allows the jacquard to have a greater figuring scope.

combined rope
See under **rope**.

combing
The straightening and parallelising of fibres and the removal of short fibres and impurities by using a comb or combs assisted by brushes and rollers.

combing roller
See **beater**.

comfort stretch
See **stretch fabric**.

commercial allowance; conventional allowance
The agreed value to be added to the clean, dry mass of a textile material (after extracting and drying it using prescribed methods) in order to obtain its commercial mass, linear density or mass per unit area. It is expressed as a percentage of the mass of the extracted and dried material.

commercial mass
The calculated mass that a consignment of textile material would have if either the commercial moisture regain were added to the dried mass of the material or the commercial allowance were added to the extracted and dried mass of the material. (See also **correct invoice mass**.)

commercial moisture regain
The agreed value to be added to the dry mass of a textile material (after drying it using prescribed methods) in order to obtain its commercial mass, linear density or mass per unit area. It is expressed as a percentage of the mass of the dried material. (See also **recommended allowance**.)

compensation (technical textiles)
The reduction in size of cutting patterns to take account of fabric stretching when **prestress** is applied to air-supported and tension-membrane structures.

composite; composite material
A product formed by intimately combining two or more discrete physical phases, usually a solid matrix and a fibrous material.

composite yarn
A yarn composed of both **staple** and **continuous-filament** components, e.g., **core spun** or **wrap spun**.

compound fabric, woven
A generic term for layered fabrics in which the separate layers or plies, each with its own warp and weft, are produced simultaneously and stitched together in one weaving process. Fabrics comprising two plies are known as double (or two-ply) fabrics and others by the number of plies they contain, e.g., three-ply fabric.

compound feed
See under **feed mechanisms (sewing)**.

compound needle (machine knitting)
See under **needle (machine knitting)**.

compression hosiery
See **graduated compression hosiery**.

compressive shrinkage
A process in which fabric is caused to shrink in length, e.g., by compression. The process is often referred to as CCS (controlled compressive shrinkage). Commercial processes include Sanforizing® and Rigmel®.

condensation polymer
See **polymer, condensation** under **polymer**.

condensation polymerization
See **polymerization, condensation** under **polymerization**.

condense dye
A dye which, during or after application, reacts covalently with itself or other compounds, other than the substrate, to form a molecule of greatly increased size.

condenser (ring-doffer or tape)
The last section of a **condenser card**: it divides a broad thin web of fibres into narrow strips, which are then consolidated by rubbing into **slubbings**.

condenser card
A roller-and-clearer type of card, as distinct from a flat card, which converts fibrous raw materials to **slubbings**, by means of a **condenser**.

condenser spun
Descriptive of yarn spun from **slubbing**.

condition
1. The moisture present in textiles (see **regain** and **moisture content**).
2. To allow textile materials (raw materials, slivers, yarns, and fabrics) to come to hygroscopic equilibrium with the surrounding atmosphere or with the **standard atmosphere for testing**.
3. To add relatively small quantities of water to textile materials (raw materials, slivers, yarns, and fabrics).
Note: The object of conditioning is to prepare for testing, or to bring textiles to an agreed moisture content for sale or to facilitate later processing. Among methods used for applying water are: mechanical means during gilling or winding; the use of conditioning machines; and storing in an atmosphere of high relative humidity.

conditioner tube
A tube supplied with steam or hot air surrounding a melt-spun threadline and located between extrusion and wind-up, whose purpose is to control the fine structure of the yarn.

cone
1. A conical support on which yarn is wound.
2. A conical package of yarn wound on a conical support.

confidence interval
The range within which a value can be expected to lie with a given probability.

conjugate (bicomponent) fibre
See under **bicomponent fibre**.

constant rate of crosshead movement
A machine setting used in materials testing in which the rate of travel of the moving crosshead is uniform.
Note: Constant rate of crosshead movement is not equivalent to **constant rate of extension** unless the specimen is held in grips which do not allow the specimen to extend around the grip.

constant rate of extension; CRE
A machine setting used in materials testing in which the rate of increase in the length of the specimen is uniform with time.
Note: The rate of increase of force or elongation is dependent upon the extension characteristics of the specimen.

constant rate of loading; CRL
A machine setting used in materials testing in which the rate of increase of the force being applied to the specimen is uniform with time.

constant rate of traverse; CRT
A machine setting used in materials testing in which the straining clamp moves at a uniform rate and the force applied is measured via a force balancing system through the small movement of the nominally fixed jaw.
Note: The rate of increase of force or elongation is dependent upon the extension characteristics of the specimen.

continuous polymerization
See **polymerization, continuous** under **polymerization**.

continuous process
A process in which material passes in sequence through a series of stages to give a continuous output of processed material. (See also **batchwise process**.)

continuous yarn felting
A process whereby **slivers**, **rovings**, **slubbings**, or **yarns** are felted on a continuous basis. This is achieved by passing wool-rich material through a unit where it is agitated in an aqueous medium where **felting** takes place. The process is used to produce a yarn, or to consolidate a spun yarn. (See also **spinning**.)

continuous-filament yarn; filament yarn
A yarn composed of one or more **filaments** that run essentially the whole length of the yarn.
Note: Yarns of one or more filaments are usually referred to as 'monofilament' or 'multifilament', respectively.

control specimens
Set(s) of specimens from a batch, kept under specified storage conditions, and tested as required to provide comparison with changes in physical properties of other specimens from the same batch after exposure to defined conditions.

conventional allowance
See **commercial allowance.**

conventional moisture regain
The agreed value applying to a textile material, which is used to represent the mass of water in any form which that material contains when, after preconditioning, it comes into equilibrium with the standard atmosphere. It is expressed as a percentage of the mass of the dried material.

converter
See **merchant converter.**

converting; conversion (tow)
The production, from a filament tow or tows, of a **staple sliver** in such a way that the essentially parallel arrangement of the filaments is maintained. If subsequently a **top** is required, further processes of **re-breaking** and/or **gilling** may be necessary and the whole operation is then often referred to as tow-to-top converting or conversion.
Note: The two methods of converting most commonly employed are: (i) crush cutting, in which the filaments of the tow are severed by crushing between an anvil roller and a cutting roller with raised 'blades' helically disposed around its surface; and (ii) stretch breaking, in which the filaments of the tow are broken by progressive stretching between successive sets of rollers.

cooling cylinder
An open cylinder, or alternatively a closed cylinder filled with cold water, over which hot fabric is passed to accelerate cooling.

cop
1. A type of yarn package spun on a mule spindle.
2. A ring tube.
3. A small cylindrical flangeless cardboard or plastic tube on to which sewing thread is wound. (See also **supercop**, **rocket package** and **pirn.**)

cop-end effect
The increase in yarn tension that occurs when unwinding from the base of some kinds of cop, bobbin, or pirn.
Note: One result of this is the gradual narrowing of the fabric as the pirn empties during weaving with a single shuttle, forming the characteristic dog-legged selvedge (see **selvedge, uneven**).

copolymer
A polymer in which the repeating units are not all the same. Usually, but not always, copolymers are formed from two or more different starting materials. For example, chloroethene (vinyl chloride) and l,l-dichloroethene (vinylidene chloride) form a copolymer that contains the repeating units

$$—CH_2—CHCl— \text{ and } —CH_2—CCl_2—.$$

Note: The different classes of **copolymer** include random copolymers, alternating copolymers, **block copolymers**, and **graft copolymers**.

copolymer, block
A **copolymer** in which the repeating units in the main chain occur in blocks, e.g.,

$$-(A)_m-(B)_n-(A)_p-(B)_q-,$$

where A and B represent different repeating units, and m,n,p,q... are integers which may be different.

copolymer, graft
A **copolymer** formed when sequences of one repeating unit are built as side branches on to a backbone polymer derived from another repeating unit, e.g,

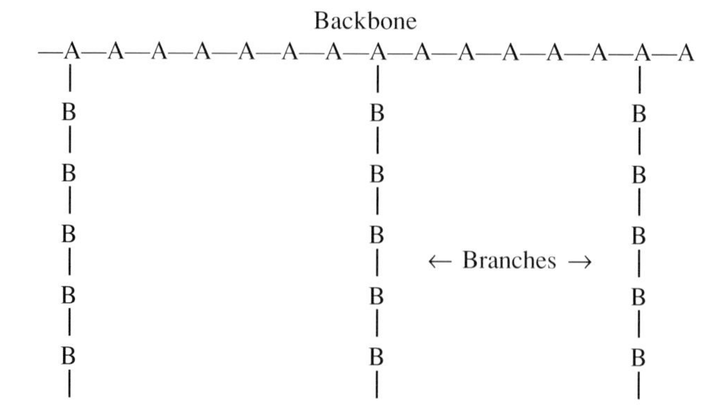

where A and B represent the different repeating units.

cord
A term applied to a variety of textile strands including: (i) **cabled yarns**, (ii) **plied yarns**, and (iii) structures made by **braiding**, **knitting** or **weaving**. (See also **bullion cord**, **cable cord**, **case cord**, **crêpe cord**, **upholstery cord** and **welting cord**.)

cord carpet
A low-level loop-pile carpet with pronounced rows of loops in the weft direction.

cord fabric, woven
See **rib fabric, woven**.

cord mail heald
See under **heald**.

cordage
Any product, regardless of size, made by twisting or braiding textile yarns, which is generally round in cross section and capable of sustaining loads.

cordon yarn
A two-ply union yarn made from a single cotton yarn and a single worsted or woollen yarn.

cordonnet
Silk sewing yarn with Z-twist, usually three thread.

corduroy
A cut-weft-pile fabric in which the cut fibres form the surface. The binding points of the pile wefts are arranged so that after the pile has been cut, cords or ribs are formed in the direction of the warp. (See also **fustian**.)
Note: Velveteen fabrics are sometimes cut to give a corduroy appearance.

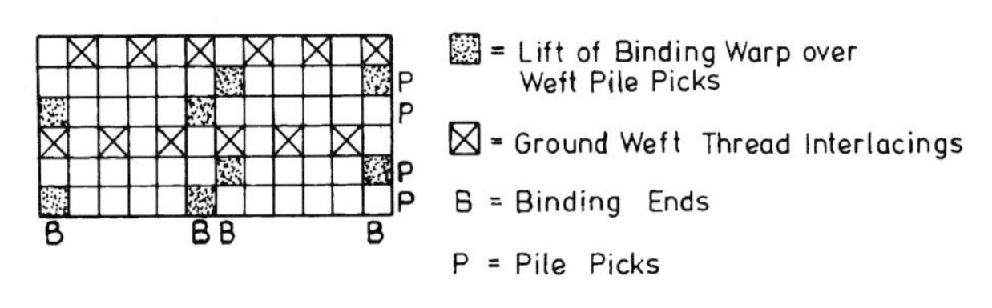

Corduroy weave

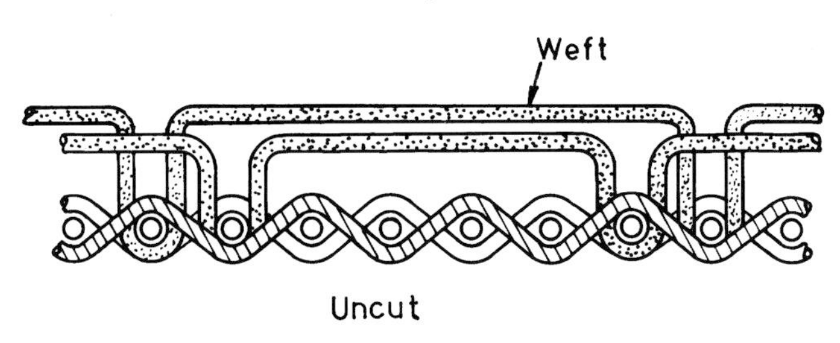

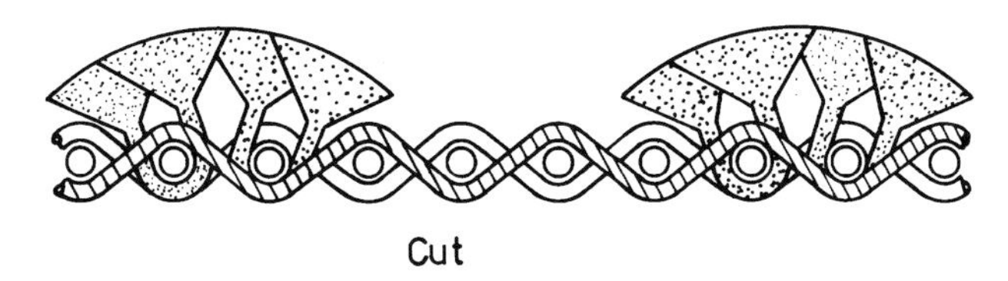

Corduroy: section through warp

core
The central portion of a **gimp**, **cord** or **rope** which may consist of parallel, twisted, cabled or knitted strands but which is not combined structurally with the gimp, cord or rope.

core sampling
A method of taking representative samples from bales or packs of textile fibres obtained by inserting a coring tube driven by hand or machine into each package.
Note 1: Core samples can be used for the determination of yield and fineness, but not fibre length.
Note 2: The term 'mini-core sampling' is applied to small-scale sampling.

core yarn
A yarn consisting of a number of component yarns, of which one or more are constrained to lie permanently at the central axis of the final thread, whilst the remaining yarns act as covering yarns.

core-sheath bicomponent fibre
See under **bicomponent fibre**.

core-spun yarn
Yarn consisting of a yarn surrounded by staple fibres. The yarn has the strength and/or elongation of the central thread whilst exhibiting most of the other characteristics of the surface staple fibres. Examples are:
(i) a sewing thread consisting of a central synthetic continuous-filament yarn surrounded by cotton fibres;
(ii) worsted yarn with bulked-nylon core, e.g., typically 1/24s worsted count (37 tex) with

core-spun yarn *(continued)*
approximately 33% of nylon. These yarns are normally produced to give strength and elasticity to the fabric;
(iii) a spun yarn from either natural or manufactured fibres incorporating an elastomeric core. These yarns are normally used in stretch fabrics.
Note: The term core-spun yarn is also associated with a central yarn or yarns in a braid (see *Note* under **braiding**).

coredon
A 2x1 woollen dress fabric produced with fine warp and coarse weft.

corkscrew weave
A warp-faced fabric produced in a steep twill weave.

corkscrew yarn
1. See **spiral yarn** under **fancy yarn**.
2. A term which, when applied to conventional folded yarns, indicates that the yarn is faulty due to spiralling of one or more of the component ends. This may be caused by a difference in linear density, twist direction, twist level, or tension between the ends.

coroa
See **caroa**.

corona discharge treatment
An electrical discharge treatment carried out at atmospheric pressure which in textiles is used to modify the surfaces of fibres.

coronation gimp
A woven trimming, about 12.5mm wide, consisting of a continuous-filament ground warp with plain weave and a **gimp** cord weft. Down each side, two two-ply cords of opposing twist are woven three up one down to produce a 'grain of wheat' effect. The centre of the fabric has a continuous-filament figure warp, the weave of which **mocks** the elbowing cords.

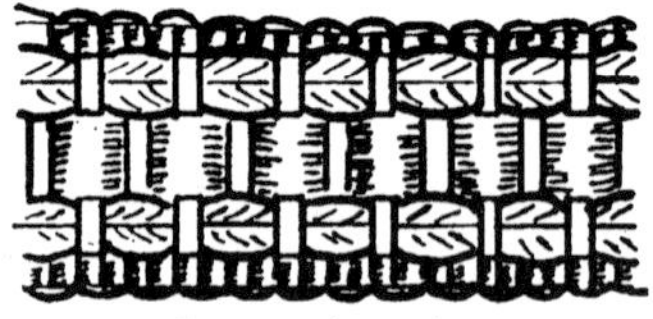

Coronation gimp

corporate clothing
See **workwear**.

correct invoice mass
The mass of material calculated from either the clean, oven-dry mass or the oven-dry mass and the **recommended allowance**.

cortex
The inner portion of most animal hair fibres. It consists of so-called 'spindle-cells' which are about 100 microns (μm) in length and are built from longitudinally oriented **fibrils**. The cortex usually represents more than 90% of the total mass of the fibre.
Note: In crimped, fine wools the cortex is divided into two hemicylinders, the orthocortex and the paracortex, which are wound around each other helically in phase with the fibre crimp. The dividing line between ortho and paracortex generally corresponds to the major axis of the elliptical cross section of the fibre. The two components have different physical and chemical

properties with the orthocortex being more reactive, more accessible to dyestuffs, and more sensitive to swelling agents than the paracortex. The different physical properties also give rise to crimp. (See also **bicomponent fibre**.)

cotted wool
Wool from a fleece that is felted or matted together so that it will not tear apart freely when pulled by hand.

cotton
The **seed hair** of a wide variety of plants of the *Gossypium* family.

cotton, coloured
Cotton which has a natural (genetic) colour, i.e., not white or cream. The most common colours are brown, green, and yellow.

cotton classing
A procedure by which a sample of cotton is ascribed various numbers or descriptors to indicate its relative market value.
Note: The most common numbers and descriptors refer to **staple length**, **micronaire value**, colour, **trash content**, and fibre strength. Classing is a combination of subjective and objective quality assessment. Increasing use is being made of **high volume instrumentation** to assist or replace manual classing.

cotton dust
See **trash (cotton)**.

cotton lap (warp knitting)
A lapping movement made on a Milanese machine in which the yarn traverses one wale per course.

cotton maturity
See **maturity (cotton)**.

cotton waste
See **waste**.

cotton wool; cotton
A fibrous product used for medical or cosmetic purposes which is made from cotton and/or viscose.

cotton-spun
A term applied to staple yarn produced on machinery originally developed for processing cotton into yarn. (See also **short-staple spinning**.)

count of reed; reed number; sett
The number of **dents** per unit width of reed.
Note 1: There have been many units in common use, e.g., (i) the number of dents per inch, (ii) the number of dents per 2 inches, (iii) the number of groups of 20 dents per 36 inches, (iv) the number of dents per 10cm.
Note 2: The recommended unit is dents/cm.

count of yarn; yarn count; yarn number; yarn linear density; grist
Methods of variously expressing the mass per unit length or the length per unit mass of a yarn. (See also **tex system**, **linear density**, **titre** and Tables, p.396-397.)
Note: The term 'size' is used in North America for count of yarn.

counting glass; linen prover; piece glass
A small mounted magnifying glass for examining fabric. The base of the mount generally contains a unit of measurement having an aperture one centimetre square, one inch square or cross-shaped with various dimensions, convenient for counting ends and picks, or courses and wales in a fabric.

couple
To combine a suitable organic component, usually a phenol or an arylamine, with a diazonium salt to form an azo compound as in the manufacture of azo colorants, in **azoic dyeing** or in the after treatment of direct dyeings.

course (knitted fabric)
A row of loops (i) across the width of a flat fabric, or (ii) around the circumference of a circular fabric.
Note 1: In weft knitting the course may be formed by one or more traverses of a feeder.
Note 2: Some fabrics knitted on two needle beds have a different number of courses on one side of the fabric from the other.

course density (knitted fabric)
The number of visible loops per unit length measured along a **wale**.
Note 1: The traditional unit has been courses per inch but the value is now expressed as courses per cm.
Note 2: In certain constructions, the number of visible loops in one wale may be different from that in another, and there may also be different results on the back and front of the fabric. Consequently, in such constructions, it is necessary to specify where the count is made.

course length, weft-knitted
The length of yarn in a knitted **course**.

course spirality (weft knitting)
The inclination of **courses** from the horizontal, i.e., across the fabric, caused by multiple feeds on a circular knitting machine.
Note: Fabrics made on circular knitting machines have an inherent inclination of the courses to the wales. This should not be confused with spirality.

coutil; coutille
A strong fabric, bleached or piece-dyed, woven in 2/1 warp-faced twill, usually in herringbone stripes, which is used for corsets, typically: 21x30; 46x27tex; 175g/m^2; K=21.0+14.8.

cover
1. The degree of evenness of thread spacing. Good cover gives the effect of a uniform plane surface and cannot be obtained with hard-twisted yarns.
2. The degree to which, in fabric finishing, the underlying structure is concealed by the finishing materials or treatments.

cover factor (knitted fabrics); tightness factor (weft-knitted fabrics)
A number that indicates the extent to which the area of a knitted fabric is covered by the yarn. It is also an indication of the relative looseness or tightness of the knitting. (See also Table 2, p.400.)

cover factor (woven fabrics)
A number that indicates the extent to which the area of a fabric is covered by one set of threads. By introducing suitable numerical constants, its evaluation can be made in accordance with any system of counting. For any fabric there are two cover factors: warp cover factor and weft cover

factor. (See Table 2, p.400.)
Note: The traditional cover factor in the cotton system (sometimes known as Peirce's cover factor) is the ratio of the number of threads per inch to the square root of the cotton yarn count.

cover roller
A roller engraved with a delicate pattern, used for over-printing an existing print to obtain a distinct subdued patterned ground. The operation is termed 'covering'.

covered yarn
A yarn made by feeding one yarn under a controlled degree of tension through the axis or axes of one or more revolving spindles carrying the other (wrapping) yarn(s) (see, for example, **elastomeric yarn** and **bourdon cord**).
Note: Covered yarn may also be produced by utilising air-jet technology.

coverstock
A permeable fabric used in hygiene products to cover and contain an absorbent medium.

covert cloth
A warp-faced fabric, usually of twill weave, having a characteristic mottled appearance, that is obtained by the use of a grandrelle or mock-grandrelle warp and a solid-coloured weft.

cow fibre (hair)
Fibre from the common ox (*Bos taurus*).

crabbing
A process used in the worsted trade to set fabric in a smooth flat state so that it will not cockle, pucker, or wrinkle during subsequent wet processing. The fabric is treated in open width and under warp-way tension in a hot or boiling aqueous medium, the tension being maintained while the fabric is cooling (see **setting**).

crack; split (defect)
A narrow streak, running parallel with the warp or weft threads, characterized mainly by the existence of a marked space between two adjacent threads. Such streaks may be caused by mechanical defects on the loom, such as a loose crank-arm or crank-shaft bearing, banging-off, a bent reed wire (see **reed mark**), etc.

cracked selvedge
See **selvedge, cracked**.

cracky weft
See **weft crackiness**.

crammed pick
See **pick, dead**.

cramming motion; retarding motion
A mechanism which makes the **take-up motion** inoperative for a defined period of time. (See also **pick, dead**.)

cramping; cold flatting
The pressing of a pile of fabric into a convenient thickness for transport or storage without adversely affecting the finish. The process is used after cuttling (see **cuttle**) and **rigging** woollen or worsted fabrics as the last process in the finishing routine.

crank shedding
A shedding mechanism in which the healds are driven from a continuously rotating eccentric or crank and complete their movement in a two-pick cycle.

crash
A fabric, originally made of linen, which has an irregular appearance arising from the use of thick, uneven yarns, particularly in the weft. Fabrics woven in plain or fancy crêpe weaves are now made of linen, cotton, spun viscose rayon or other suitable manufactured fibre, and unions of these. Typical linen crash fabrics for towels are plain-woven traditionally from 65-100 tex (16-25 lea) flax-tow yarns.

cratch; back rack; warp beam back frame (narrow fabrics)
A frame behind a narrow fabric weaving machine, which holds the warps and their tensioning devices.

craua
See **caroa**.

CRE
See **constant rate of extension**.

crease (fabric defect)
A fold in a fabric introduced unintentionally at some stage in processing. (See also **crease mark** and **rope marks**.)

crease mark
A mark left in a fabric after a crease has been removed. It may be caused by mechanical damage to the fibres at the fold, by variation in treatment due to the constriction along the fold, or by disturbance of the fabric structure.

crease-recovery
The measure of **crease-resistance** specified quantitatively in terms of crease-recovery angle.

crease-resist finish
A finish, usually for cellulosic-fibre fabrics or their blends, that improves the crease recovery and smooth-drying properties. In the process used most extensively, the fabric is impregnated with a solution of a reagent that penetrates the fibres, and, after drying and curing, cross-links the fibre structure under the influence of a catalyst and heat. The crease resistant effect is durable to washing and using the fabric.

crease-resistance; wrinkle resistance
Resistance to, and/or recovery from, creasing of a textile material during use.
Note 1: It may still be possible to introduce defined creases in a crease-resistance fabric by the application of heat and pressure. (See also **durable press**.)
Note 2: The early use of the term 'non-crush' finish was replaced by 'anti-crease' finish and **crease-resist finish**; **easy-care** fabric is a more general term.

Credit rain simulation tester
See **water repellency**.

creel
A structure for holding supply packages in textile processing.

creel, beam
See **beam creel**.

creel, warping
See **warping creel**.

creep
The time-dependent increase in strain resulting from the continuous application of a force.
Note: Creep tests are usually carried out at constant load and constant temperature.

creep recovery
The time-dependent decrease in strain following removal of stress.

crêpe, warp-knitted
See under **crêpe fabric**.

crêpe, weft-knitted
See under **crêpe fabric**.

crêpe cord; moss cord (narrow fabric)
A cord comprising two to four strands, each in turn comprising a core covered by several fine threads in an acute helix, over-wrapped in the opposite direction by a strong thread, giving a soft crêpe or spiral effect. Two or more strands are laid together to complete the cord.

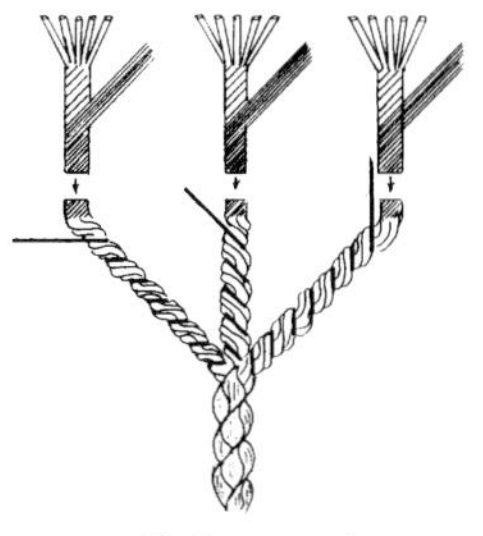

Crêpe cord

crêpe de chine
A light-weight crêpe fabric of plain weave, made with two S and Z highly twisted continuous-filament yarns alternating in the weft, and with a normally twisted continuous-filament warp.

crêpe embossing
1. The embossing of a fabric with a pattern resembling a true crêpe.
Note: The effect may be either permanent or not according to: the nature of the fibre; the conditions of embossing; and/or the accompanying finishing treatment.
2. See **precrêping**.

crêpe fabric
A fabric characterized by a crinkled or puckered surface. The effect may be produced in a variety of ways, for example, by the use of S and Z high twist yarns, by the use of a particular construction, or by chemical or thermal treatment to provide differential shrinkage in the finished fabric.

crêpe, warp-knitted
A double-faced warp-knitted fabric which contains more rows of cleared stitches per unit of length on one side than on the other.

crêpe, weft-knitted
An irregular, surfaced fabric either plain or rib-based usually constructed from knit-float or knit-tuck loops introduced in a predetermined random order.

crêpe fabric *(continued)*

crêpe, woven

A fabric produced by the use of highly twisted S and Z yarns, e.g., **crêpe de chine**, **crepon**, **marocain**, **georgette**, **crêpe suzette**, or by the use of **crêpe**, **oatmeal crêpe** or **moss-crêpe** weaves.

embossed crêpe

A fabric with a crêpe appearance imparted by embossing rollers (see **precrêping**).
Note: The permanence of the effect is governed by: (i) the fibre used; (ii) the finish accompanying the embossing process.

crêpe sizing

See **sizing, crêpe**.

crêpe suzette

See **crepon georgette**.

crêpe weave

A weave having a random distribution of floats so as to produce an 'all-over' effect in the fabric to disguise the repeat.

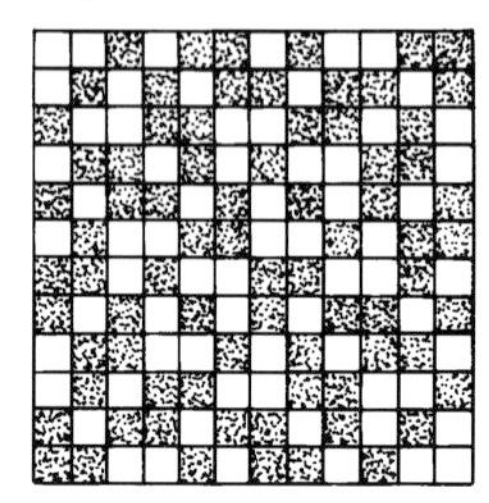

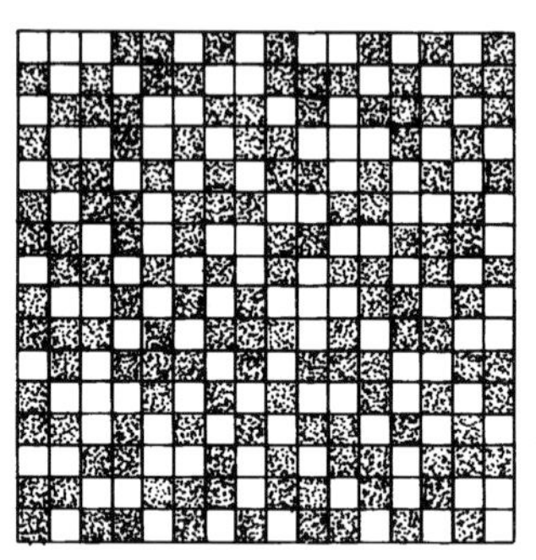

Crêpe weaves

crêpe yarn

1. A twisted and highly twist-lively yarn (see **twist liveliness**) which may be used in the production of crêpe fabrics.
Note: The twist liveliness may be achieved through the insertion of a very high twist or through the insertion of a moderate twist with an additional, thermally-set false twist.
2. A balanced handknitting yarn with a plaited appearance produced by twisting together 3 or 4 two-fold yarns normally using Z, S and Z twists (see **twist direction**).

crêping

1. A wet treatment that allows the relaxation of the strain of highly twisted crêpe yarns in fabric and so produces a characteristic crêpe effect.
2. A chemical treatment designed to produce an effect similar to 1.

crepon

A **crêpe fabric**, more rugged than the average crêpe, with a fluted or crinkled effect in the warp direction. It may be produced: (i) with crêpe-twist weft yarn, all with the same direction of twist; (ii) with weft yarns of different amounts of crêpe twist in the same or different directions; or (iii) by chemical or thermal means.

crepon georgette; crêpe suzette

A **georgette** in which all the weft yarn has the same direction of twist.

cretonne
A printed fabric originally and usually of cotton and of heavier weight than a chintz.

crimp
1. (Fibre) The waviness of a fibre.
Note: This fibre characteristic may be expressed numerically as the **crimp frequency** or as the difference between the lengths of the straightened and crimped fibre, expressed as a percentage of the straightened length.
2. (Yarn) The waviness or distortion of a yarn that is due to interlacing in the fabric.
Note 1: In woven fabrics, the crimp is measured by the relation between the length of the fabric sample and the corresponding length of yarn when it is removed therefrom and straightened under suitable tension.
Note 2: Crimp may be expressed numerically as (i) percentage crimp, which is 100 divided by the fabric length and multiplied by the difference between the yarn length and the fabric length, and (ii) crimp ratio, which is the ratio of yarn length to fabric length. In both methods, the fabric length is the denominator.

crimp, latent
A crimp that is potentially present in specially prepared fibres or filaments and that can be developed by a specific treatment such as by thermal relaxation or by tensioning and subsequent relaxation.

crimp contraction
See **crimp retraction**.

crimp frequency
The number of full waves or crimps in a length of fibre divided by the straightened length.

crimp recovery
A measure of the ability of a yarn to return to its original crimped state after being subjected to tension.

crimp retraction; crimp contraction
The contraction in length of a previously textured yarn from the fully extended state (i.e., where the filaments are substantially straightened), owing to the formation of crimp in individual filaments under specified conditions of crimp development. It is expressed as a percentage of the extended length.

crimp rigidity test
A form of **crimp retraction** test, used in the U.K. for the testing of false-twist textured nylon yarns (see **textured yarn**, *Note 1* (i)).

crimp stability
The ability of a **textured yarn** to resist the reduction of its crimp by mechanical and/or thermal stress.
Note: Crimp stability is normally expressed as the ratio of values of **crimp retraction** measured before and after a specified mechanical and/or thermal treatment of the yarn.

crimped length
See under **fibre length**.

crimped loop ruche
See under **ruche**.

crimped yarn
See **textured yarn**.

crinoline
1. A stiff fabric similar to **buckram**.
2. A stiff fabric made with a cotton warp and a horsehair weft.

critical application value; CAV
In a low **wet pick-up** easy-care finishing system, the CAV is the amount of finishing liquor which must be applied to a given fabric to avoid a non-uniform distribution of **cross-linking** after drying and **curing**.

CRL
See **constant rate of loading**.

croa
See **caroa**.

crochet-knitting machine
A warp-knitting machine, using latch or carbine needles, generally mounted horizontally in a needle bed. The fabric is removed at 90° to the needles' movement (vertically downwards) and is controlled by the top of the needle bed and a special bar placed in front of the fabric. The laying-in bars on a crochet machine make a special movement and do not swing in the same manner as the ground bars or the guide bars on a tricot or raschel machine. Because this type of machine finds wide-spread application in the production of various types of edgings or trimmings, it is sometimes described as a 'trimming machine'. (See also **needle (machine knitting)** and **knitting machine**.)

crocking; rubbing
A transfer of colour from the surface of a coloured fabric to an adjacent area of the same fabric or to another surface principally by rubbing action. (See also **colour fastness**.)

crockmeter
An apparatus for evaluating the colour fastness to rubbing of dyed or printed textiles.

crofting
See **grass bleaching**.

crop
See **shear**.

cropped terry pile
See **velour, woven** 3.

cross border dobby
A **dobby** that may be controlled by any one of two or more pattern chains for the purpose of weaving borders or hems across the fabric.

cross dyeing
The dyeing of one component of a mixture of fibres of which at least one is already coloured.

cross lapping; cross laying
The production of a nonwoven **web** or **batt** from a fibre web by traversing it to and fro across a conveyor moving at right angles to the direction of traverse.

cross-ball
See **warp**.

cross-ball warping
See **warp**.

cross-linking
The creation of chemical bonds between polymer molecules to form a three-dimensional polymeric network, e.g., in a fibre or in a pigment binder. This generally restricts swelling, inhibits solubility and alters elastic recovery. (See also **crease-resist finish**.)

cross-over tufting
See under **tufting machine**.

cross-plated fabric, knitted
See **plated fabric, weft-knitted**.

cross-tuck, weft-knitted
A generic name used to describe either plain or rib weft-knitted fabrics in which the construction repeats on a minimum of two courses and where tuck loops alternate with knitted loops within a course and between one course and another.

cross-wound package
A package characterized by the large crossing angle of the helices of sliver or yarn.

crossbred
1. A sheep that is the progeny of parents from two different breeds.
2. (Common usage) Wool coarser than 25 microns (μm) in diameter.

crossed shed (leno weaving)
The shed formed when the crossing end is lifted over the weft in its crossed position when using bottom douping (see **leno weaving**, *Note 3*).

crossing
See **cobwebbing**.

crossing heald
See **leno weaving**.

crotch; crutch; fork
The region in a bifurcated garment where the legs of the garment join, normally the junction of the inside leg seams.

> **depth of crotch; depth of crutch; body rise**
> The measurement from the side waist level to the fork level where the legs join the trunk. *Note:* This measurement is taken from a seated figure: from the side waist level to the horizontal surface.
>
> **width of crotch; width of crutch; fork quantity**
> The measurement in bifurcated garments of the distance through the trunk, from front to back, measured through a point where the legs join the trunk.

crotch; crutch; fork *(continued)*

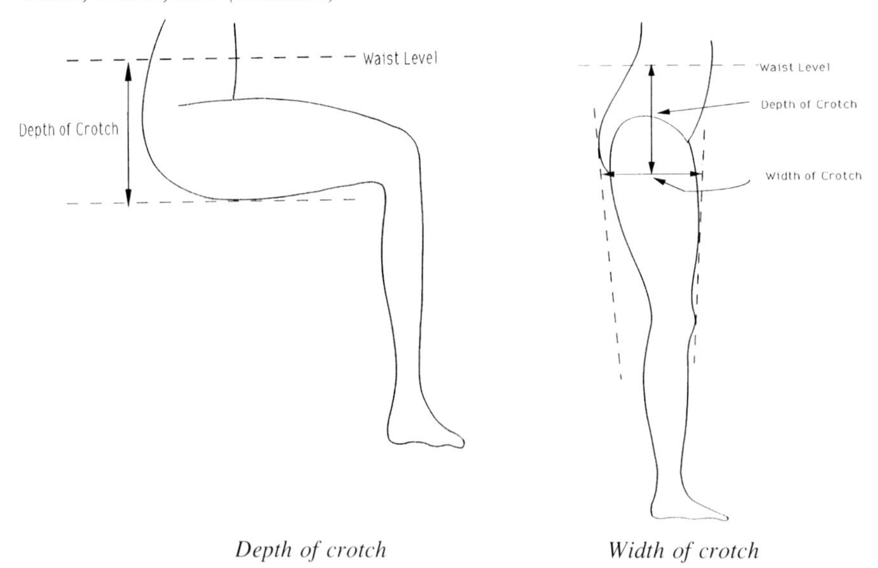

Depth of crotch *Width of crotch*

crow twill; broken crow; crow weave
A 3/1 warp **twill** generally used on wool and worsted fabrics.

crow's feet (defect)
Fabric breaks or wrinkles of varying degrees of intensity and size, resembling birds' footprints in shape, and occurring during the wet processing of fabrics.

crow's feet (knitting)
A puckering effect, usually in the heel or toe of circular-knitted hosiery and generally associated with the **suture line**.

crow's foot (weaving)
A small colour-and-weave effect in a fabric, produced by a combination of a matt weave and a specific order of colouring in warp and weft (e.g., 2/2 matt weave and 4 and 4 order of colouring; 3/3 matt weave and 6 and 6 order of colouring etc.).

croze marks
Crease marks which may be produced during wet treatments of **hoods** or **bodies**, e.g., in dyeing.

CRT
See **constant rate of traverse**.

crumbs
Shredded **alkali-cellulose**.

crush cutting
See **converting**.

crutch
See **crotch**.

crutchings
Wool from around the crutch of the sheep, often stained.

Note: This is often removed some time before lambing or shearing to reduce the attack of blowfly.

crystallinity
Three-dimensional order in the arrangement of atoms and molecules within a chemical phase.
Note: Most chemical compounds of low molecular weight may be obtained in a state of virtually complete three-dimensional order. When polymers crystallize, in general the product consists of regions of high order (crystallites), regions of low order (amorphous regions), and regions of intermediate order. Different methods of measuring the degree of crystallinity (e.g., density, wide-angle X-ray scattering, enthalpy measurement) emphasize different aspects and therefore lead to quantitatively different values. In recent years the simple distinction of crystalline and amorphous regions has been questioned and terms such as para-crystalline have been introduced.

CSP
See **lea count-strength product**.

cuff
The lower part of the **sleeve** of a garment, and also the corresponding parts of trousers and gloves. It may consist of a simple turned-in or turned-up section of the sleeve material, usually secured by stitching, or it may be a separate section, sometimes of different material, attached to the lower edge of the sleeve. In knitted garments the cuff is often the **rib fabric** knitted integrally with the sleeve.

cumber board
See **comber board**.

cup seaming
The joining together of two edges, usually knitted selvedges. The edges to be joined are positively fed to a sewing point by two cup-like wheels. (See also **linking**.)

cuprammonium rayon (fibre)
See **cupro (fibre)**.

cupro (fibre) (generic name); cupra fibre (fibre) (generic name U.S.A.)
A manufactured fibre of cellulose obtained by the cuprammonium process. (See also Classification Table, p.401.)

curing
1. A process following application of a finish to textile fabrics in which appropriate conditions are used to effect a chemical reaction. Heat treatment for several minutes is usual, but higher temperatures for short times (flash-curing) and long times at low temperatures and higher regain (moist curing) are also used.
2. A chemical reaction causing cross-linking of a thermoset resin in a fibre composite.

curing tape
A woven **narrow fabric**, usually a twill, used in the rubber hose industry to stabilize the dimensions of flexible tubes during hot processing. To facilitate helical wrapping, most curing tapes have a coloured central guideline. The diameter of the hose determines the width of the tape used.

curled pile
See under **pile (carpet)**.

curled yarn
A yarn so constructed or treated, or both, as to produce a pile with a curled effect when used in a suitable fabric construction.

curtain machine
See **lace furnishing machine** under **lace machines**.

curvature
See **bow**.

cut
1. A length of fabric in the grey or loom state, or the length of warp required to produce it. A cut of fabric was usually of the order of 45-90 metres (50-100 yards), probably because it was a convenient length to cut from the cloth roller during the weaving operation. (See also **piece**.)
2. See Table, p.397.

cut (knitting)
An American term for **gauge (knitting)**.

cut crimped ruche
See under **ruche**.

cut mark
An indication on a weaver's warp of a precise length of material, generally a piece length or a fraction of a piece length.

cut pile
See under **pile (carpet)**.

cut presser (warp knitting)
A **presser** with an edge castellated so that only selected bearded needles are pressed during a particular knitting cycle.

cut ruche
See under **ruche**.

cut weft
See **chopped weft**.

cut-loop pile
See under **pile (carpet)**.

cut-off; binder
One or more ends, in plain or rib weave, which separate and highlight the division between the edge and body weave in a ribbon.

cuticle
The surface layer of animal hair fibres, consisting of flat overlapping scales.

cutting (knitting)
A defect in knitted fabric caused by the cutting of yarn by the knitting elements during stitch formation. (See also **bursting**.)

cuttle
To place fabric in loose transverse folds, usually in open width. (See also **plaiting 2.**)

Note: Cuttled fabric is sometimes known as 'lapped' fabric.

cyclic test
A test in which a sequence of operations is repeated between defined limits.

cylinder; card cylinder; swift
Usually the largest diameter wire-covered cylinder of the card, often referred to as the main cylinder of a card. (See also **workers** and **card flats**.)

cylinder, drying
See **drying cylinder**.

cylinder and dial knitting machine
A circular weft-knitting machine in which knitting is carried out on two sets of needles, one set being mounted vertically and in parallel around a needle cylinder and the other mounted horizontally and radially on a dial, concentric with the needle cylinder. (See also **knitting machine**.)
Note: In some types of machine the dial may contain elements such as **pelerine** points or **jacks** as well as or instead of needles.

cylinder bed
See under **bed (sewing machine)**.

D and K
The term stands for 'damaged and kept', and denotes lengths of fabrics spoilt in processing.

dag; dags; daggings (wool)
Soiled and tangled wool from the back end of the sheep.
Note: This is not suitable for use in textile processing.

damaged length
The maximum extent, in a specified direction, of the damaged area of a material under specified combustion test conditions.
Note: The deprecated term 'char length' may be used synonymously for damaged length.

damask
A figured fabric made with one warp and one weft in which, generally, warp-satin and weft-sateen weaves may sometimes be introduced.

Damask

damasquette
A **damask** woven with more than one weft to provide extra colour.

dart
A wedge or diamond-shaped section removed from the surface area of a garment part by stitching or cutting and stitching. (See also **fish**.)

deacetylated acetate (fibre)
A manufactured fibre of regenerated cellulose obtained by almost complete de-ethanoylation (deacetylation) of a cellulose ethanoate (acetate).
Note: This name was formerly an ISO **generic name**. (See also Classification Table, p.401.)

dead fibre (cotton)
See **maturity (cotton)**.

dead frame yarn (carpet)
In a Wilton carpet, a complete frame of pile yarn which lies embedded in the substrate.

dead pick
See **pick, dead**.

dead twist
See **twist set**.

dead wool; fallen wool
Wool taken from sheep that have died from natural causes. (See also **skin wool**.)

dead yarn (carpet)
The pile yarn in a multiframe Wilton carpet which lies completely flat in the substrate and is not raised by the jacquard mechanism. It excludes all yarn forming the effective pile and the pile root.

de-aeration
The removal of undissolved gases and part of the dissolved gases (chiefly air) from spinning solutions prior to extrusion.

deburring
A process in wool yarn manufacture for extracting burrs, seeds and vegetable matter from raw wool.

decatizing; decating
A finishing process used chiefly to improve fabric handle and appearance. In batch decatizing, the fabric, interleaved with a smooth cotton wrapper, is wound tightly on a perforated roller through which steam is blown. In continuous decatizing, the steam is blown through fabric which passes continuously between one or more perforated drum rollers and a smooth cotton wrapper forming an endless belt.
Note: The process is used mainly for fabrics containing wool.

decitex
See **tex system**.

deck-chair canvas
A fabric in plain, repp, or twill weave, made of vegetable or manufactured fibres, or a combination of any of these fibres. Traditionally, widths have not exceeded 500mm (20 inches) and weights have not been less than 185g/sq m (6oz/sq yd).

declared composition
The nominal composition of a mixture of textile materials, used for labelling and invoicing purposes and, where appropriate, for the calculation of the weighted **commercial allowance** or weighted moisture **regain** of the mixture.

deep dyeing
A property of fibres modified so as to have greater uptake of dyes than normal fibres.

deer fibre (hair)
Fibre from the deer (Genus *Cervus*).

degradation
A deleterious change in properties of a textile.

degreasing
1. The removal of grease, **suint**, and extraneous matter from wool by an aqueous or solvent process.
2. The removal of natural fats, waxes, grease, oil, and dirt from any textile material by means of an organic solvent.

degree of orientation
The extent to which the macromolecules composing a fibre or film lie predominantly in one direction. In the case of fibres the predominant direction is usually the fibre axis.
Note 1: There are several methods for assessment of the degree of orientation, of which measurement of **birefringence** is one of the most usual.
Note 2: The degrees of orientation of crystalline and non-crystalline regions may be evaluated separately.

degree of polymerization; DP
The average number of repeating units in the individual macromolecules in a polymer.
Note: In general, this average will depend on the basis on which it is calculated, which should be stated. For example, it may be based upon a mass (weight) or a number average.

degrees Baumé
See **Baumé, degrees**.

degrees Twaddell
See **Twaddell**.

degumming; boiling off
The removal of sericin (silk gum) from silk yarns or fabrics, or from silk waste prior to spinning, by a controlled, hot, mildly alkaline treatment intended to have little or no effect on the underlying **fibroin**.

delaine
A light-weight all wool fabric in plain weave, ornamented by printing.

delayed needle timing (weft knitting)
See **timing (weft knitting)**.

delustrant; delusterant
A material added to a spinning composition before extrusion to reduce the lustre of a manufactured fibre.
Note 1: The anatase form of titanium dioxide is commonly used for this purpose.

delustrant; delusterant *(continued)*
Note 2: Terms used to indicate the level of delustrant in manufactured fibres include: clear, **bright**, semi-dull, **dull**, **matt**, extra dull, and super-dull.

delustring
1. The use of a **delustrant**.
2. A process for reducing the lustre of fabrics, e.g., cellulose acetate fabrics may be delustred by treatment with alkali, and viscose fabrics by the deposition of urea formaldehyde polymer.

denier
The mass in grams of 9000 metres of a fibre, filament, or yarn (see **tex system** and Table, p.396).

denim
Traditionally a 3/1 warp-faced twill fabric made from yarn-dyed warp and undyed weft yarn. Typical cotton construction: 32x19; 45x54tex; 310 g/m^2; K=21.5+14.0. More recently, other weaves have been used in lighter constructions.

dent
The unit of a reed comprising a reed wire and the space between adjacent wires.

denting plan; reeding plan; sleying plan
An indication of the position of the reed wires in relation to the ends in one **weave repeat**.

depitching
The removal of tar or other branding substances from wool, usually, though not necessarily, by solvent-extraction.

depth (coloration)
That **colour quality**, an increase in which is associated with an increase in the quantity of colorant present, all other conditions (viewing, etc.) remaining the same.

depth of crotch; depth of crutch
See under **crotch**.

Derby rib
See under **rib fabric, weft-knitted**.

desi cotton; deshi cotton
Coarse, short cotton of the *Gossypium arboreum* species produced in India and Pakistan. Used for waddings, medical, and sanitary products.

design paper; point paper
1. (Weaving) Paper ruled with vertical and horizontal lines in a manner suitable for showing weaves and designs.
Note: Generally, each space between vertical lines represents one end and each space between horizontal lines represents one pick. The design paper commonly used has equally spaced fine rulings, with heavy over ruling in blocks of eight by eight. For figured designs, other rulings may be used: for example, in jacquard designs, it is convenient to use these rulings according to (i) the number of needles in the short row, and (ii) the ratio of numbers of ends to picks per length, the objects being (a) to facilitate card cutting, and (b) to ensure that the design is represented in the correct proportion, width to length. (See also **point-paper design (woven fabrics)**.)
2. (Knitting) Paper printed with a series of horizontal and vertical lines, dots or other shapes representing a ground structure on which the draft of a design is plotted.

designers' blanket
See **pattern blanket**.

desizing
The removal of **size** from fabrics.

detergent
A substance, normally having surface-active properties, specifically intended for cleansing a substrate.

detwisted
Descriptive of a yarn of fibres or filaments from which twist has been removed.

developing
A step in a dyeing or printing process in which an intermediate form of the colorant is converted to the final form (e.g., oxidation of a vat leuco ester).

devoré style
See **burn-out style**.

dew point
The temperature at which the saturation pressure is equal to the actual pressure of the water vapour in air.
Note: When air is cooled at constant pressure, condensation begins at this temperature.

deweighting
The controlled alkaline hydrolysis of the surface of polyester fibres to bring about a decrease in fibre decitex and a decrease in the weight of the material. Weight losses of 10-30% may typically be brought about to confer enhanced silk-like properties on the textile material. (See also **saponification** and **causticizing**.)

diacetate (fibre)
A manufactured fibre made from propanone-soluble (acetone-soluble) cellulose ethanoate (acetate).
Note: The ISO **generic name** is **acetate**. (See also Classification Table, p.401.)

dial and cylinder knitting machine
See **cylinder and dial knitting machine**.

diamond barring
weft streaks that are distributed in characteristic pattern in a woven or flat weft-knitted fabric. It is the result of a periodic variation in the diameter, twist, tension, crimp, colour, or shade of the weft yarn.
Note: The dimensions of the fault depend on the ratio of the length of the periodic variation to the width of the fabric and only rarely is it seen in a clearly diamond form.

Diamond barring in fully fashioned silk stocking

diaper

1. The original diaper was made of linen and based on a 5-end sateen, woven on the damask principle. The cloth has a smooth even surface. The preferred method of producing this group of weaves is shown at A. It is based on a diced weave e.g., the 5-end sateen. Another form is shown at B where 2 basic weaves are used. Many weaves within this group produce effects which are of a diamond character as shown at C. A very ancient form of diaper is the **huckaback**. Diapers are produced in cotton, linen and flax, and used principally for towel and toilet purposes.
2. A baby's napkin.

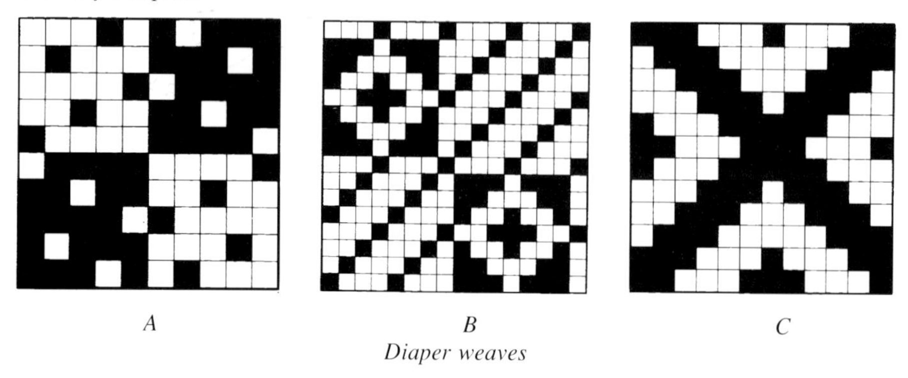

Diaper weaves

diazotize

To convert a primary aromatic amine into the corresponding diazonium salt, by treatment with nitrous acid.

diced weaves

A group of weaves produced by quartering and reversing a weave element, thus forming opposite surfaces and directions in each quarter and clean cuts between the sections. Weaves tend to repeat on relatively small areas, e.g., 8x8 or 10x10, but larger versions are possible. Also referred to as **diaper** weaves.

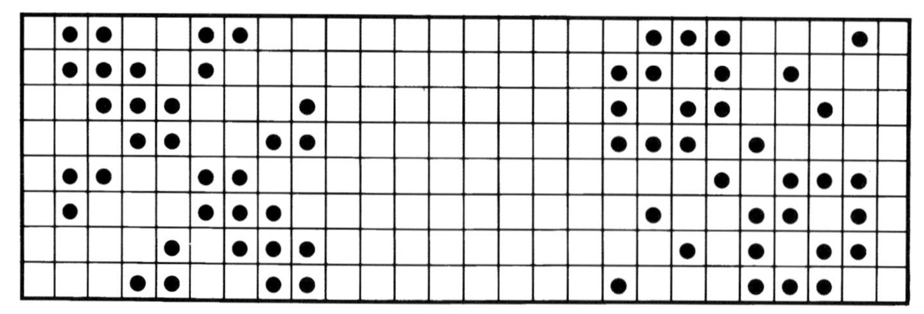

Diced weaves

die swell

The increase in diameter that occurs as a visco-elastic melt or solution emerges from a die or spinneret hole. (See also **swell ratio**.)

differential bottom feed

See **differential drop feed** under **feed mechanisms (sewing)**.

differential drop feed

See under **feed mechanisms (sewing)**.

differential dyeing
Usually descriptive of fibres of the same generic class, which have potential dyeing properties different from those of the standard fibre. (See also **deep dyeing**.)

diffusion
The movement of substances owing to the existence of a concentration gradient.

dimensional stability
1. The ability of a fabric or floorcovering to retain its dimensions when exposed to use and/or an ageing process, to water, washing, steaming, drying or other process.
2. Changes in length and/or width of a textile when subjected to specified conditions.
Note 1: The changes may be positive or negative.
Note 2: The changes may be reversible or irreversible.

dimity
A fabric, usually of cotton, that is checked or striped by corded effects, made by weaving two or more threads as one.

dip
1. An immersion of relatively short duration of a textile in a liquid.
2. The depth of liquid in the inner cylinder of a rotary washing machine.
3. A laboratory dyeing, usually to develop a dye formula.

direct beaming
The winding of the total number of warp ends in full width in a single operation from creeled bobbins, either on to a weaver's beam, or on to a sectional beam.

direct cabling
A system of producing a twist-balanced folded yarn from balanced (or twistless) single yarns. The process is usually carried out on a modified **two-for-one twisting** machine.

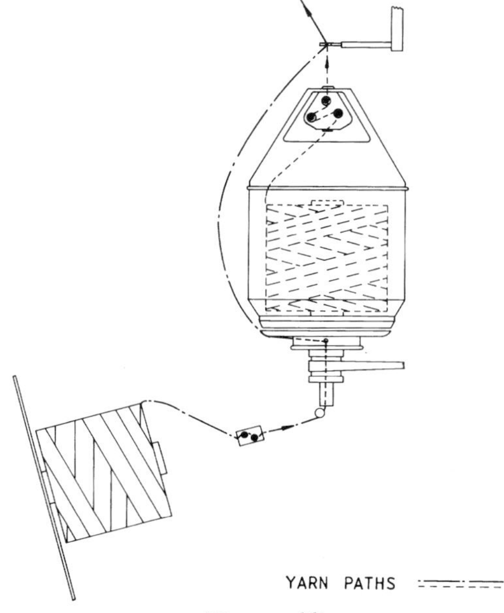

Direct cabling

direct dye
An **anionic dye** having substantivity for cellulosic fibres, normally applied from an aqueous dyebath containing an electrolyte.

direct spinning
1. (Manufactured fibre production) Integrated polymerization and fibre extrusion without intervening isolation or storage of the polymer.
2. (Manufactured fibre yarn production) The process whereby **tow** is converted to staple fibre and spun into yarn in an integrated operation.
3. (Bast fibre production) A method of dry-spinning bast fibres whereby untwisted slivers are drafted with suitable controls and directly twisted into yarn.
4. (Short-staple yarn production) The production of yarn directly from card sliver.

direct style
A style of printing in one or several colours where the dyes are applied and then fixed by ageing or other appropriate means. The fabric is usually initially white but may sometimes have been previously dyed (see **discharge (printing)** and **resist style**).

direct warping
The transference of yarn from a package creel directly on to a beam. (See also **beaming** and **section warping**.)

direct-spun
1. A term used to describe filaments or yarn produced by **direct spinning**.
2. Descriptive of woollen yarns spun on a mule on to weft bobbins.

discharge (printing)
To destroy by chemical means a dye or mordant already present on a substrate to leave a white or differently coloured design.

discharging
The destruction by chemical means of a dye or mordant already present on a material to leave a white or differently coloured pattern.
Note: This term is also used to describe the removal of gum from silk (see **degumming**).

disperse dye
A substantially water-insoluble dye having substantivity for one or more hydrophobic fibres, e.g., cellulose acetate, and usually applied from fine aqueous dispersion.

dispersion
See under **fibre length**.

dispersion spinning
A process in which polymers that tend to an infusible, insoluble, and generally intractable character (e.g., polytetrafluoroethylene) are dispersed as fine particles in a carrier such as sodium alginate or sodium cellulose xanthate solutions, that permit extrusion into fibres, after which the dispersed polymer is caused to coalesce by a heating process, the carrier being removed either by a heating or by a dissolving process.

dissolving pulp
A specially purified form of cellulose made from wood tissue.

distorted selvedge
See **selvedge, distorted**.

distorted thread effect
A **mock-leno** weave where threads can be displaced from their normal line in either the warp direction (see diagram), or the weft direction.

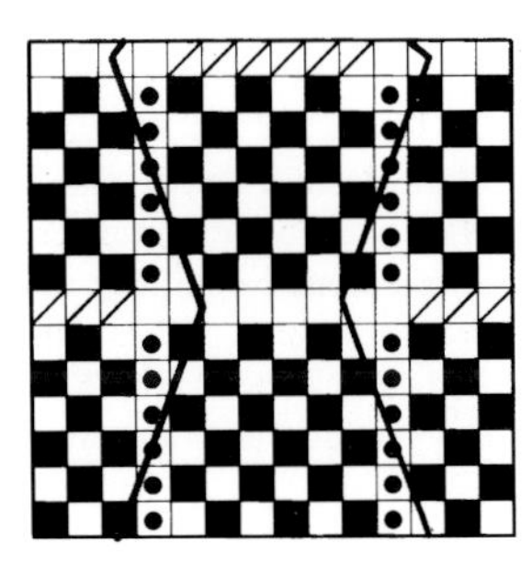

Distorted thread effect

district checks
Bold distinctive woollen checks originally made in Scotland, usually in 2x2 twill weave.

divided harness
A **jacquard** in which the odd numbered ends in a fabric are controlled by the hooks in the first length-wise or vertical rows in the jacquard, and the even numbered ends are controlled by the other rows.
Note: This type of harness is used when weaving jacquard fabrics having two distinct sets of figuring threads, e.g., double cloths figured by interchange or repps. It makes card cutting easier and simplifies fault tracing.

dobby
A mechanism for controlling the movement of the **heald shaft** of a loom. It is required when the number of heald shafts or the number of picks in a repeat of the pattern or both are beyond the capacity of tappet shedding.

dobby fabric
Any one of a variety of fabrics or a variety of weaves which require a **dobby**.

doctor
A straight-edged metallic blade mounted parallel to a moving surface, e.g., a printing roller, to remove excess of, or unwanted, material.

doctor streak
A white or coloured streak in the lengthwise direction on a coated or printed substrate caused by a damaged or incorrectly set **doctor**.

doeskin fabric
A five-end satin or other warp-faced fabric with dress-face finish.
Note: Today, other weaves, such as 2/1 warp twill and 3/1 broken crow, are very often used and given a dressed finish, and the name doeskin is applied. It is often the effect and the kid-glove handle due to the finish that cause such a fabric to be placed in the category of a doeskin. The fabric is all wool, often all merino, or possibly blended wool including merino.

doffer
1. A wire-covered cylinder used for the removal of fibres from the main cylinder of a card. (See also **cylinder** and **carding**.)

doffer *(continued)*
Note: The fibres are condensed on to the doffer to form a card web.
2. A person who, or mechanism which, removes packages or material from a textile machine used in yarn manufacture.

doffing
1. The removal of material or packages from a textile machine.
2. The replacement of full spinning packages with empty spinning tubes (cops). This can be performed automatically by the use of automatic doffing units.

doffing tube (rotor spinning)
An extension to the **navel** to guide the withdrawn yarn from the rotor. (See also **rotor spinning**.)

dog-legged selvedge
See **selvedge, uneven**.

dogstooth check
See **houndstooth check**.

dolly
1. A machine in which fabric pieces sewn end to end are circulated repeatedly through a liquor by means of a single pair of squeeze rollers and a drawing-off roller above the liquor.
2. A machine, also known as a 'tom-tom', in which lace, hosiery, or knitwear are subjected to the action of free-falling beaters while immersed in a detergent solution and carried in a moving trough.

domet; domett
Imitation flannel made from a cotton warp and a wool or cotton-wool weft, finished with a raised surface on both sides and used for linings, skirtings, and pyjamas, having an approximate mass per unit area of 84 gm^{-2}.

Donegal
A plain-weave fabric woven from woollen-spun yarns characterized by a random distribution of brightly coloured flecks or slubs. It was originally produced as a coarse woollen suiting in County Donegal.

dope
See **spinning solution**.

dope-dyed
See **mass colouration**.

dosuti
A Hindi word, which literally means 'two threads', used to describe the operation of combining two threads together at a winding machine, in which case the operation is known as 'dosuti winding'. When applied to fabric, it means that two warp ends are working in pairs and that two weft threads are placed in the same shed. (See also **assembly winding**.)

double (yarn)
See **fold (yarn)**.

double bow
See **bow (weaving)**.

double braided rope
See under **rope**.

double cloth, woven
A **compound fabric** in which the two component fabrics are held together by one of the following: (i) centre-stitching, in which a special series of stitching threads, lying between the two fabrics, are interlaced alternately with them and thus bind them together; (ii) self-stitching, in which threads from one fabric interlace with the other (e.g., by taking a back warp thread over a face weft thread); (iii) interchanging, in which the two fabrics are so woven as to interchange with each other. In some cases, the fabrics are completely interchanged whereas in others only the warp or weft threads interchange.

double cylinder knitting machine
A circular weft-knitting machine with two cylinders, one superimposed above the other. It is equipped with one set of double-ended needles that can be caused to knit in either cylinder as required to make **plain**, **rib**, or **purl** structures. This arrangement is commonly used in **hosiery knitting machines** and **garment-length knitting machines**. (See also **knitting machine**.)

double end (defect)
1. See **spinners' double**.
2. Two ends unintentionally weaving as one. The fault may appear as a line down the length of a fabric.

double jersey, weft-knitted
A generic name applied to a range of knitted fabrics made on a **rib** or **interlock** basis, the construction of which is often designed to reduce the natural extensibility of the structure. The term is generally confined to fabrics knitted on machines of E10 **gauge** or finer and it may be classified as either non-jacquard or jacquard double jersey. For examples of non-jacquard fabrics see **bourrelet, double piqué, Swiss double piqué, French double piqué, eight-lock, interlock, Milano rib, half-Milano rib, piquette, punto di Roma, single piqué,** and **texipiqué**; and for jacquard fabric see **rib jacquard (weft knitting)**.

double knitting yarn
A four-ply handknitting yarn with a resultant yarn number of approximately 440tex.

double lift (weaving)
A term applied to dobbies (see **dobby**) and **jacquard mechanisms** in which there are two knives or **griffes**, one operating on odd picks, the other on even picks, to effect the lift (see **lift (weaving)**).

double locker machine
See under **lace machines**.

double London
A worsted twill tape, generally with a two-fold warp.

double marl yarn
See **worsted yarns, colour terms**.

double pick (defect)
The insertion of two picks in a shed during weaving where only one is intended. The fault may appear as a line across the width of the fabric.

double piqué, weft-knitted
A non-jacquard **double jersey** fabric made on a rib basis, using a selection of knitted loops and floats. The two most important sequences are known as Swiss and French double piqué respectively, and the knitting sequences for each are shown. Double piqué is also known as 'wevenit', 'rodier', and 'overnit'. (See also **double jersey, weft-knitted**.)

double piqué, weft-knitted *(continued)*

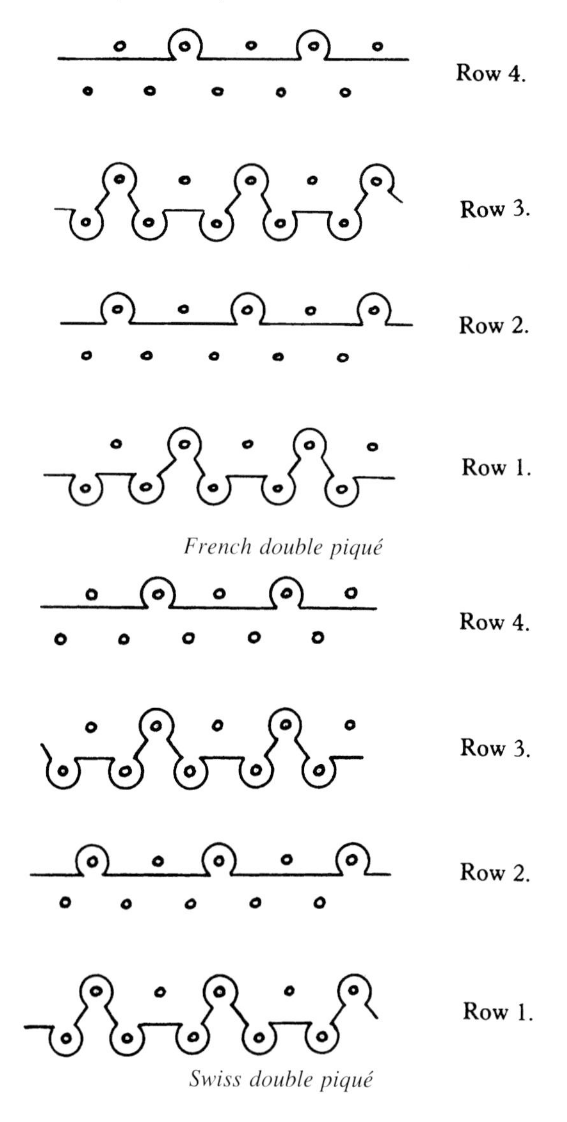

French double piqué

Swiss double piqué

double plain
A form of woven double fabric. (See also **double cloth, woven**.)

double plush, warp-knitted
See **plush, double, warp-knitted**.

double roving spinning
A system in which two **rovings** are fed to each spindle of a ring frame so producing a pseudo two-fold or two-strand yarn. The rovings are separated in the **drafting** system by means of special guides and the two drafted strands then combined after the drafting system. (See also **spinning**.)
Note: The term was originally used to describe the spinning of a yarn from a two-roving feed in order to achieve improved regularity.

double satin
A warp-backed satin, frequently used for ribbons, with the backing ends stitched in sateen order to present a warp-satin surface on both sides of the fabric.

double satin ribbon
Originally a ribbon in which every other warp end floats on one surface and the other warp ends on the other surface in a satin weave. In recent times, it is more usually produced in a 4/4 weave, and thus every end floats alternately on both surfaces. It is produced from continuous-filament yarns.

double shot (narrow fabrics)
The construction of a **ribbon** or **galloon** in which the body of the fabric is in 2/2 weave, providing a pronounced weft-way rib. A plain weave **binder** separates the body from the selvedge.

double-V twill
A narrow fabric weave in which there are two repeats of a herringbone pattern in the width of the fabric.

double-ended needle (machine knitting)
See under **needle (machine knitting)**.

double-faced jacquard, weft-knitted
A patterned rib-based fabric in which a different jacquard design appears on both sides of the fabric. (See also **reverse jacquard, weft-knitted**.)

double-plush carpets
See **face-to-face carpets**.

doubled yarn
See **folded yarn**.

doubling (knitting)
In the process of joining a rib border to the body of a garment, the action of joining two rib loops to one body loop, at spaced intervals. This is necessary when, in order to give a better fit, a rib border has more loops than the body. At these spaced intervals, two rib loops may be run into one **point** during **bar-filling**.

doublings (drawing)
The number of laps, slivers, slubbings, or rovings fed simultaneously into a machine for drafting into a single end.
Note: Doubling is employed to promote blending and regularity.

doup
See **leno weaving**.

douping heald
See **leno weaving**.

doupion; dupion
A silk-breeding term meaning double-cocoon; hence, an irregular, raw, rough silk reeled from double cocoons.
Note: Doupion is often used as weft for furnishing fabrics.

dowlas
1. Originally, a plain-woven coarse linen fabric used for clothing.

dowlas *(continued)*
2. A low-quality cotton fabric made of coarse rough-spun yarn, finished to imitate linen and used for towels, aprons, etc.

downtwisting
See **ring twisting**.

DP
See **degree of polymerization**.

draft
1. To reduce the linear density of a fibrous assembly by drawing.
2. When drafting, the degree of attenuation calculated either as the ratio of the input to output linear densities, or as the ratio of the surface speeds of the output and input machine components which bring about drafting.

draft (clothing)
A scale plan of a garment or garment section. Drafting is the process of creating flat patterns from 3-dimensional measurements.

draft board; draft box
A device which forms part of a rope-scouring machine (**dolly** scouring machine or a rotary milling machine) and keeps separate the ropes of fabrics scoured at the same time. In a rotary milling machine or a combined scouring and milling machine, the draft board may also serve as a stop motion should the fabric become entangled in the machine.

drafting
1. The process of drawing out laps, slivers, slubbings, and rovings to decrease the linear density (see **drawing**).
2. The order in which threads are drawn through heald eyes before weaving (see **drawing-in**).
3. (Lace) See **draughting**.

drafting plan
See **drawing-in plan**.

drape
1. The ability of a fabric to hang in graceful folds, e.g., the sinusoidal-type folds of a curtain or skirt.
2. The direct application of fabric to a stand/dummy or body and the manipulation of the fabric to develop a design or as a means of producing a pattern.

drapes (surgical)
Sterilised fabric sheets used to cover a person undergoing surgery or sterilised instruments.
Note: Patient drapes are cut to different shapes according to the surgical procedure and may include an opening over the location of the incision. Drapes traditionally have been made from cotton, but increasingly disposable nonwoven materials are now used.

draughting; drafting (lace)
The conversion of a design into diagrammatic form preparatory to punching of jacquard cards, or assembly of pattern chains, or preparation of magnetic tape.

draw (sampling)
See **pull**.

draw cord
A cord which passes through a **channel** and allows the garment to be drawn into gathers.

draw mechanism (knitting)
A mechanism on a straight-bar knitting machine for converting rotary motion into reciprocating motion for the purpose of laying the yarn and kinking it round the needles.

draw pin
A stationary pin or guide that, by inducing a localised change in yarn tension and/or temperature, may be used to stabilize the position of the draw-point or neck in some processes of drawing of manufactured-fibre yarns. (See also **snubber pin**.)
Note: For the drawing of some fibre types, e.g., polyester, a heated pin may be used; with other types, e.g., nylon, the pin is normally not heated.

draw ratio; stretch ratio

machine draw ratio
In a drawing process, the ratio of the peripheral speed of the draw roller to that of the feed roller.

natural draw ratio
The ratio of the cross-sectional areas of a filament before and after the **neck** when a synthetic filament or film draws at a neck.

residual draw ratio
The draw ratio required in draw-winding, draw-texturing and draw-warping to convert a partially oriented yarn into a commercially acceptable product.

true draw ratio
In a drawing process, the ratio of the linear densities of the yarn before and after drawing.

draw roller; draw roll
The output roller of a zone in which drawing takes place.

draw thread (knitting)
A thread introduced during knitting which, on removal, separates articles that have been knitted as a succession of connected units.

draw threads (lace)
Removable threads included in the construction of lace either to act as a temporary support for certain parts of the pattern or to hold together banded laces, that are separated subsequently by their removal.

draw-beaming
See **draw-warping**.

draw-down
See **spin-stretch ratio**.

draw-spinning
A process for spinning partially or highly oriented filaments in which the orientation is introduced prior to the first forwarding or collecting device. (See also **spin-drawing** and **high-speed spinning**.)

draw-texturing
A process in which the **drawing** stage of synthetic yarn manufacture is combined with the texturing process.

draw-texturing *(continued)*
Note 1: The drawing and texturing stages may take place in separate, usually consecutive, zones of a machine (sequential draw-texturing) or together in the same zone (simultaneous draw-texturing).
Note 2: The usual feedstock for simultaneous draw-texturing is **partially oriented yarn**.

draw-twist
To orient a filament yarn by drawing it and then to twist it in integrated sequential stages.

draw-warping; draw-beaming; warp drawing
A process for the preparation of warp beams or section beams from a creel of packages of **partially oriented yarn** in which the stages of drawing and beaming are combined sequentially on one machine.

draw-wind
To orient a filament yarn by **drawing** it, and then to wind it to form a **package** in an integrated process without imparting twist.

drawing (staple yarn)
Operations by which slivers are blended, doubled, or levelled, and by **drafting** reduced to a **sliver** or a **roving** suitable for spinning.
Note: In short-staple spinning the term is only applied to the process at a drawframe. Various systems of **drawing** are practised in modern worsted spinning, but with machinery development, and the greater use of manufactured staple fibres, the differences are becoming less distinct. Most modern worsted drawing sets comprise three passages of pin drafting (see **gilling**) and a roving process.

drawing (synthetic filaments and films)
The stretching to near the limit of plastic flow of synthetic filaments or films of relatively low molecular orientation.
Note: This process orients the molecular chains in the direction of stretching.

drawing, cold (synthetic filaments and films)
The **drawing** of synthetic filaments or films without the intentional application of external heat.
Note: Free drawing of filaments or films at a **neck** is also referred to as cold drawing even though this may be carried out in a heated environment.

drawing, hot (synthetic filaments and films)
The **drawing** of synthetic filaments or films with the intentional application of external heat.

drawing-in
The process of drawing the threads of a warp through the eyes of a heald and the dents of a reed. The operation thus includes that of reeding (see **reed**).

drawing-in plan; drafting plan
An indication of the order in which ends are controlled by specified heald shafts in one **weave repeat**. (See also **drafting**.)

drawn piece
A piece of fabric in which, as a result of distortion during some process subsequent to weaving, the warp yarns are not at right angles to the weft yarns (see **bow** and **off-grain**).

drawn yarn
Extruded yarn that has been subjected to a stretching or **drawing** process to orient the long-chain molecules of which it is composed.

drawn-pile finish; laid-pile finish
A finish given to textile fabrics to produce a surface **nap** or **pile** that is laid in one direction. The effect is usually produced by raising the wet or damp fabric. (See also **dress-face finish**.)

dress
1. Clothing, particularly outerwear.
2. A type of adjustment made to one side of trousers to improve fit (bespoke).

dress-face finish (wool fabric)
A finish characterized by a closely cropped surface and high lustre.
Note: This is obtained partly by raising and cropping and partly by the high degree of regularity of the lie of the fibres.

dresser sizing
See **Scotch dressing** 2 under **dressing (warp preparation)**.

dressing (flax)
A combing process applied to **stricks** or pieces of **line flax** fibre to parallelise the strands, remove **naps** or bunches of entangled fibres, and square the ends of the pieces by pulling or breaking fibre strands that protrude from the ends.

dressing (lace)
The operation of stretching lace, net, or lace-furnishing products to size, then drying, after the application of stiffening or softening agents. The stretching and drying may be carried out on either a running **stenter** or a stationary frame.

dressing (warp preparation)
The operation of assembling yarns from a ball warp, beam, or chain on a beam immediately prior to weaving.

Scotch dressing
1. (dry taping; Scotch beaming) A method of preparing striped warps for weaving, suitable for use when long lengths of any one pattern are to be woven. Three operations are involved:
(i) splitting-off from stock **ball warps** (bleached or dyed, and sized) the required number of threads of the required colours;
(ii) the winding of the differently coloured warps, each on to a separate 'back' or warper's **beam**; and
(iii) the simultaneous winding of the threads from a set of back beams through a coarse **reed** on to a loom beam.
2. (dresser sizing; Scotch warp dressing) A method of warp preparation, used particularly in the linen industry, which incorporates **sizing**. Yarn in sheet form is withdrawn from two sets of warper's beams (one set at each end of the machine) and wound on to a loom beam at a central headstock. Each half of the machine has its own size box and hot-air-drying arrangement.

Yorkshire dressing
A method of preparing a striped warp beam for a loom. Four operations are involved:
(i) splitting-off from stock **ball warps** (bleached or dyed, and sized) the required number of threads of the required colours;
(ii) the disposition of these threads to pattern in the **reed** with or without ends from stock grey warps;
(iii) the slow and intermittent winding of the threads on to the loom beam, during which process they are tensioned by means of rods and rollers, brushed by hand, and kept parallel and in correct position and if, as is usual, there are two or four ends per reed dent, these are further separated by means of a rod; and

dressing (warp preparation) *(continued)*

(iv) the picking of an end-and-end **lease**. The process ensures that in the warp all threads will be kept parallel, separated one from another, in their correct position, and correctly tensioned.

dressing, surgical
See **wound dressing materials**.

drill
A twill fabric of construction similar to a denim, but usually piece-dyed. Typical construction: 39x18; 37x50tex; 11.5x9.0%; 260 g/m^2; K=23.1+13.4; 3/1 twill weave, usually made in 0.7m widths.
Note: Drills made with a five-end satin weave are known as satin drills.

Actual size

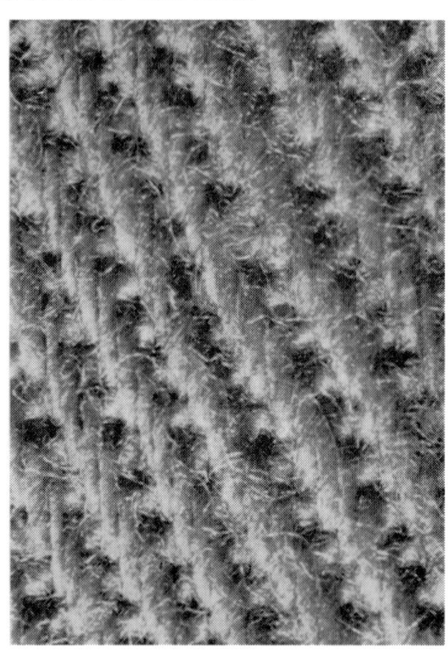

Magnification x 5

Drill

drip-dry
See **easy-care**.

driving bar (lace machines)
A bar running the net-making width of the double locker bobbinet machine. There are two driving bars, one each side of the **well** of the machine, situated above or below the combs. The action of these bars propels the carriages through the combs towards the well and, with the **locker bars**, is responsible for the motion of the carriages through the well from the front to the back of the machine and *vice versa*.

drop feed
See under **feed mechanisms (sewing)**.

drop wire; dropper
One of a series of wire or metal strips suspended on individual warp threads during warping or weaving. When the thread breaks, the drop wire falls, causing the machine to stop.

droplea
The substrate of a woven carpet with no pile on the surface. It consists essentially of only warp and weft yarns in the case of Axminster droplea, but Wilton droplea will usually include dead yarns (see **dead yarn (carpet)**).

dropper pinning
The placing of a **drop wire** on each **end** of the warp.

dry beating (finishing)
A process consisting of gently brushing fabrics on a teazle gig (see **teazle**). (See also **beating (finishing)**.)

dry clean
To remove grease, oil, and dirt from garments or fabrics by treating them in an organic solvent, as distinct from aqueous liquors. Examples of suitable solvents are white spirit, trichloroethylene and tetrachloroethylene (perchlorethylene). The process was originally known as 'French cleaning'.

dry laying
A process for forming a **web** or **batt** of staple fibres by **carding** and/or **air laying**.

dry spinning (manufactured fibre production)
The conversion of a dissolved polymer into filaments by **extrusion** and evaporation of the solvent from the extrudate.

dry taping
See **Scotch dressing** 1 under **dressing (warp preparation)** and **assembly beaming**.

dry-combed top
A wool **top** containing not more than 1% of fatty matter based on the oven-dry fat-free weight as tested by the International Wool Textile Organisation's method which specifies Soxhlet extraction with dichloromethane.
Note: The standard **regain** of a dry-combed top is 18.25% based on the combined weight of oven-dry fat-free wool and the fatty matter.

dry-jet wet spinning
See **wet spinning (manufactured fibre production)**.

dry-laid nonwoven fabric
A fabric made from a **web** or **batt** by **dry laying**, followed by any type of bonding process.

dry-spun
1. Descriptive of a worsted yarn produced from a **dry-combed top** or of synthetic or blended yarns spun on similar machinery.
2. Descriptive of coarse flax yarn spun from air-dry roving. (See also **wet-spun**.)
3. Descriptive of manufactured filaments produced by **dry-spinning**.

dryer fabric
A fabric which transports the paper sheet through the steam heated cylinder dryer section of a paper-making machine. These fabrics are generally woven flat and seamed in various ways. For some applications seamed needled fabrics or fabrics formed by linking together monofilament spirals are used.

drying cylinder
Heated, rotating, hollow cylinder(s) around which textile material or paper is passed in contact

drying cylinder *(continued)*
to dry it.
Note: A cylinder dryer may consist of a series of cylinders with which the material is in contact on alternate faces or, by the use of auxiliary rollers, one face only of the material is arranged to touch the surface of the cylinder. Large single cylinders are also used for faced cloths, to avoid flattening the surface or raised threads on one side, and for drying backfilled cloths.

duchess satin
A heavyweight yarn-dyed silk fabric woven in a satin weave, the warp crossing over at least 7 weft threads at a time. This produces a pearl-like lustre on the satin face. The fabric is woven from net mulberry silk yarn. Typical end-uses are wedding, evening and cocktail dresses.

duck
1. A term used in Scotland to describe a degree of bleaching.
2. A closely woven, plain-weave fabric similar to **canvas** usually made from cotton or linen yarns. The names canvas and duck have become almost generic and are usually qualified by terms that indicate the use of the fabric, e.g., Royal Navy canvas, artist's canvas, duck suiting, belting duck.

duffel; duffle
A heavy low-grade fabric, napped on both sides, made from woollen yarn. Generally it is made up for short coats referred to as duffel coats (Duffel, near Antwerp). (See also **flushing**.)

dull
Lacking in **lustre**.

dullness (of a colour)
That **colour quality**, an increase in which is comparable to the effect of the addition of a small quantity of neutral grey **colorant**: it is the opposite of brightness.

dummy needle; blank needle (knitting)
An element for the filling of otherwise empty needle spaces. It may perform other functions, such as holding down fabric and/or preservation of selvedge regularity.

dummy slider (knitting)
An element inserted into an otherwise empty **trick** to protect the trick and act as a latch guard.

dungaree
A 3/1 or 2/1 twill fabric used for overalls. Some are piece-dyed, but better qualities are made from dyed warp and weft yarns. (See also **denim**.)

duo
A pair of skewed **godets** or other rollers (often heated) used in continuous-filament yarn production to transmit the yarn between extrusion and wind-up. Duos are often found at the beginning and/or end of the draw-zone in **spin-drawing**. The axes of the two rollers are invariably set at a small angle to each other to ensure that successive wraps of yarn remain separate.

dupion
See **doupion**.

dupion fabric
Originally a silk fabric woven from **doupion** yarns. The term is nowadays applied to imitations woven from manufactured fibre yarns, but it is recommended that in such contexts the name of the fibre is indicated.

durability
The ability of a textile to perform its required function until an agreed limiting state is reached.

durable finish
Any type of finish reasonably resistant to normal usage, washing, and/or dry-cleaning.

durable press; permanent press
A finishing treatment designed to impart to a textile material or garment the retention of specific contours including defined creases and pleats resistant to normal usage, washing, and/or dry-cleaning.
Note: The treatment may involve the use of a cross-linking agent, which may be applied and cured either before or after fabrication of a garment, or, in the case of textiles composed of heat-settable fibres, may involve high-temperature pressing.

duration of afterglow
See **afterglow time**.

duration of flaming
See **afterflame time**.

Dutch tape
A **tape** made from linen warp and cotton weft, sometimes with cotton selvedges.

dye
A **colorant**, usually organic, soluble or dispersed in its medium of application and which is designed to be absorbed or adsorbed by, made to react with, or deposited within a substrate in order to impart **colour** to that substrate with some degree of permanence.

dye affinity
See **affinity**.

dye-boarding
See **boarding**.

dye-fixing agent
A substance, generally organic, applied to a dyed or printed material to improve its fastness to wet treatments.

dyeing
The application and fixing of a **dye** to a **substrate**, normally with the intention of obtaining an even distribution throughout the substrate.
Note: In certain instances this term is used to describe processes resulting in non-uniform coloration, e.g., **space dyeing**, **tie-and-dye**.

dynamic loading test
1. A test which measures the thickness loss which occurs in a floorcovering when it is exposed to walking traffic. A specimen of floorcovering is exposed to a specified number of impacts from a standard metal block.
2. A test in which a load is repeatedly imposed on a textile fabric for a given number of cycles.

ease; ease allowance; tolerance; body tolerance
An addition to body measurement, primarily to facilitate movement, for inclusion in the pattern calculations. This addition is determined by a number of factors, for example, fabric and style. The body measurement plus ease gives the finished garment measurement.

easer motion
See **leno weaving**.

easy-care; drip-dry; wash-and-wear; minimum-care; smooth-drying
Descriptive of textile materials that are reasonably resistant to disturbance of fabric structure and appearance during wear and washing and require a minimum of ironing or pressing.

eccentric disc (tufting)
See under **tufting machine**.

eccentric yarn
See under **fancy yarn**.

ecru
Descriptive of fibres, yarns, or fabrics that have not been subjected to processes affecting their natural colour.

edge; edgings; trimmings (lace)
Narrow laces used for trimming, with one edge straight and the other usually scalloped or indented.

edge, leno
See under **selvedge, woven**.

edge, sealed
See under **selvedge, woven**.

edge cable (technical textiles)
See **boundary cable**.

edge wire; looper wire
A length of wire, or monofilament yarn, drawn in through a heald and working as a warp thread at one or both edges of the warp in a weaving machine. It is held permanently so that it does not move forward with the woven fabric, which slides off it as weaving proceeds. Its forward end, which is often slightly tapered, is arranged to terminate a fixed distance beyond the **fell**. Edge wire1s are used:
(i) to provide soft, regular selvedges;
(ii) to form **picot**, **scallop**, or other loop-pattern selvedge effects on narrow fabrics;
(iii) to provide a resistance to weft tension, so as to maintain a fabric at full width during weaving;
(iv) to avoid dog-legged selvedges.
Note: Edge wires may be used for similar purposes in the production of certain flat braids.

edge-crimped yarn
See **textured yarn**, *Note 1* (v).

edging
A narrow fabric or strip of knitted fabric, lace or embroidery, attached to another fabric or made-up article by one edge, usually for decorative purposes. (See also **binding**.)

edging (seaming)
Overstitching along the edge of a fabric to prevent fraying or for ornamentation.

edgings (lace)
See **edge (lace)** and **banded laces**.

effect side (weft-knitted fabric)
See **plain fabric, weft-knitted**.

effect threads
Yarns inserted in a fabric that are sufficiently different in fibre, count, or construction to form or enhance a pattern.

effective length
See under **fibre length**.

effective pile (carpet)
See **pile, effective (carpet)**.

Egyptian tape
A **tape** of Egyptian-type cotton, typically R20/2tex x 9.8tex: 31 ends x 16.5 double picks per cm.

eight-lock fabric, weft-knitted
A double-faced interlock-based fabric that usually repeats over four wales. (See also **double jersey, weft-knitted**.)
Note: When knitted in colour the fabric exhibits a vertical or check effect.

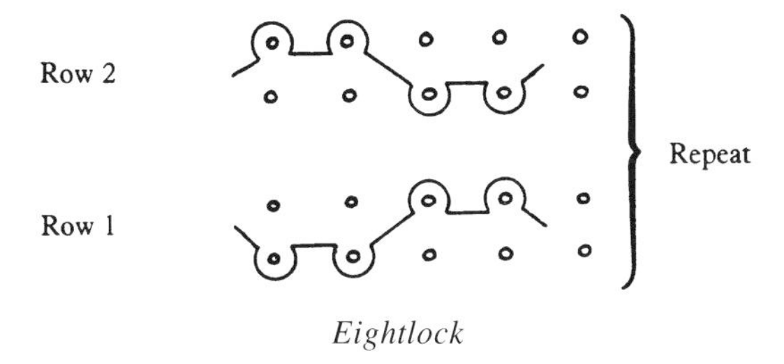

Eightlock

8-strand plaited rope
See under **rope**.

elastane (fibre) (generic name)
A manufactured fibre composed of synthetic linear macromolecules having in the chain at least 85% (by mass) of segmented polyurethane groups that rapidly reverts substantially to its unstretched length after extension to three times that length. (See also Classification Table, p.401.)
Note 1: Elastane fibres typically contain hard (highly inextensible) segments alternating with soft (highly extensible) segments in the molecular chain.
Note 2: Many elastane fibres can be extended reversibly to much more than three times their unstretched length.

elastic binding
A light-weight elastic narrow fabric often with a **scallop** or **picot** edge.

elastic fabric
A fabric containing rubber or other elastomeric fibres or threads, having high recoverable extensibility. (See also **stretch fabric**.)

elastic limit
The greatest strain which a material is capable of sustaining without any permanent strain remaining after complete release of the stress.

elastic narrow fabric
A term used to describe **narrow fabrics** incorporating elastomeric fibres, which extend when stretched and recover their original dimensions when the stretching load is removed. The principal types of elastic are:
(i) elastic **web** or narrow woven elastic;
(ii) elastic **braid**, flat or tubular;
(iii) crochet-knit elastic;
(iv) narrow elastic **lace**.

elastic recovery
The immediate reduction in extension observed in a material when, after being held at a defined elongation for a given time, the applied force is removed.

elasticate; elasticise
1. To apply an **elastomeric yarn** under tension to the surface of a fabric.
2. To incorporate an elastomeric yarn under tension into part of a fabric. (See also **elastic fabric** and **stretch fabric**.)

elasticity
That property of a material by virtue of which it tends to recover its original size and shape immediately after the removal of the force causing deformation.

elastique
A high warp **sett** fabric with a steep double twill line used for trousers, jackets and coats. This fabric is also referred to under a series of other names, e.g., **cavalry twill**, **whipcord**, etc. In the 19th century the fabric was made in fine merino wool for over coating.

elastodiene (fibre) (generic name)
A manufactured fibre composed of natural or synthetic polyisoprene, or composed of one or more dienes polymerized with or without one or more vinyl comonomers, that rapidly reverts substantially to its unstretched length after extension to three times that length. (See also Classification Table, p.401.)

elastomer
Any polymer having high extensibility together with rapid and substantially complete elastic recovery.
Note: Most fibres formed from elastomers have breaking elongations in excess of one hundred per cent.

elastomeric yarn
A yarn formed from an **elastomer**.
Note 1: Elastomeric yarn may either be incorporated into fabric in the bare state or wrapped with relatively inextensible fibres. Wrapping is done by covering (see **covered yarn**), core spinning (see **core-spun yarn**) or **uptwisting**.
Note 2: Examples are **elastane** and **elastodiene** yarns.

electromagnetic warp protector
A mechanism to prevent the loom sley from making a beat-up if the shuttle has not passed a sensor in the raceboard by a specific time in the weft insertion process.

electrostatic flocking
The process of applying **flock** to an adhesive-coated substrate in a high-voltage electrostatic field.

elongation
The increase in length of a specimen during a tensile test, expressed in units of length. (See also **extension percentage**.)

elysian fabric
A thick, soft woollen fabric in which extra weft is floated to the surface and subsequently burst in the finishing treatment. The floats may be planned for various patterns, such as twills and waved effects. The ground structure may be a single (weave A) or double (weave B) fabric.

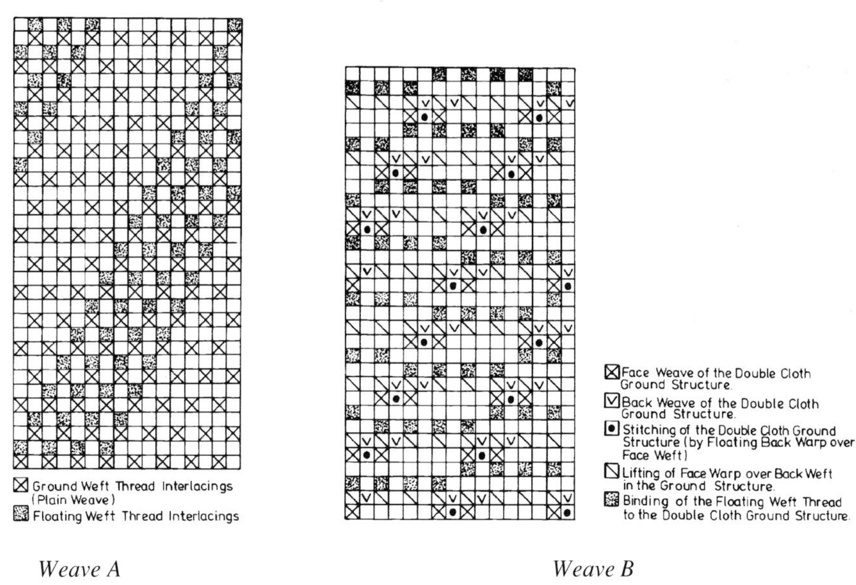

Weave A *Weave B*

Elysian weaves

emboss
To produce a pattern in relief by passing fabric through a **calender** in which a heated metal bowl engraved with the pattern works against a relatively soft bowl built up of compressed paper or cotton on a metal centre.

embossed crêpe
See under **crêpe fabric**.

embroidery
A decorative pattern superimposed on an existing fabric by machine stitching or hand needlework.

embroidery lace
A lace construction obtained by working with any suitable stitching thread on a pre-existing ground of bobbinet, tulle, net, or lace, in order to produce an ornamental effect on that ground.

embroidery-plated fabric, knitted
See **plated fabric, weft-knitted**.

emerizing
A process in which fabric is passed over a series of rotating emery-covered rollers to produce a suede-like finish.
Note: A similar process is known as ‘sueding’.

empress fabric
A 1x2 twill-face and finely ribbed backed double-faced woollen dress fabric.

end
1. (Spinning, braiding) An individual strand.
2. (Weaving) An individual warp thread.
3. (Fabric) A length of finished fabric less than a customary unit (piece) in length. (In certain districts a half-piece.)
4. (Finishing) Each passage of a length of fabric through a machine, for example, in jig-dyeing.
5. (Finishing) A joint between pieces of fabric due, for example, to damage or short lengths in weaving or damage in bleaching, dyeing, or finishing.

end brasses (lace machines)
Brass spacing-pieces set vertically between two supporting bars attached to a Leavers lace machine at each end. They serve to guide and separate the **steel bars** where they fan out between the **well** and the jacquard at one end and between the well and the spring frame at the other end.

end-down
A situation that exists when a warp thread has broken in a loom. Failure to correct leads to a fault (see **end-out**).

end-fent
A short length of finished fabric from the end of a piece that is not usable for the same purpose as the rest. (See also *Note* under **leader fabric**.)
Note: End-fents are not necessarily imperfect. At least three classes of end-fents exist:
(i) those formed in dressing pieces of fabric in the final stages of manufacture. Such end-fents usually consist of ragged imperfect pieces of fabric;
(ii) in garment manufacture, those formed in preparing pieces for cutting out. Such end-fents usually contain length markings and **truth marks**;
(iii) in garment manufacture, those formed as end residues. Such end-fents are of good material but may exhibit the 'cutting-out' contour.

end-group
A chemical group that forms the end of a polymer chain. Linear polymers possess two end-groups per molecule and branched polymers with n branch points possess $n + 2$ end-groups per molecule.

end-out (defect)
A line, running warp-way through part or all of a piece, caused by the absence of a warp thread.

ending
A dyeing fault consisting of a change in colour from one end of a length of fabric to the other or a difference in colour between the bulk and the end of a length of fabric. (See also **tailing**.)
Note: This term is commonly used with reference to batch-dyed material.

endless woven belting
See under **belting (industrial and mechanical)**.

English rib
See under **rib fabric, weft-knitted**.

English welt (knitting)
See **roll welt** under **welt (knitting)**.

entering lease
See **lease, entering**.

epitropic fibre
A fibre whose surface contains partially or wholly embedded particles that modify one or more of its properties, e.g., its electrical conductivity.

eri
See **silk, wild**.

ermine fibre (hair)
Fibre from the ermine (*Mustela erminea*).

eskimo
A piece-dyed, double-faced, all-wool overcoating with a 5-end satin face and twill back.

estamenes
A loosely woven 2x2 twill worsted dress fabric with a rough finish.

exfoliation
See **lousiness**.

exhaustion
The proportion of dye or other substance taken up by a substrate at any stage of a process to the amount originally available.

expression (per cent)
The weight of liquid retained by textile material after mangling or **hydro extraction**, calculated as a percentage of the air-dry weight of the goods. (If the dry weight is used, this should be indicated.)

extension at break; breaking extension
The **extension percentage** of a test specimen at breaking point.

extension percentage
The increase in length of a specimen during a tensile test, expressed as a percentage of the **gauge length** or the **nominal gauge length**. (See also **elongation**.)

extractable matter
The **non-fibrous matter** that can be removed by liquid treatment of a textile, e.g., the amount of oil or grease based material, extracted with organic solvents.

extrusion (fibre production)
The process of forming fibres by forcing materials through orifices (see **spinning**, *Note 2*).

extrusion ratio
See **spin-stretch ratio**.

eye of buttonhole
That part of the buttonhole into which the shank or neck of the button will sit when the garment is fastened.

eyelet (weft knitting)
An openwork effect produced by transferring sinker loops usually to two adjacent needles. Generally, two consecutive sinker loops are collected and transferred.

eyelet (weft knitting) *(continued)*

Eyelet

eyeletting; beading (lace machines)
A flattened-spring wire coil used as a multiple guide to separate and determine the spacing of beam threads on a Leavers machine.

fabric (textile)
A manufactured assembly of **fibres** and/or **yarns** that has substantial surface area in relation to its thickness and sufficient cohesion to give the assembly useful mechanical strength.
Note: Fabrics are most commonly woven or knitted, but the term includes assemblies produced by felting, lace-making, net-making, nonwoven processes, and tufting.

fabric field (technical textiles)
A region of fabric in an air-supported or tension-membrane structure surrounded by **ridge cables** or **boundary cables**. It usually consists of more than one strip of fabric welded, sewn or glued together.

fabric length
See **length, fabric**.

fabric slippage
1. (Sewing) The non-uniform passage of one or two plies of fabric past the needle whilst stitching. Slippage is attributable to the frictional properties of the fabric or to the machining conditions.
2. (Wear) Movement of fabric in a garment against another garment or over the skin of a wearer.
Note: Fabric slippage should not be confused with **seam slippage** or **yarn slippage**.

fabric width
See **width, fabric**.

face (fabric)
The side of a fabric which is intended to be the use surface or which is to be visible in an end product.

face (weft-knitted fabric)
See **plain fabric, weft-knitted**.

face loop (weft knitting)
See under **knitted loop (weft knitting)**.

face-finished (fabric)
Descriptive of a finish , for example, to wool fabrics, in which the face side is treated selectively, as in raising.

face-to-face carpets; double-plush carpets
Carpets manufactured as a "sandwich" in which the pile is attached alternately to two substrates; two cut pile carpets are made by cutting the pile yarns between the two substrates.

facing ribbon
See **faille ribbon**.

facing silk
A fine lustrous fabric of silk (usually of cord, satin, twill weave, or barathea) used for facing, e.g., lapels in men's evening wear.
Note: Fabrics of other fibres are used for facing purposes but are not properly described as 'facing silk'.

facings
Components cut to the same shape as garment parts and applied primarily to neaten raw edges. Facings are usually applied to the inside of a garment but may be applied to the face of the garment in order to produce a visible and/or decorative finish.

façonné; faconne
The French word for 'figured'. It is used in relation to textiles to describe jacquard fabrics with a pattern of small scattered figures.

fade; fading
1. In fastness testing, any change in the colour of a textile caused by light or contaminants in the atmosphere, e.g., burnt-gas fumes.
Note: The change in colour may be in hue, depth or brightness or any combination of these.
2. Colloquially, a reduction in the depth of colour of a textile, irrespective of cause.

faille
A fine, soft fabric, woven from continuous-filament yarn, made in a plain weave with weft-way ribs formed by the intersection of a fine, close-set warp with a coarser weft. It was originally made of silk with a warp of the order of 5 tex and a coarser weft of about 13 tex.
Note: Faille belongs to a group of fabrics having ribs in the weft direction. Examples of this group arranged in increasing order of prominence of the rib are taffeta, poult, faille, and grosgrain.

faille ribbon; facing ribbon
A **double shot** or ribbed plain woven ribbon commonly used to reinforce the button line of cardigans.

fall plate fabric, warp-knitted
A patterned fabric made on a **raschel warp-knitting machine** using one needle bar, involving the use of a solid metal plate (fall plate or chopper bar) to push the newly formed laps of the pattern bars out of the needle hooks, to be cast off with the ground loops of the previous course.

fallen wool
See **dead wool**.

fallers
1. Straight, pinned bars employed in the control of fibres between drafting rollers.
2. Curved arms fixed to two shafts on a mule carriage and carrying the faller wires (see **mule spinning** diagram).

false reed
Used in addition to an ordinary reed to keep the threads of a fibrous or high-sett warp separated. A false reed is composed of short wires with a loop at the top which are threaded on a band of wire. The wires of the false reed are passed through the warp behind the ordinary reed and divide the warp into groups of three or four threads.

false-twist direction
The direction, S or Z, of twist generated by a false-twisting device upstream of itself.

false-twist level
See *Note 1* under **false-twisting**.

false-twist textured yarn
See **textured yarn**, *Note 1* (i).

false-twisting
A twisting operation applied at an intermediate position on a yarn or other continuous assembly of fibres, so that no net twist can be inserted, as distinct from twisting at the end of a yarn where real twist is inserted.
Note 1: Real twisting necessarily involves either rotation of a yarn end, as in **uptwisting** or in downtwisting (see **ring twisting**), or the repeated passage of a thread loop around an end, as in **two-for-one-twisting**. In false-twisting, a yarn normally runs continuously over or through a false-twisting device which may act at either a constant or varying rate. When the twisting rate is constant and equilibrium has been established, the yarn passes through a zone of added twist, then on leaving the twisting device, returns to its original twist level. The added (false) twist level is equal to the ratio of the rotational and axial speeds of the yarn.
Note 2: Equilibrium false twisting is used in one method of yarn texturing where thermal setting is carried out in the zone of temporary twist; it is also used to provide temporary cohesion and thus strength in some **staple-fibre** processing systems. (See also **pin-twisting** and **friction-twisting**.) The **self-twist spinning** (e.g., Repco) process is an example of the use of a varying false-twisting rate.
Note 3: Some devices that are usually used for false-twisting have been used to generate real twist in, for example, certain **open-end spinning** processes (see **pin-twisting** and **friction-twisting**).
Note 4: Static elements such as yarn guides may, in certain circumstances, generate either equilibrium or varying false-twist in running yarns.

fancy
A wire-covered roller used for lifting fibres from the base to the top of the clothing on a **swift** prior to transfer to the **doffer** on a roller and clearer card.
Note: The clothing on the swift is usually fillet wire.

fancy atlas fabric, warp-knitted
A warp-knitted fabric similar in construction to a single or two bar atlas fabric but in which the direction of traverse may change more than once within the repeat and the movement of the threads is not restricted to one wale per course. (See also **atlas fabric, single bar** and **atlas fabric, two bar**.)

fancy purl
See **purl fabric, weft-knitted**.

fancy yarn
A yarn that differs from the normal construction of single and folded yarns by way of deliberately produced irregularities in its construction. These irregularities relate to an increased input of one or more of its components, or to the inclusion of periodic effects, such as knops, loops, curls, slubs, or the like.

bouclé yarn
A compound yarn comprising a twisted core with an effect yarn wrapped around it so as to produce wavy projections on its surface.
Note: Bouclé yarns belong to a group of which the other members are gimp yarns and loop yarns. The effect is achieved by differential delivery of the effect component as compared with the core yarn, the former wrapping around the latter either tightly or loosely according to the amount of excess delivery and the doubling twist inserted. Generally speaking, bouclé yarns exhibit an irregular pattern of semi-circular loops and sigmoid spirals; gimp yarns display fairly regular semi-circular projections and loop yarns have well-formed circular loops.

chenille yarn
1. A yarn consisting of a cut pile which may be of one or more of a variety of fibres helically disposed around axial threads which secure it. Chenille yarns are traditionally used in the manufacture of furnishing fabrics and as decorative threads in many types of broad and narrow fabrics.
2. A tufted, weft yarn made by weaving in a loom (known as a weft loom) in which the warp threads are arranged in small groups of 2 to 6 ends, which interlace in a gauze or cross-weaving manner, the groups being a definite distance apart to suit the length of pile. The weft is inserted in the normal way, each pick representing a potential tuft. The woven piece is cut into warp-way strips, which are then used as weft yarn in the production of chenille fabrics. (See also **mock chenille yarn** below.)

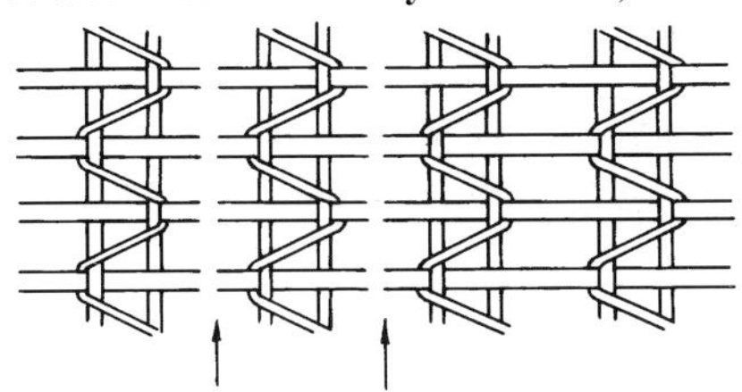

Cutting points to produce chenille yarn

3. The product of a chenille machine. Pile yarn is introduced between and at right angles to a pair of axial threads at the point at which these axial threads engage as they are twisted together. The pile yarn is then cut.
Note: A yarn resembling chenille is produced by electrostatic **flocking** an axial yarn prepared with an adhesive.

cloud yarn
A type of yarn using two threads of different colours in such a manner that each thread alternately forms the base and cover to 'cloud' the opposing thread. It is made by alternate fast and slow deliveries from two pairs of rollers.

eccentric yarn
An undulating gimp yarn.
Note: Generally, it is produced by binding an irregular yarn, such as a stripe or slub, in the direction opposite to the initial stage, to create graduated half-circular loops along the compound yarn.

fancy yarn *(continued)*

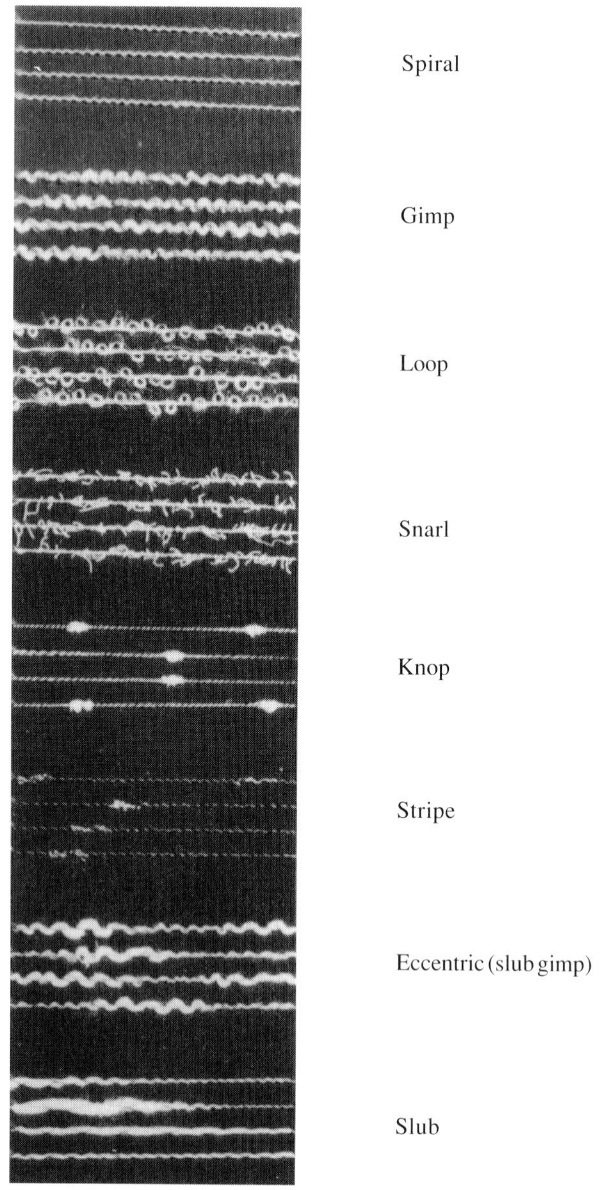

Fancy yarns

fleck yarn
A mixture yarn of spotted and short streaky appearance, due to the introduction of a minority of fibres of different colour and/or lustre.

knickerbocker yarn; nepp yarn; knicker yarn
A yarn made on the woollen system and showing strongly contrasting spots on its surface that are made either by dropping in small balls of wool at the latter part of the carding process or by incorporating them in the blend and so setting the carding machine that these small lumps are not carded out.

knop yarn
A yarn that contains prominent bunches of one or more of its component threads, arranged at regular or irregular intervals along its length.
Note 1: The yarn is usually made by using two pairs of rollers, capable of being operated independently, as follows: (i) foundation threads-intermittent delivery; (ii) knopping threads-continuous delivery. The knopping thread(s) join(s) the foundation threads below the knopping bar and is (are) gathered into a bunch or knop by the insertion of twist.
Note 2: The knop yarn may be bound with a thread in the direction opposite to the initial stage to secure the knops and/or to produce an additional spiral between the knops.

loop yarn
A compound yarn comprising a twisted core with an effect yarn wrapped around it so as to produce wavy projections on its surface (see *Note* under **bouclé yarn** above).

mock-chenille yarn
A doubled corkscrew yarn (see **spiral yarn** below).
Note: It is made by doubling together two or more unbalanced corkscrew yarns in the reverse direction with sufficient twist to form a balanced structure.

slub yarn
A yarn in which **slubs** are deliberately created to produce a desired effect.
Note: Generally, slub yarns are divided into two classes: (i) spun slub yarns, and (ii) plucked (or inserted) slub yarns.
Note: Spun slubs may be produced by an intermittent acceleration of one pair of rollers during spinning or by the blending of fibres of different dimensions. Plucked slub yarns are composed of two foundation threads and periodic short lengths of straight-fibre materials that have been plucked from a twistless roving by roller action. (See also **slubby yarn**.)

snarl yarn
A compound yarn that displays snarls or kinks projecting from the core.
Note: It is made by the same procedure as a **loop yarn**, but, instead of a resilient thread, a lively highly twisted yarn is used. Thus, snarls are formed in place of loops when the tension is released at the front rollers. The snarls may be controlled to vary in size and frequency, either continuously or in groups at places along the yarn.

spiral yarn; corkscrew yarn
A plied yarn displaying a characteristic smooth spiralling of one component around the other.
Note: Spiral yarns include: (i) a plied yarn made up of two single ends or groups of ends of equal length containing S and Z twists, respectively; (ii) a plied yarn produced by delivering one or more of its components at a greater rate. The shorter length forms the base, while the greater length of its companion(s) creates a spiral round it; (iii) a plied yarn made from two ends of equal length, one coarser than the other.

stripe yarn
A yarn that contains elongated knops (see **knop yarn** above).
Note: It can be made by either of two methods as follows: (i) as a knop with a moving knopping bar to spread the surplus thread or knop; (ii) by alternate fast and slow delivery of one or more of its component threads and a constant rate of delivery of the base threads. The threads join below a stationary bar to form the intermittent stripes.

fasciated yarn
A staple fibre yarn that, by virtue of its manufacturing technique, consists of a core of essentially parallel fibres bound together by **wrapper fibres**. The current technique of manufacture is often referred to as **jet spinning**.

fashioning (weft knitting)
See **shaping (weft knitting)**.

fast reed (warp protector)
1. A **reed** rigidly mounted in a loom **sley**. (See also **loose reed**.)
2. A mechanism that will halt the advancement of the sley if the shuttle fails to reach the receiving box. A **trap** results and if the mechanism fails to operate a **smash** occurs.

fastness
See **colour fastness**.

FBA
See **fluorescent brightening agent**.

FDY
See **fully drawn yarn**.

fearnought (fabric)
A stout, thick, woollen fabric with a heavy appearance, used chiefly as a covering for port-holes and the doors of powder magazines and also for scarves and coats.

feather-edged slivers
See **clicking top sliver**.

feed; feeder (circular weft-knitting machines)
1. The final element which guides the yarn(s) to the needles.
2. A device which supplies yarn to the feeder (see **positive feed (weft knitting)**).
Note: A large diameter machine with a multiplicity of feeders may be referred to as a **multifeeder machine**.

feed dog (sewing machine)
A toothed element which acts on the surface of material being sewn to move it in a controlled way during stitching. The action is normally on the lower surface of the material, but mechanisms which act on the upper surface are also in use. For most of the time, the operative motion is forwards (away from the machinist), but in specific cases (such as back tacking) the motion is reversed. (See also **feed mechanisms (sewing)**.)

feed mechanisms (sewing)
The means for moving the material being stitched from one stitch position to the next.

> **compound feed**
> A feed mechanism comprising synchronised **drop feed** and **needle feed**.
>
> **drop feed**
> A feed mechanism in which the **feed dog** alternately engages and disengages the underside of the material to transport it forward.
>
> **differential drop feed; differential bottom feed**
> A type of **drop feed** with two **feed dogs** arranged in tandem to move differentially. One feed dog is in front of the needle and the other is behind. Material may be gathered or stretched by adjusting the stroke of the front feed dog so that it is longer or shorter respectively than the stroke of the rear feed dog.
>
> **needle feed**
> A feed mechanism in which the feeding of the material is accomplished or assisted by the needle which moves forward by one stitch whilst penetrating the material.

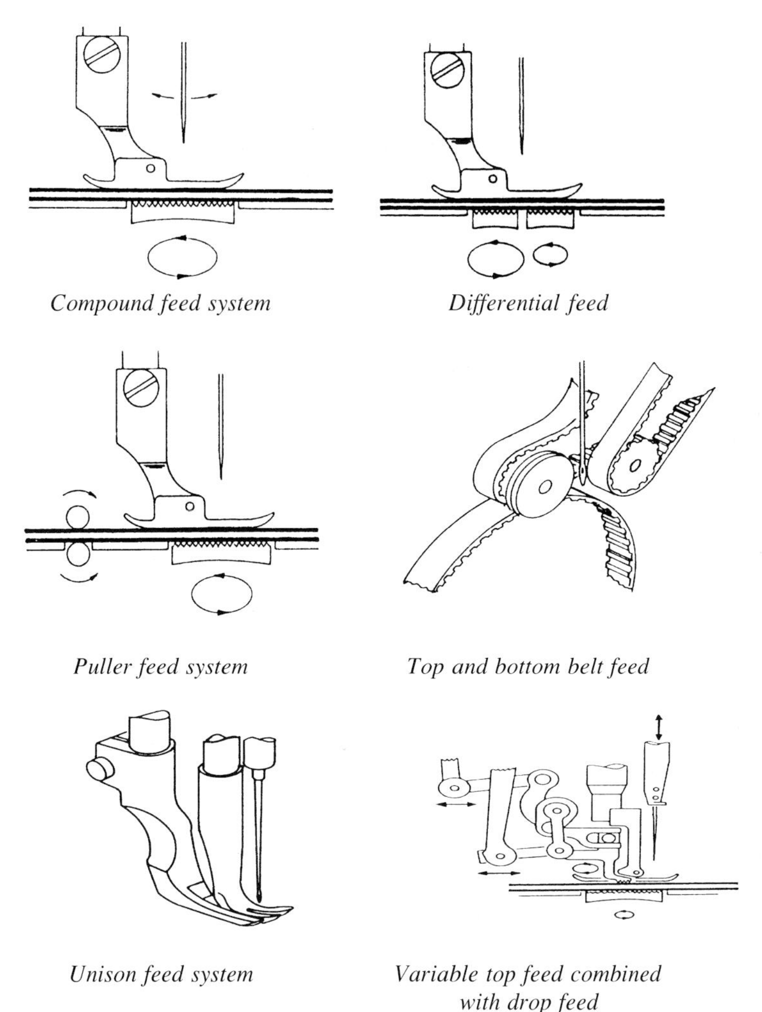

Compound feed system *Differential feed*

Puller feed system *Top and bottom belt feed*

Unison feed system *Variable top feed combined with drop feed*

(Source: H. Carr and B. Latham. *The Technology of Clothing Manufacture*, 1994)

puller feed
A feed mechanism situated behind the **presser foot** and consisting of one or more rollers, at least one of which is driven, in contact with the material. Puller feeds may be either continuous or intermittent, and may be used in conjunction with other feed mechanisms.

top feed
A feed mechanism in which a **presser foot**, known as a 'feeding foot', alternately engages and disengages the upper surface of the material to transport it forward.

adjustable top feed; variable top feed
A **top feed** system with the **presser foot** in two sections: one of which holds the material in position during needle penetration whilst the other, with its teeth on the lower side, engages the upper ply to feed the material forward when the needle is raised.

top and bottom feed
A feed mechanism in which **top feed** and **drop feed** work in unison.

feed mechanisms (sewing) *(continued)*

unison feed
A feed mechanism providing **needle feed** in addition to the **top and bottom feed**.

wheel feed
A feed mechanism in which a driven wheel or roller, in constant engagement with the material, replaces all or part of the normal presser foot or feed dog. It is described as 'upper-feed' or 'under-feed' wheel according to its position relative to the material. The feed may be continuous or intermittent.

feed roller; feed roll
A roller that forwards a yarn to a subsequent processing or take-up stage.

feed-off-the-arm bed
See under **bed (sewing machine)**.

feeder variation
A fault, affecting all the stitches in a course in a knitted fabric on a multi-feeder machine. It recurs in all the courses knitted by a particular feeder.

feeding foot
See **top feed** under **feed mechanisms (sewing)**.

feeler motion
A mechanical device used to detect when the weft on a pirn in a shuttle is becoming exhausted.

fell (of the cloth)
The line of termination of the fabric in the loom formed by the last weft thread.

felling (making-up)
The flattening of an overseam by stitching it down, usually with a single chainstitch.

felling simili
See **simili binding** under **binding**.

fellmongering
The process of pulling wool from sheep skins. (See also **skin wool**.)

felt
A textile fabric characterized by the entangled condition of many or all of its component fibres. Three classes of felt can be distinguished:
(i) pressed felt (mechanical or sheet felt), which is formed from a **web** or **batt** containing animal hair or wool, consolidated by the application of moisture, mechanical action and heat which cause the constituent fibres to mat together;
(ii) woven or knitted felts formed from staple fibre fabrics having some wool or animal hair content. These are subjected to the processes identified in (i) to such a degree that the original fabric construction is completely obscured by the smooth felted surface;
(iii) **needlefelt**.

felted yarn
A wool-rich yarn produced from **sliver**, **slubbings**, **rovings**, **yarn**, or by **felting**. (See also **continuous yarn felting**.)

felter; fray
A faulty area in fabric caused by local interference with the shed during weaving, that results

in a concentration of stitches or floats (see **stitch (defect)**), and is sometimes accompanied by broken ends.
Note: One common cause of a felter is a broken end that becomes entangled with adjacent ends.

felting
The matting together of fibres during processing or in use (see **milling (fabric finishing)** and **felt**).

felting needle
see **needle (needlefelt)**.

fents; bribe
Short lengths of fabric cut from an end, piece, or lump of fabric. They may or may not be of imperfect material.
Note: The term 'remnants' has been used as a synonym for fents. These are usually short lengths of fabric that accumulate in the marketing of textile material.

fettling
1. The maintenance of textile machines.
2. The process of cleaning fibres and other material from the clothing of a carding machine. This was necessary with flexible card clothing, which tended to readily accumulate material, but is less common with metallic card clothing.

FIBC
See **flexible intermediate bulk container**.

fibre; fiber
Textile raw material, generally characterised by flexibility, fineness and high ratio of length to thickness. (See also Classification Table, p.401.)

fibre, chemical
A literal translation of various non-English terms having the same meaning as manufactured fibre or man-made fibre (see **fibre, manufactured**).

fibre, manufactured; fibre, man-made
A fibre that does not occur in nature, although the material of which it is composed may occur naturally. (See also **fibre, natural**.)
Note: The raw materials for the manufacture of fibres may be derived from:
(i) naturally occurring, non-fibrous materials, for example, **metal fibres** from a variety of metals and their ores, **glass fibre** from silica and other minerals;
(ii) natural polymers, for example, **rubber fibre** from latex, **viscose** from wood cellulose, **azlon** from natural proteins (see **fibre, regenerated**);
(iii) synthesised polymers such as the polyamides and polyesters (see **fibre, synthetic**);
(iv) other manufactured fibres which then undergo further significant physical or chemical modification, for example, **carbon fibre** from acrylic or pitch fibres.

fibre, natural
A fibre occurring in nature.
Note: Fibres are found in all three sectors of the natural world, for example: animal (silk, wool); vegetable (cotton, jute); mineral (asbestos).

fibre, regenerated
A fibre formed from a solution of a natural polymer or of a chemical derivative of a natural polymer and having the same chemical constitution as the natural polymer from which the solution or derivative was made.

fibre; fiber *(continued)*

fibre, synthetic

A fibre manufactured from a polymer built up from chemical elements or compounds, in contrast to fibres made from naturally occurring fibre-forming polymers.

fibre cohesion

See **cohesion (fibre)**.

fibre diagram

See under **fibre length**.

fibre extent

See under **fibre length**.

fibre fineness

The fineness of cotton, silk and manufactured fibres is usually expressed in terms of average linear density. The fineness of animal fibres is usually expressed as the mean fibre diameter.

fibre length

The distance between the ends of a fibre when measured under specified conditions. In commercial practice the following definitions apply:

barbe

The mean length of the fibres of a sliver or of a roving calculated from the proportions by mass of the fibres in the sliver or roving.

Note 1: This term is used for long staple fibres, particularly wool.

Note 2: Barbe is always greater than **hauteur** (see below) for a given **top**; the following relation holds and may be used to convert hauteur to barbe and *vice versa*:

$$\text{Barbe} = \text{Hauteur}\ (1 + V^2)$$

where V is the fractional coefficient of variation of hauteur, i.e.

$$\frac{\text{Coefficient of variation}}{100}$$

crimped length

The distance between the ends of a fibre when substantially freed from external restraint, measured with respect to its general axis of orientation.

dispersion

A measure of length variation in cotton fibres. Statistically it is the inter quartile range of the fibres greater than half the maximum length when determined using a comb sorter.

effective length

A measure of the characteristic length of a sample of cotton fibres. Statistically it is the upper quartile length of fibres longer than half the maximum length when determined using a comb sorter.

fibre diagram; staple diagram

A graphical representation of the length characteristics of a sample of staple fibres. It is a graph of length against cumulative frequency.

fibre extent

The distance between two planes which just enclose a fibre without intercepting it, each plane being perpendicular to the general direction of the yarn or other assembly of which the fibre forms a part.

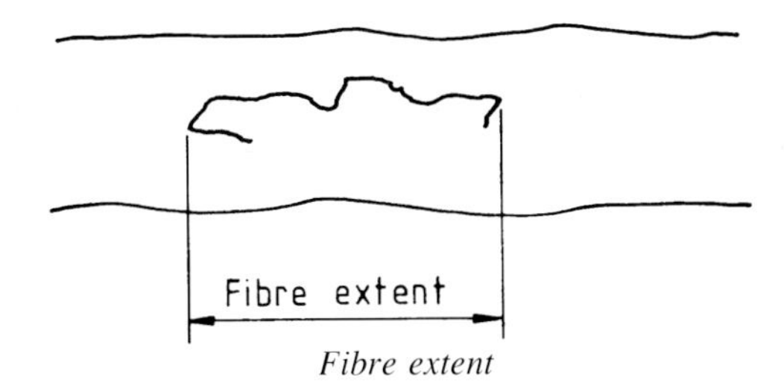

Fibre extent

fibrogram
A graph showing the length distribution of a sample of cotton fibres as determined using the Fibrograph instrument. Statistically it is the curve representing the second cumulation (integral) of the frequency distribution.

floating fibre index
The percentage of fibres not gripped by either the front or back rollers of a drafting system. It is determined by the Fibrograph instrument which is used to test samples of cotton fibre. The floating fibre index is given by:

$$\left(\frac{S}{L} - 0.975\right) \text{x } 100$$

where S = 2.5% **span length,** L = mean length.

hauteur
The mean length of the fibres of a sliver or of a roving, calculated from the proportions by **titre** of the fibres in the sliver or the roving. (See also **barbe** above.)
Note: This term is used for long staple fibres, particularly wool.

span length
The fibre extent exceeded by a stated percentage of cotton fibres by number as determined using the Fibrograph instrument (e.g., 2.5% span length is the extent exceeded by only 2.5% of the fibres).

staple length
The characteristic fibre length of a sample of staple fibres (usually estimated by subjective visual assessment for natural fibres).
Note: The staple length of wool is usually taken as the length of the longer fibres in a hand-prepared tuft or 'staple' in its naturally crimped and wavy condition (see **crimp**). The staple length of cotton corresponds very closely to the modal or most frequent length of the fibres when measured in a straightened condition.

uniformity index
A measure of length variation in cotton fibres determined using the Fibrograph instrument. It is the ratio of the mean length to the upper half mean length expressed as a percentage.

uniformity ratio
A measure of length variation in cotton fibres determined using the Fibrograph instrument. It is the ratio between two span lengths (50% and 2.5%) expressed as a percentage of the longer span length.

upper half mean length
The mean length by number of the longer one half of the fibres by weight as determined using the Fibrograph instrument for cotton.

fibre migration
The change in distance of a fibre or filament from the axis of a yarn during production.

fibre quality index
A numerical value indicating the processability of cotton calculated from its fineness, length and tenacity values. Quality index is given by:

$$\frac{2.5 \times \text{uniformity ratio} \times \text{tenacity (gf / tex)}}{\text{micronaire value}}$$

fibre ultimate; ultimate fibre
That unit cell beyond which subdivision is not possible without loss of a fibre's identity.

fibre wall thickness
See **maturity (cotton)**.

fibre-bonded floorcovering
A textile floorcovering which is composed of entangled textile materials bonded together by a mechanical, physical, or chemical process (e.g.,**needle-punching**, stitch-bonding, heat treatment, or resin impregnation) or by a combination of two or more of these processes.

fibrid
A netted filamentary or fibrillar structure, substantially longer in one dimension than in the other two, that exhibits a capacity for mechanical entanglement with other structures and much higher water-holding capacity than fibres produced by conventional spinning means. Fibrids are used as binding elements in the production of wet-laid synthetic papers.

fibril
1. A small, fine fibre.
2. A fine, fibre-like element embedded within a fibre and usually consisting of a different material to its matrix (see **matrix-fibril bicomponent fibre** under **bicomponent fibre**).
3. Structural sub-units of fibres, formed from bundles of linear polymer molecules in crystalline or semi-crystalline organisation.

fibril-matrix bicomponent fibre
See under **bicomponent fibre**.

fibrillae
Whitish specks often visible on the surface of dyed silk yarns. (See also **lousiness**.)

fibrillated yarn
A yarn produced by the process of **fibrillation**.

fibrillated-film fibre
Staple fibre produced by cutting, chopping or stretch-breaking **fibrillated yarn** or **fibrillated-film tow**.

fibrillated-film tow
An assembly of fibrillated textile films.

fibrillated-film yarn
Yarn produced from **fibrillating film** that has been converted into a longitudinally fibrillated structure. (See also **polymer tape**.)

fibrillating film
A polymer film in which molecular orientation has been induced by stretching to such a degree that it is capable of being converted into yarn or twine by manipulation, e.g., by twisting under tension or by rubbing laterally, which results in the formation of a longitudinally split structure (split fibre).

fibrillating roller
A pinned roller used for **fibrillation**.

fibrillation
The process of splitting a longitudinally oriented **fibre**, **textile film** or **tape** into a network of interconnected fibres. Fibrillation can also refer to a fibrillar failure of a fibre or film structure.
Note: Processes for producing fibrillation may be divided into two groups: (i) those producing random splitting to give a relatively coarse network, e.g., by twisting or rubbing; and (ii) those producing controlled splitting to give a relatively fine network, e.g., by rapidly rotating pinned rollers.

fibrogram
See under **fibre length**.

fibroin
The part of a silk thread remaining after the gum has been discharged.

fiddle string
A particular form of **tight end (defect)** or **tight pick (defect)** that becomes evident only after wet processing.

figured casement
A **casement cloth** in which a pattern has been introduced by weaving.

figured fabric
A fabric in which patterns or motifs are produced by a combination of distinct weaves usually requiring a **dobby** or **jacquard** mechanism.

filament
A fibre of indefinite length

filament blend yarn; biconstituent yarn; co-spun yarn
A filament yarn containing separate filaments of two distinct types, the filaments being more or less randomly blended over the cross-section of the yarn. (See also **bicomponent yarn**.)

filament yarn
See **continuous-filament yarn**.

filamentation
See *Note* under **broken filaments**.

filet net (lace)
1. (Furnishing and Leavers lace) A lace construction in which a square-mesh net consists of parallel warp threads bound by one or more bobbin threads, and mesh threads that alternately pillar and throw at right angles to the warp threads. Pattern may be added by more frequent throwing of the mesh threads or by throws of a further set of patterning threads that pillar when not patterning.
2. (Warp-knitted) A lace construction similar to the above except that a knitted chain of loops in the warp thread binds the mesh threads and patterning threads (if any).

filet net (lace) *(continued)*

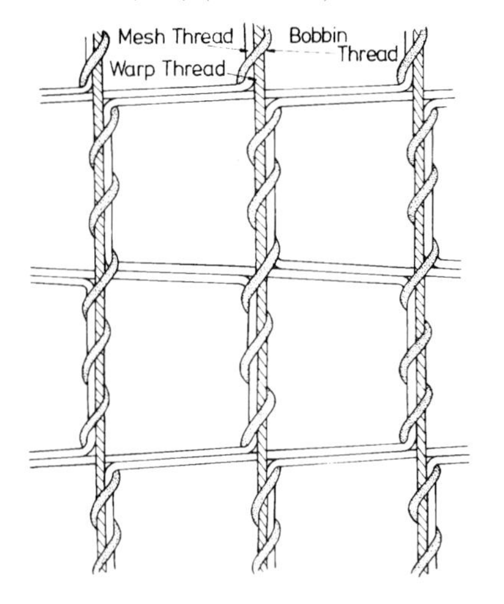

Filet net 1 (furnishings and Leavers lace)

Filet net 2 (warp-knitted)

filet net, woven
A net woven in such a way that the yarns are locked at the intersections.

filler
See **filling**.

filler fabric
See under **tyre textiles**.

filling; filler
1. Generally insoluble materials, such as China clay, gypsum, etc., added to fabrics together with starches, gums or polymers during finishing to add weight and/or to modify their appearance and handle. (See also **weighting**.)
Note 1: This term is usually applied only to cellulosic textiles. (See also **loading**.) Finishes in which starches or gums are used without the addition of insoluble materials are sometimes referred to as 'fillings' but are more correctly described as 'assisted finishes'.
2. A synonym for weft yarns (see **weft**).
3. See **wadding thread**.

filling-in (knitting)
A means of providing a loop for a needle otherwise left empty. Its chief application is in preventing a hole forming in the fabric during the widening process of fully fashioned knitting.

fingering yarn
A plied handknitting yarn normally produced in 2, 3 or 4-ply to a specific diameter. The count of the 4-ply yarn is half that of a double knitting yarn.

finings, bobbin (lace)
See **bobbin finings (lace)**.

finish
A term used broadly in the textile industries to include:
1. A substance or mixture of substances added to a **substrate** at any stage in the process to impart

desired properties.
2. A process, physical or chemical, applied to a substrate to produce a desired effect.
3. Properties, e.g., smoothness, drape, lustre, gloss or crease resistance produced by 1 and/or 2 above.
4. To apply or produce a finish.

finishing
Descriptive of processes, physical or chemical, applied to a **substrate** to produce a desired effect.

fique
A fibre from the leaf of the plant *Furcraea macrophylla*.

fire retardant
A substance added, or a treatment applied to, a material in order to suppress, significantly reduce or delay the combustion of the material.

firmness factor
A term derived from cloth-setting theories: it takes account both of the thread-spacing relative to the yarn diameter (**cover factor**) and of the frequency of the interlacings. It may be referred to as a percentage of the maximum possible cover factor for a particular weave structure (**percentage cover**).
Note: For plain-weave, it is identical with the cover factor; for other weaves, e.g., for twill weaves, it is the cover factor multiplied by a value characteristic of the weave and indicative of the frequency of the interlacings.

fish dart
A fish-shaped **dart** with a small cut at the widest part. It is generally used in the waist of a close-fitting garment without a waist seam.

fisheye (knitting)
See **snag (knitting)**.

fishnet, warp-knitted
A range of structures generally having a large diamond net appearance often used for stockings and tights.

fishnet, weft-knitted
See **float-plated fishnet, weft-knitted**.

fixation accelerator
A product added to a finishing formulation to speed up, or lower the temperature required for, chemical reaction.

flame resistance; flame retardance
The property of a material whereby flaming combustion is slowed, terminated or prevented.
Note 1: This definition is published in ISO 4880-1984.
Note 2: Flame resistance can be an inherent property of the basic material or it may be imparted by specific treatment. The degree of flame resistance exhibited by a material during testing may vary with test conditions.

flame retardant
A substance used to impart improved flame resistance to a material.
Note: This definition is published in ISO 4880-1984. (See also **flame resistance**.)

flame retardant treatment
Chemical process or treatment whereby improved flame resistance is imparted to a material.
Note 1: This definition is published in ISO 4880-1984.
Note 2: One or more flame retardants may be applied during textile finishing or coating, or in the case of manufactured fibres, introduced into the fibre during fibre production.

flame retarded
Treated with a flame retardant.
Note: This definition is published in ISO 4880-1984.

flame spread
Propogation of a flame front.
Note: This definition is published in ISO 4880-1984.

flame spread rate
Distance travelled, per unit time, by a flame during its propagation under specified test conditions. (See also **flame spread time**.)
Note: This definition is published in ISO 4880-1984.

flame spread time
The time taken by a flame on a burning material to travel a specified distance under specified test conditions. (See also **flame spread rate**.)
Note: This definition is published in ISO 4880-1984.

flammability
The ability of a material or product to burn with a flame under specified test conditions.

flammable
Capable of burning with a flame under specified test conditions.
Note: The use of the term 'inflammable' is deprecated.

flammé
1. Woollen dress fabric made from printed yarns.
2. Plain-weave fabric produced from yarn-dyed linen warp and cotton weft, used for tablecloths and curtains.

flannel
An all-wool fabric of plain or twill weave with a soft handle. It may be slightly milled and raised.
Note: When fibres other than wool are present, the proper qualification is, e.g., union flannel.

flannelette
A fabric made from cotton warp and soft-spun cotton weft, the fabric being subsequently raised on both sides to give an imitation of the true woollen flannel. The weave may be plain, plain with double-end warp, or twill.
Note 1: It may be woven **grey** and dyed or printed, or it may be woven from dyed yarns.
Note 2: Fibres other than cotton are sometimes present in the weft yarn. If these exceed 7% they are named in the description, e.g., cotton-viscose flannelette.

flap
A shaped piece of material which provides a covering for a pocket mouth or is for ornamentation.

flash spinning
A modification of the **dry-spinning** method in which a solution of a polymer is extruded at a temperature well above its boiling point, such that, on its emergence from the **spinneret**, evaporation occurs so rapidly that the individual filaments are disrupted into fine **fibrils**.

flash-spun fabric
A **nonwoven** formed by the bonding of a **web** made by the fibrillation of an extruded film by the rapid evaporation of solvent.

flat
A place where warp ends are not leased in the correct order. For instance, two adjacent ends that pass together on the same side of the two lease rods in an otherwise end-and-end lease form a flat.

flat bed
See under **bed (sewing machine)**.

flat card
A card which has flats, revolving and/or stationary, and which disentangles short-staple or cotton-length fibres to produce thin webs or slivers for further processing. (See also **carding**.)

flat fabric
A two-dimensional woven or knitted fabric that has no pile loops.

flat knitting machine
A weft-knitting machine having straight needle beds carrying independently operated, usually latch needles. (See also **knitting machine**.)
Note 1: Rib machines (V-type), usually, have two needle beds, which are opposed to each other in inverted-V formation.
Note 2: Purl machines, usually, have two needle beds horizontally opposed in the same plane.
Note 3: Reciprocating carriage flat knitting machines are flat knitting machines, usually, provided with one or two carriages containing 1,2,3 or 4 cam systems that traverse to and fro across the needle beds.
Note 4: Circulating carriage flat knitting machines are flat knitting machines in which several carriages each with its own cam system move continuously across the needle beds in the same direction.

flat metal yarn
A yarn consisting of one or more continuous lengths of metal strip or incorporating one or more continuous length(s) as a major component.
Note 1: A notable example is a single metal yarn in banknotes.
Note 2: Twist inserted in flat metal yarns may form irregular facets which reflect light accordingly to give decorative effects in fabrics. (See also **metallized yarn**.)

flat ruche, knitted
See under **ruche**.

flat ruche, woven
See under **ruche**.

flat seam
See **plain seam** under **seam type**.

flat setting
The **setting** of fabric at open-width. The term is particularly used in the finishing of woven wool fabrics, where setting is usually effected by steaming under pressure.

flat yarn
1. A yarn consisting of fully drawn continuous-filaments substantially without twist and untextured. (See also **twistless yarn**.)
2. See **straw yarn**.

flatlock seam
See under **seam type**.

flax
1. Plants of the species *Linum usitatissimum* cultivated for the production of fibre, or seed and fibre.
2. Fibre extracted from flax plants.

flax, green; natural flax
Scutched flax produced from deseeded straw without any intermediate treatment such as retting.

flax, line
Hackled flax (see **hackling**).

flax fibre bundle
One of the aggregates of ultimate fibre (see **fibre ultimate**) that run from the base of the stem up to the top of the branches of flax straw. They are each composed of large numbers of ultimate fibres overlapping each other.

flax fibre strands
Flax fibres after removal from the plant, consisting of more than one ultimate fibre in the cross-section.

flax tow
Short flax fibres that are removed during the **scutching** or **hackling** processes:
(i) rug tow: short flax fibre removed during scutching and containing extraneous woody material;
(ii) re-scutched tow: short fibre which has been cleaned in a tow-scutching apparatus;
(iii) machine tow: short fibre which has been removed from scutched long flax during the hackling process.

flax yarn bundle
The standard length by which wet-spun flax yarns are bought and sold. The bundle traditionally contained 60,000yds (about 55,000m) of yarn.

flax-spun
A term applied to staple yarn that has been prepared and spun on machinery originally designed for spinning yarns from flax.

fleck yarn
See under **fancy yarn**.

fleece
The fibrous covering of a sheep or similar animal.

fleece wool
Any **wool** as shorn from a living sheep. The term is in use to distinguish this wool from other forms such as **skin wool**.

fleecy
Resembling a wool fleece in appearance and handle, or descriptive of fabrics having a fine, soft,

open, and raised structure.

fleecy fabric, weft-knitted
1. A general term for any weft-knitted fabric which has been brushed or raised on one or both sides.
2. A specific term for a plain, weft-knitted fabric with a ground yarn in which a yarn of low twist, laid-in and secured by a **binder**, appears on the back of the fabric and may be brushed or raised.
Note: Fleecy fabrics may be produced with only two yarns. (See also **laid-in fabric, weft-knitted**.)

flexible intermediate bulk container; FIBC
A fabric container used for transporting and storing quantities (approximately 1 to 5 tonnes) of bulk commodities, such as powders and granular materials.
Note: The containers are characterised by webbing or fabric lifting loops that enable them to be handled mechanically by crane or fork-lift. They may be made from coated or uncoated woven fabric and may be fitted with an inner liner made of polymer film and filling and emptying tubes. Uncoated fabrics are normally constructed from tape yarns woven at high sett to give a fabric of low permeability. FIBCs may be designed to be single trip or multiple trip.

flexural rigidity
A measure of the resistance of materials to bending by external forces. It is related to stiffness and is one of the factors sensed when a fabric is handled.
Note: Flexural rigidity is one of the factors that determine the manner in which a fabric **drapes**.

flipper fabric
See **filler fabric** under **tyre textiles**.

float (defect)
See **stitch (defect)**.

float (lace)
1. A pattern thread traversed over more than one wale and not tied to the intermediate warp thread or threads.
2. See **clips**.

float (warp knitting)
A length of yarn not received by a needle and connecting two loops of nonconsecutive courses.

float (weave)
A length of yarn on the surface of a fabric between adjacent intersections. This corresponds to the number of threads over or under which the intersecting yarn passes in a woven structure.

float loop; missed loop (weft knitting)
A length (or lengths) of yarn not received by a needle and connecting two loops of the same course that are not in adjacent wales.

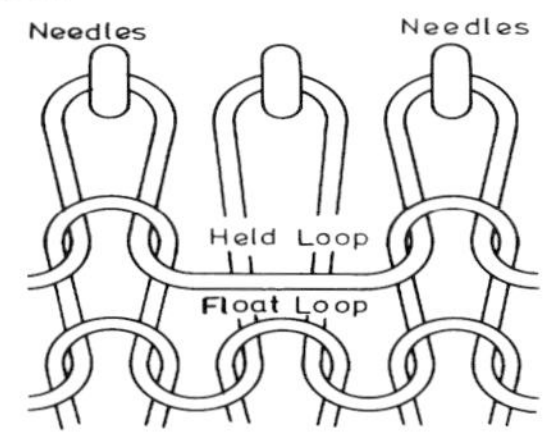

Float or missed loop (shown from back of fabric)

float-plated fabric, knitted
See **plated fabric, weft-knitted**.

float-plated fishnet, weft-knitted
A knitted fabric resembling a fine-meshed net construction generally made by **plating** a thick and thin yarn. The thick yarn is floated across the thin yarn to give either an all over or patterned open-work effect. (See also **float-plated fabric, knitted** under **plated fabric, weft-knitted**.)

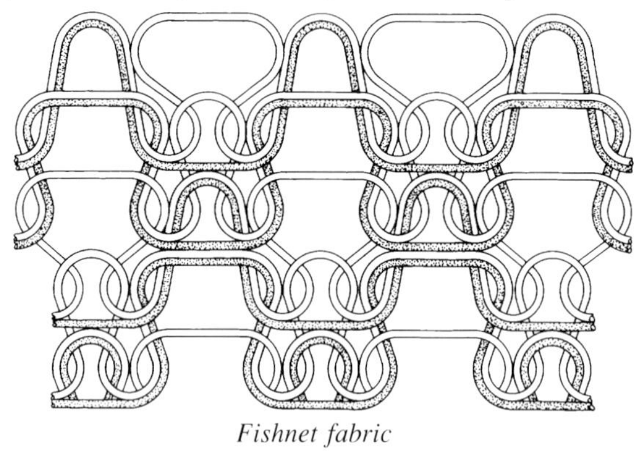

Fishnet fabric

floating breaker
See **breaker fabric**.

floating fibre index
See under **fibre length**.

flock
A material obtained by reducing textile fibres to fragments by, for example, cutting, tearing, or grinding. There are two main types:
(i) stuffing flock: fibres in entangled small masses or beads, usually of irregular broken fibres, obtained as a by-product from, for example, milling, cropping, or raising of wool fabric, and mainly used for stuffing, padding, or upholstery;
(ii) coating flock: cut or ground fibres used for application to yarn, fabric, paper, wood, metal, or wall surfaces prepared with an adhesive. (See also **electrostatic flocking**.)

flock printing
A method of fabric ornamentation. The fabric is printed with an adhesive and then finely chopped fibres are applied all over by means of dusting-on, an air-blast, or electrostratic attraction. The fibres adhere only to the printed areas and are removed from the unprinted areas by mechanical action.

flocked carpets
Carpets manufactured by applying short chopped lengths of fibre (flock) to an adhesive-coated backing fabric. The application is usually carried out electrostatically.

flocks
1. (Cotton) Bunches of cotton fibres produced in the intermediate preparation stages of a spinning process, between bale opening and carding.
2. (Wool) Waste fibres obtained from wool during the different finishing processes.

floorcovering
See **textile floorcovering**.

florentine
Heavy grey woven 3/1 twill cotton fabric, used for overalls and uniforms, having approximately 38 ends x 19 picks/cm and 37x49 tex cotton yarn.

floss silk
1. See **cocoon strippings**.
2. A degummed silk yarn or singles without twist used for embroidery.

flouncings (Leavers lace)
Wide dress lace having one edge scalloped and the other usually straight. The width was traditionally between 30 and 180 cm, across the width of the machine.

fluidity
The ease with which a fluid flows; numerically the reciprocal of **viscosity**. The unit of fluidity is the reciprocal pascal second (Pa^{-1} s^{-1}).
Note: The fluidity of dilute solutions of polymers is inversely related to the polymer molecular weight and, for certain fibre-solvent systems, may be used as an indicator of polymer degradation.

fluorescent brightening agent; FBA; optical brightening agent; OBA
A substance that when added to a substrate increases the apparent reflectance in the visible region by converting ultra-violet radiation into visible light and so increases the whiteness or brightness.

fluorochemical finish
A polymeric **finish** containing combined fluorine applied to the surface of textile materials to confer water and oil repellency and improved soil and stain release properties.

fluorofibre (fibre) (generic name)
A term used to describe fibres composed of linear macromolecules made from fluorocarbon aliphatic monomers. (See also **polytetrafluoroethylene (fibre)** and Classification Table, p.401.)

flushing
A heavy woollen coating cloth originally made in Flushing, Holland. (See also **duffel**).

fluted roller (lace machines)
A shaft running the net-making width of a roller locker **bobbinet machine** and fluted in the form of a gear wheel. There are four fluted rollers, two on each side of the **well** of the machine, under the combs. The flutes of the rollers engage with the teeth of the carriages, and their alternate forward and backward rotation propels the carriages through the well from the front to the back of the machine and *vice versa*.

fly; lint
Fibres that fly out into the atmosphere during processing.

fly-beam
See *Note* under **sley** 1.

flyer spinning
A spinning system in which yarn passes through a revolving flyer leg guide on to the package. The yarn is wound-on by making the flyer and spinning package rotate at slightly different speeds. (See also **spinning**.)

flyer spinning *(continued)*

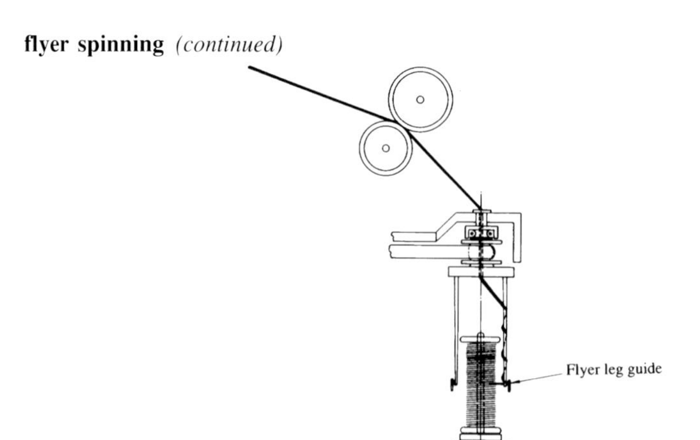

Flyer spinning

flyshot loom
A multi-piece loom for weaving narrow fabrics in which each shuttle is knocked through the open shed by means of a peg fixed in a slide. The term is also sometimes applied to single-head narrow-fabric looms.

foam backing (carpet)
See under **substrate (carpet)**.

foam bonding
A method of making **nonwoven fabrics** in which a fibre **web** or **batt** is treated by the application of a foamed adhesive material. (See **adhesive-bonded nonwoven fabric**.)

foam finishing
The application of one or more liquid chemical finishes in the form of a foam to a textile material with the advantage of a low **wet pick-up**.

foambacked fabric
A **combined fabric** usually having two layers, one of which is of cellular plastics material.

fogmarking
The soiling of textiles during processing by deposition of atmospheric dirt. It is almost invariably associated with charges of static electricity. The stains are characterized by their resistance to removal by normal scouring processes. The term derives from the belief that the presence of fog during processing was the primary cause of the fault, but low relative humidity is now recognised as the prime cause.
Note 1: In spinning and winding processes, exposure of a package in a dirty atmosphere may result in severe marking. The contaminated yarn produces a series of dirty streaks or bars in fabric.
Note 2: In weaving and warp-knitting, dust particles are attracted to charged yarns in the exposed part of the warp sheet during a pause in processing; they appear as a series of dirty streaks or bars across the width of the fabric.
Note 3: The fault occurs notably with yarns of low electrical conductivity, such as those composed of the hydrophobic manufactured fibres.

fold; double; twist; ply (yarn)
To combine by twisting together two or more **single yarns** to form a **folded yarn**.

folded yarn; doubled yarn; plied yarn
A yarn in which two or more single yarns are twisted together in one operation, e.g., two-fold yarn, three-fold yarn, etc.
Note: In some sections of the textile industry, e.g., the marketing of handknitting yarns, these yarns are referred to as two-ply, three-ply, etc.

folder
A sewing machine attachment which folds one or more materials into the desired configuration for sewing. Folders are constructed for specific applications, which may provide for the insertion of tapes, trim fabrics, elastic, etc., at the same time as the material is folded. (See also **hemmer**.)

folk weave
A term applied to any construction which, when used in loosely woven fabrics made from coarse yarns, gives a rough and irregular surface effect. Coloured yarns are commonly used to produce weftway and/or warpway stripes.

footing
See **tab.**

fork
See **crotch**.

fork quantity
See **width of crotch** under **crotch**.

forked needle
A needle with a U-shaped end used to produce surface-structured products by **needling**.

End of a forked needle

form (hat manufacture)
The production of a loosely constructed cone of fur by drawing a weighed quantity of blown fur on to a revolving perforated metal cone, the adhering layer of fur being moistened by spraying with water and subsequently carefully removed from the cone by hand.

forming fabric
A continuous belt of woven fabric, normally made from polyester monofilament, which transports the paper sheet through the drainage section of a paper-making machine. These fabrics are woven flat and seamed by yarn interlacing to give an invisible join.

foulard
1. A light-weight fabric, frequently printed, originally of silk, and of 2/2 twill weave.
2. See **padding mangle**.

foundation bar (lace machines)
The combination of **jack bar** and **trick bar** in a lace furnishing machine. A shogging motion is imparted to the foundation bar in unison with the pillar motions of the guide-bar cams to ensure that each jack always controls threads from the same pillar.

foundation net (lace)
See **ground (lace)**.

FOY
See **fully oriented yarn**.

frame (Wilton or gripper Axminster carpets)
The pile warp yarns coming from one beam or in one section of the creel, normally of one colour and controlled by the same operation.
Note: The terms 2 frame, 3 frame etc. are often used to indicate the number of colours used in Wilton or Axminster carpets.

fray
See **felter**.

fraying
Descriptive of the loss of yarns from, or unravelling at, the raw edge of fabric. This term is most commonly used to describe the loss of yarns that are more or less parallel to a raw edge of a woven fabric, resulting in the projection of short, loose lengths.

free-fibre-end yarn
An air-jet textured yarn (see **textured yarn**, *Note l* (viii)) in which the protruding filament loops are cut or broken.
Note: Because of its hairy surface, such a yarn resembles a spun staple yarn.

French clean
See **dry clean**.

French crêpe cord
See **cable cord**.

French double piqué
See **double piqué, weft-knitted**.

French seam
See under **seam type**.

French welt (knitting)
See **tubular welt** under **welt (knitting)**.

friction calendering
The process of passing fabric through a calender in which a highly polished, usually heated, steel **bowl** rotates at a higher surface speed than the softer (for example, cotton or paper-filled) bowl against which it works, thus producing a glaze on the face of the fabric that is in contact with the steel bowl. The friction ratio is the ratio of the peripheral speed of the faster steel bowl to that of the slower bowl and is normally in the range 1.5 to 3.0.

friction spinning
A method of **open-end spinning** which uses two surfaces moving in opposing directions to insert twist into an assembly of fibres positioned between them.
Note 1: Usually the external surfaces of two rollers are used, at least one of which is perforated so that air can be drawn through it to facilitate fibre collection. Twisting occurs near the line of closest proximity of the rollers where the fibre assembly is rotated by frictional contact with the roller surfaces.
Note 2: High rates of twist insertion can be achieved due to the large difference between the yarn and roller diameters. (See also **spinning**.)

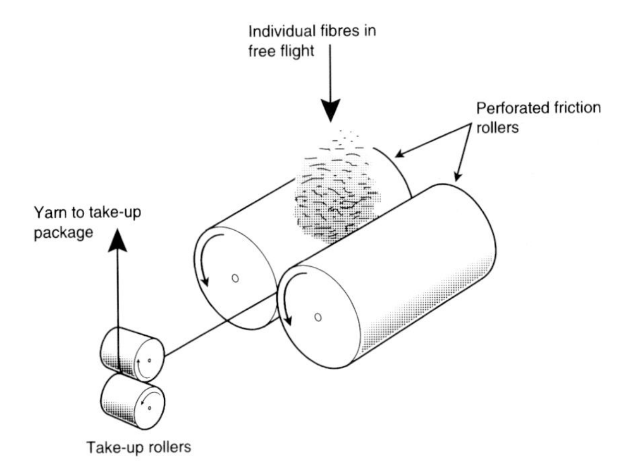

Friction spinning

friction tests

Tests which measure frictional properties during relative movement between fibre/fibre, yarn/yarn or fabric/fabric, or between fibres, yarns and fabrics and other surfaces such as metals, ceramics, etc.

Note: The frictional properties of textile materials may show variations from those predicted by the classical laws of solid friction.

friction-twisting

The generation of false-twist (see **false-twisting**) by a device in which the yarn lies in contact with one or more surfaces of high friction driven in a direction at a substantial angle to the yarn axis.

Note: In practice, belts, bushes, cylinders or disks are used, one rotation of which generates many turns of twist. (See also **pin-twisting**.)

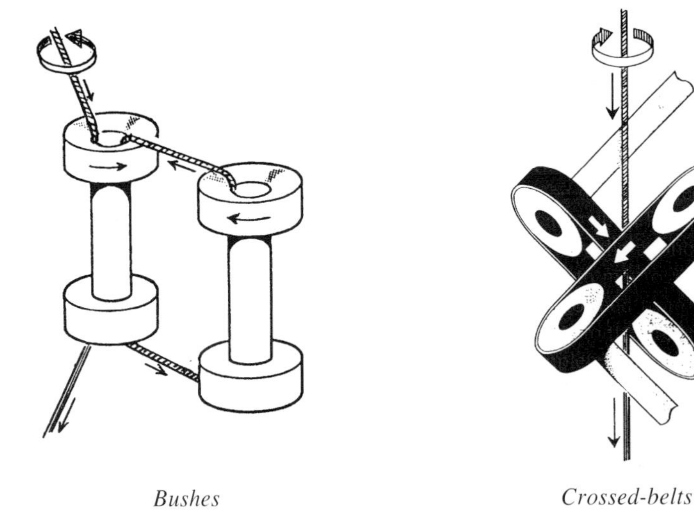

Bushes *Crossed-belts*

friction-twisting *(continued)*

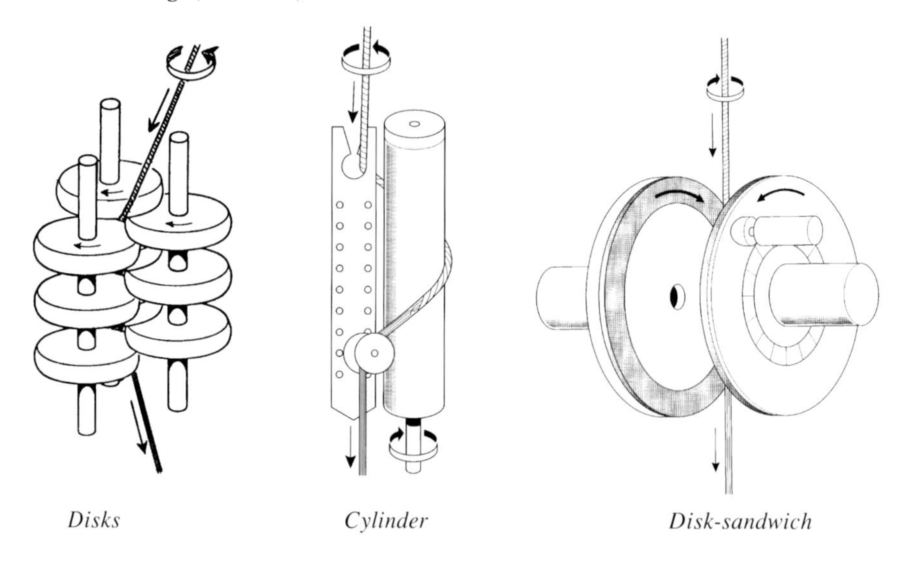

Disks *Cylinder* *Disk-sandwich*

friezé
1. A heavy woollen overcoating made from a coarse wool. The fabric is consolidated by very heavy milling and then raised to give a rough fibrous surface with the nap laid in one direction. The weave structure is hidden by the milling and raising.
2. nap friezé: a friezé in which the nap is rubbed into small pills or balls.

fringe
1. An edging or border of loose threads, tassels, or loops.
Note: These may be produced by the constituent threads or by threads added to a fabric after weaving or knitting.
2. (Narrow fabric) A trimming having, on one or both edges, cut or looped weft threads, which extend substantially beyond the width of the warp threads to form a decorative edge. The threads forming the fringe are sometimes bunched or knotted together to increase the decorative effect.
Note 1: Tassels, balls, or other adornments may be added.
Note 2: That part of a fringe comprising both warp and weft is known as the 'heading'.
Note 3: That part of a fringe containing only weft is known as the 'skirt'. (See also **bullion fringe**.)

frisé pile
See **curled pile** under **pile (carpet)**.

frisons; kibisu; knubbs; strusa
The first waste obtained in the process of reeling silk cocoons, composed of the tangled first lengths of the silk filament which are removed by the reeler up to the point when the filament begins to reel properly.

front crossing heald
See **leno weaving**.

front loop (weft knitting)
See **face loop** under **knitted loop (weft knitting)**.

front rise
The distance, measured along the surface of a bifurcated garment, from the **crotch** to the centre front of the waistline. (See also **back rise**.)

front standard
See **leno weaving**.

frosting
A whitish appearance of coloured textiles, normally regarded as detrimental, caused by the presence of lightly coloured surface fibres.

fud
Droppings or fettlings from woollen cards consisting of very short fibres that may be heavily charged with oil and dirt. (See also **fettling**.)

fugitive tint; sighting colour
A colorant applied to textile materials for their identification during handling. The colorant must be removed easily during normal textile scouring or dyeing procedures.

full cardigan rib; polka rib (knitting)
1x1 rib fabric, every stitch of which consists of a held loop and a tuck loop.

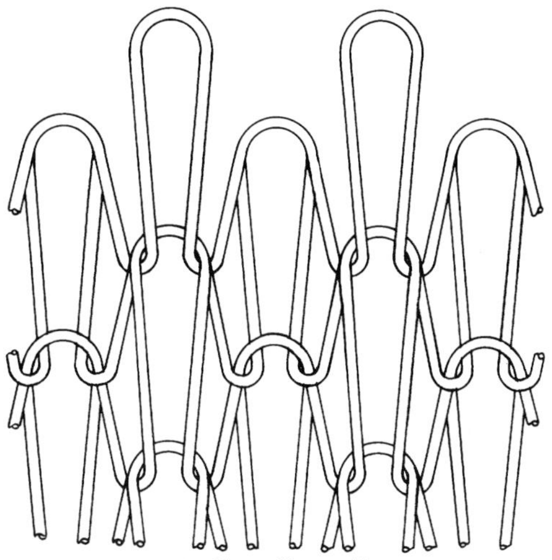

Full cardigan rib

full-fashioned; fully fashioned
See under **shaping (weft knitting)**.

fulling
See **milling (fabric finishing)**.

fullness
An extension to the length on one of two sections of a garment joined by a seam, used to create volume or shape in the garment, e.g., in a **sleeve head**.

fully drawn yarn; FDY; fully oriented yarn; FOY; highly oriented yarn; HOY
A melt-spun continuous-filament yarn that has been highly oriented either by drawing at a high draw ratio (preferred term, fully drawn yarn) or by spinning at a high wind-up speed such that

fully drawn yarn; FDY; fully oriented yarn; FOY; highly oriented yarn; HOY *(continued)*
little residual drawability remains (preferred term, highly oriented yarn). (See also **draw-twist**, **draw-wind**, **spin-drawing** and **draw-spinning**.)
Note: These terms are used in contrast to **low orientation yarn** and **partially oriented yarn**.

furniture cord
See **upholstery cord**.

fustian
A hard-wearing type of clothing fabric containing a large amount of weft yarn. At different times, the term has been used to describe a considerable variety of structures made from different natural fibres. It is now used to describe a class of heavily wefted fabrics (usually made from cotton) of which **swansdown**, **imperial sateen**, **beaverteen**, **moleskin**, **velveteen**, and **corduroy** are examples.

fuzz fibres
Short, relatively coarse fibres which remain on the cotton seed after **ginning**.
Note: Not all varieties of cotton seed possess fuzz fibres. *Gossypium barbadense* varieties generally have naked seeds; *Gossypium hirsutum* varieties generally have fuzzy seeds. Fuzz fibres are distinguished from **lint** fibres as follows: they begin to develop several days after the lint; they grow to a length of only 2 to 5mm; they are often a darker colour; they are more firmly anchored to the seed due to the fact that their cell walls are not thinned at the point where the fibre meets the seed surface. Fuzz fibres may be separated from the seeds, to form the major part of **linters**, by means of a second and third ginning operation

fuzzing
The roughening of the surface fibres and/or teasing out of fibres from a fabric which produces an unwanted change in its appearance. This change can occur during washing, dry cleaning, testing or in wear.

gaberdine
A firmly woven, clear-finished, warp-faced fabric in which the end density considerably exceeds the pick density, the twill line thus being produced at a steep angle. It is usually woven in 2/1 and 2/2 twills and largely used for raincoats and sportswear.

Gaberdine (magnification x 5).

gait (flax)
A large handful of loose, pulled flax, stood up on end in a cone form to dry. (See also **retting**.)

gait (lace machines)
1. The distance between the centres of adjacent comb blades.

2. A measure of the distance over which a thread is moved, e.g., 'two gaits' means 'across two comb spaces'.

gait (weaving)
A full repeat of the **draft** in the **healds**, or in the case of **jacquard**, in one complete row of the harness.

gait; gaiting; gait up (weaving)
General terms used to describe the positioning of the **warp**, **healds**, and **reed** in a loom, in readiness for weaving. Where **drop wires** are mounted on the warp during warp preparation, gaiting also includes the positioning of the drop wires. (See also **looming**.)

gaiting (knitting)
See **gating (knitting)**.

galloon (lace)
Lace having both edges scalloped.

galloon (narrow fabric)
1. Ribbon used as a band on men's hats and as a binding on ladies' court shoes originally with 2/2 **warp rib** weave with a plain weave **binder** separating the **selvedge** and known as **double shot**. A plain woven construction using thicker weft is often substituted.
2. Any woven narrow fabric used for trimming.

galloons
See **banded laces**.

gamma value
The mean number of xanthate groups per 100 glucose residues in **cellulose xanthate**.

garment-length knitting machine
A **knitting machine** built for the production of individual garment panels in series, rather than for the continuous production of fabric. The term is most commonly used to specify this type of circular weft-knitting machine (see **circular knitting machine**).

garnett machine
A type of **carding** machine, containing rollers and cylinders covered with metallic teeth, similar in shape to the teeth of a saw, which is used to open up soft and hard wastes.

garter band
See **after-welt (knitting)**.

garter webbing
An elastic narrow fabric, sometimes multicoloured, characterized by selvedges that form a frill on relaxation. It may also be a medium-strength elastic narrow fabric for supporting socks or stockings.

gas
To singe, i.e., to remove unwanted surface fibres on a fabric or yarn by passage through a flame (see **singe**).

gas fume fading
An irreversible change in hue which occurs when textiles, particularly cellulose acetate, triacetate and polyamide dyed with certain aminoanthraquinone disperse dyes, are exposed to

gas fume fading *(continued)*
oxides of nitrogen which arise, for example, from gas or storage heaters.

gassed yarn; singed yarn
A yarn that has been passed through a flame or over a heated element to remove unwanted surface fibres.
Note: A 'genappe yarn' (Genappe in Belgium) is a gassed worsted yarn.

gathering
The formation of a style feature in garments or other made-up goods by drawing material together into a succession of small folds (gathers) and retaining by stitching or other means.

> **pinching**
> The formation of small individual gathers along a **stitch line**.
>
> **ruffling**
> A method of gathering material, by feeding in faster to the needle, than it is fed away from the needle. This is accomplished manually, or by a sewing machine attachment, or by means of **differential feed**.
>
> **shirring**
> The formation of a series of parallel gatherings, normally produced using a multi-needle chainstitch machine, in which **elastomeric yarn** is included to allow subsequent stretch in use. (See also **shirring thread**.)

gating; gaiting (knitting)
The relative alignment of two sets of knitting elements, e.g., needles, on knitting machines. Two forms of needle gating (rib and interlock) are common and may be interchangeable on the same machine. Types of gating are:

> **interlock gating; interlock gaiting (knitting)**
> The opposed alignment of one set of needles with the other on a knitting machine.
>
> **purl gating; purl gaiting (weft knitting)**
> The opposed alignment of **tricks** of two needle beds lying in the same plane, on a machine equipped with double-headed needles.
>
> **rib gating; rib gaiting (knitting)**
> The alternate alignment of one set of needles with the other on a machine equipped with two sets of needles arranged to knit rib fabrics.

gauge; cut (knitting)
1. A term giving a notional indication of the number of needles per unit length, along a needle bed or needle bar, of a knitting machine. In current practice a common unit length of 1 'English' inch (25.4mm) is used for all types of warp and weft knitting machines.
Note 1: For circular knitting machines, the length referred to is measured along the circumference of the needle cylinder. Such values are generally quoted to the nearest whole number (or to the nearest 1/2 if less than 5) and, in keeping with the present national and international use of the concept, written as 'E10', say, rather than as '10 needles/inch', or the older abbreviated form '10G'.
Note 2: Gauging systems other than the above are still used. The more common systems and the types of knitting machine to which they apply are as follows:
(i) loop wheel machines: the number of needles in 1.5 inches (38.1mm) of circumference;
(ii) raschel machines: the number of needles in 2 'English' inches (i.e., 50.8mm) or, much less commonly, 2 saxon inches (i.e., 47.2mm) of a needle bed. Such gauges based on English inches are written in the form 'ER10';
(iii) sinker wheel machines: the number of needles in 1 zoll (fein) or 1.5 zoll (gross) measured

along the arc of the holes drilled in the needle ring to receive the cranked ends of the needles. (1 zoll=1/36th metre or 27.8mm);
(iv) straight-bar plain machines: the number of needles in 1.5 inches (38.1mm) of a needle bar;
(v) straight-bar rib machines: the number of needles in 1.5 inches (38.1mm) of either needle bar;
(vi) occasionally reference may also be found to a 'universal' gauging system based on the number of needles per 100mm.
Note 3: For flat knitting machines needle spacing is sometimes quoted as an alternative to the number of needles per unit length. In such cases the numerical value given is the needle spacing in mm multiplied by 10.
Note 4: For small diameter circular knitting machines a statement of gauge is replaced by one of the total number of needles and the diameter of the cylinder, in inches, e.g., 370x4.5.
Note 5: When the term gauge is applied to a knitted fabric or garment, it refers to the gauge of the knitting machine on which it was made.
2. A term specifying a dimension, usually thickness, of the needles or other loop-forming elements of a knitting machine.

gauge; pitch
The distance measured horizontally between spindles, drive drum centres, or adjacent yarns, on any multi-position machine.

gauge (lace machines)
1. A term specifying the comb spacing, usually expressed as 'x point'. Traditional values have been: (i) Leavers: the number of comb spaces in 1/2 inch (12.7mm); (ii) furnishings and bobbinet (plain net): the number of comb spaces in 1 inch (25.4mm).
Note: The number of carriages per unit length is the same in the Leavers and bobbinet machines. The bobbinet machine works two carriages in tandem in each comb space.
2. A term traditionally specifying the number of needles per inch on warp lace machines.

gauge (linking)
A term specifying the spacing of the **points** in linking and point seaming machines and expressed as the number of elements per unit of length. Gauge (English): Points in an arc of 1.5 inches (38.1mm). Gauge (American): Points in an arc of 1 inch (25.4mm). The terms 'gauge' and 'point', respectively, are used to differentiate between the two systems, e.g., an English 36-gauge linking machine has 24 points per inch and an American 24-point linking machine has 24 points per inch.

gauge length (testing)
The original length of that portion of the specimen over which strain or change of length is determined.

gauge marks (testing)
Marks of known separation applied to a specimen.

gauze
A light-weight, open-texture fabric produced in plain weave or simple leno weave.

gauze reed; leno reed
A **reed** constructed of alternate full-length and half-length dent wires, the latter having holes at the top through which the crossing ends pass.
Note: A gauze fabric may be produced by lowering the gauze reed between picks and allowing the standard end to be traversed to the other side of the dent space by a **tug reed** before the gauze reed is raised in readiness for the next pick (see **leno weaving**).

gauze weaving
A term commonly used as a synonym for **leno weaving**; strictly, a method of producing the simpler types of light-weight fabric by leno weaving.

gear
See **heald shaft**.

gear-crimped yarn
See **textured yarn**, *Note 1* (vi).

gel dyeing
A continuous tow dyeing method in which soluble dyes are applied to wet-spun fibres, (e.g., acrylic or modacrylic fibres) in the gel state, (i.e., after extrusion and coagulation, but before drawing and drying).

genappe yarn
See **gassed yarn**.

generic name (fibres)
A name used for regulatory purposes to distinguish a class of textile fibres from others.
Note 1: For natural fibres (see **fibre, natural** under **fibre**) the distinguishing attribute is the fibre source; for manufactured fibres (see **fibre, manufactured** under **fibre**) chemical difference, which often results in distinctive property differences, is the main basis for classification; other attributes are included where necessary. The attributes used for the specification of generic names of manufactured fibres are not necessarily identical with the attributes used for naming chemical molecules.
Note 2: Generic names of fibres are properly used as adjectives but are often used as nouns; they are descriptive of the nature of the fibre or filament components of the associated structure (yarn, top, sliver, fabric, garment, etc.).
Note 3: ISO 2076:1989 *Textiles - Man-made fibres - Generic names* contains a list of the generic names and definitions of the different categories of fibres manufactured on an industrial scale for textile and other purposes. In this edition of *Textile Terms and Definitions*, for the sake of presentation, the individual definitions sometimes differ slightly from the precise ISO 2076 text. Other lists of fibre names have been produced in connection with textile labelling legislation (e.g., in the EC Textile Labelling Directive and the U.S. Federal Trade Commission's Textile Fibre Products Identification Act). Names and definitions approved by the Federal Trade Commission that do not conform with the ISO list are included in this edition with the qualification '(generic name U.S.A.)'. (See also Classification Table, p.401.)

geogrid
A network of integrally connected tensile elements used to reinforce and stabilise civil engineering structures.

geomembrane
A barrier of very low permeability, which may or may not incorporate textile reinforcement, used to control the flow of fluids.

georgette
A fine, light-weight, open-texture fabric, usually in a plain weave, made from crêpe yarns, usually having two S-twisted and two Z-twisted yarns alternately in both warp and weft.

geotextile
Any permeable textile material used for filtration, drainage, separation, reinforcement and stabilisation purposes as an integral part of civil engineering structures of earth, rock or other constructional materials.

gigging
The process of raising a nap on fabrics by means of a **teazle** machine.

gilding
See **oxidized oil staining**.

gill box; pin drafter
A **drafting** machine, used in worsted processing, in which the motion of the fibres is in part controlled by pins fixed on moving bars (pinned **fallers**).

gilling; pin drafting
A system of **drafting** in which the movement of the fibres relative to one another in a sliver is controlled by pins.

gimp
1. A core helically wrapped by one or more threads, resulting in a stiff **cord**.
2. An upholstery trimming usually made wholly or partly from gimp as defined above (see **Argyle gimp**, **coronation gimp** and **scroll gimp**).

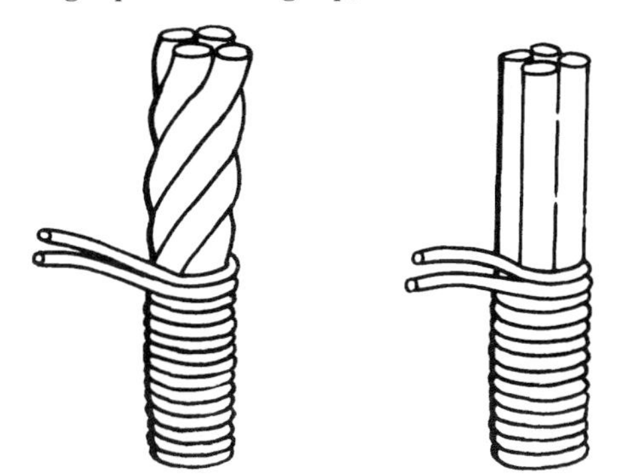

*(Both these figures illustrate the definition of **gimp** 1.)*

gimp (lace)
Threads from a beam used to fill in the **objects** of the pattern.

gimp thread
A thick thread used to support and raise buttonhole stitching.

gimp yarn
See **bouclé yarn** under **fancy yarn**.

gin cut cotton
Cotton that has been damaged in **ginning** by the cutting saws to the extent that its value is reduced.

gingham
A plain-weave, light-weight cotton fabric, approximately square in construction, in which dyed yarns, or white dyed yarns, form small checks or, less usually, narrow stripes.
Note: If fibres other than cotton are used the term should be suitably qualified, e.g., viscose gingham.

ginned lint
The main product (by value) of the cotton **ginning** process, the other products being seeds and **linters**.

ginning
A process that removes cotton fibres (lint) from the seed.

glacé binding
See under **binding**.

glass (fibre) (generic name)
Fibre, in textile form, obtained by drawing molten glass. (See also Classification Table, p.401.)
Note 1: The above is the ISO definition; in practice, glass filament or fibre is also produced by extrusion or other spinning processes.
Note 2: Glass is described in ASTM C 162-89 as: 'An inorganic product of fusion which has cooled to a rigid condition without crystallizing.'
Note 3: Glass fibres used in textiles consist of vitreous silicates or borosilicates.

glass-rubber transition temperature; Tg
Temperature region over which there is a reversible physical change from a viscous or rubbery state to a brittle glassy state.

glaze
To produce a smooth, glossy, plane surface on a fabric by heat, heavy pressure, or friction.
Note: Glazing may be produced intentionally, e.g., by **friction calendering**, or as a fault.

go-through machine
See under **lace machines**.

goat fibre (hair)
Fibre from the common goat (Genus *Capra*).

godet
1. (Fibre manufacture) A driven roller, which may be heated, around which yarn is passed in order to regulate its speed during the extrusion and further processing of certain manufactured fibres.
Note: Godets usually have a single flange and are mounted on an axle from the flange side. They may be constructed with serrations broadly parallel to the axis, and may be tapered or stepped.
2. (Clothing) Material cut in a wedge-shaped segment of a circle and let into a seam or slash to give decorative extra **fullness** to a hem.

going-part
See *Note* under **sley** 1.

goods ratio
See **liquor: goods ratio**.

gore
A shaped panel, generally in a skirt, which increases in width towards the hem or lower edge.

gossypium
The generic name of the **cotton** plant.
Note: Almost all cotton grown commercially belongs to one of only four species, *Gossypium barbadense*, *Gossypium hirsutum*, *Gossypium herbaceum*, and *Gossypium arboreum*.

grab sampling
A method of taking representative samples from packages of textile fibres. A jaw is driven by a machine into the packages and when withdrawn brings out a sample of fibres.
Note: Grab samples can be used to measure staple and fibre length.

grab test
A **tensile test** in which only the central portion of the width of the specimen is held in the jaws. (See also **tensile strength at break**.)

graduated compression hosiery
Hosiery which, when worn on the leg, exerts a specified minimum pressure at the ankle and a progressively reduced pressure at the calf and thigh.

graft copolymer
See **copolymer, graft**.

graft polymerization
The production of a graft copolymer (see **copolymer, graft**).

grain
A term used in making-up to refer to the direction of the warp threads in woven or warp-knitted fabric, and the direction of wales in knitted fabric. (See also **grain line** and **off-grain**.)

grain line; straight of grain
A straight line marked on a **pattern** piece to indicate the warp or wale direction (see **grain**) to ensure that the pattern piece is correctly positioned and cut to achieve the required appearance in the finished product.

grandrelle yarn; twist yarn
A two-ply yarn composed of singles of different colour or contrasting lustre. (See also **worsted yarns, colour terms**.)

granite weave
A weave having a **satin** base or other regular plan with adjoining ends lifted in regular order to give small broken effects. They are largely used as ground weaves for jacquard designs.

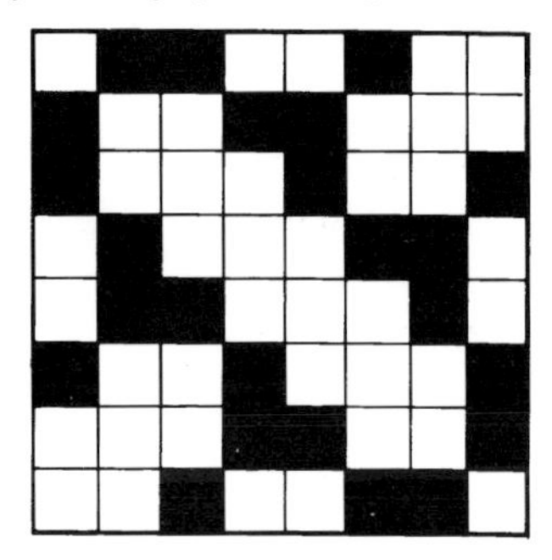

Example of a Granite weave

grass bleaching; grassing; crofting
A process for bleaching linen cloth after it has been washed by exposing it, while spread out on a grass lawn or field known as a green, to the action of the elements.

greasy piece
A piece of woollen fabric as it comes from the loom.

greasy wool; grease wool
Unscoured wool as shorn from the sheep.

Grecian alhambra
A figured quilting fabric in which **Grecian weaves** are largely used.

Grecian weave
A weave based on the counter-change principle and having floats of warp and weft that produce either a rough surface or a cellular effect on both sides of the fabric. Modifications of the basic Grecian weave are also made, in some of which the warp and weft floats appear on the face side of the fabric only.

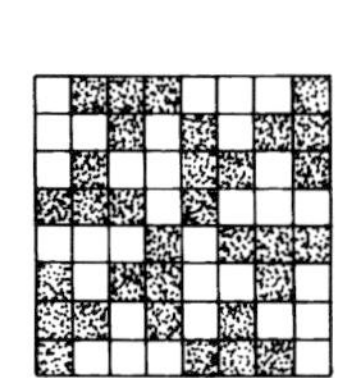
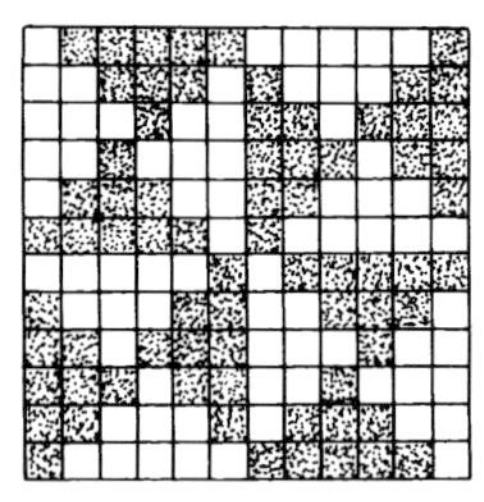

Grecian weaves (simple)

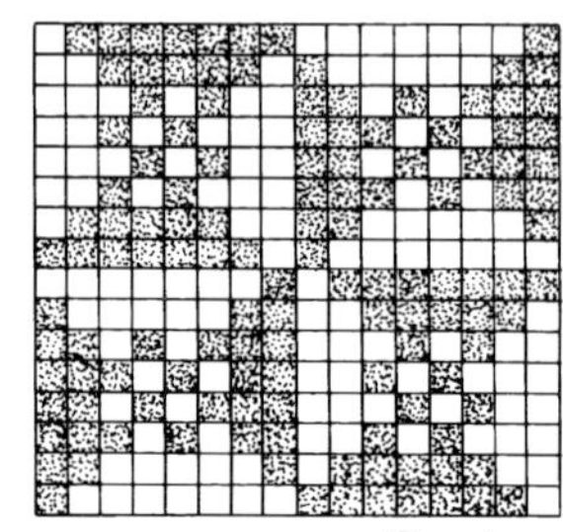
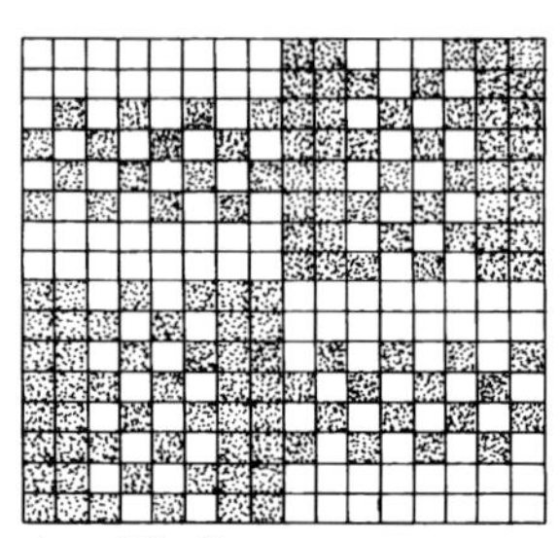

Grecian weaves (modified)

green flax
See **flax, green**.

grenadine
A thrown silk yarn similar to **organzine**, but having a higher twist.

grey; gray; greige
Descriptive of textile products before being bleached, dyed or finished. Some, however, may contain dyed or finished yarns. (See also **loom-state**.)
Note: In the linen and lace trades, the term brown goods (see **brown lace**) is sometimes used.

grey scale
A series of pairs of neutrally coloured chips, showing increasing contrast within pairs, used visually to assess contrasts between other pairs of patterns: for example the ISO Grey Scales comprise two series of chips against which the magnitude of the change in colour of a specimen submitted to a fastness test and of staining of adjacent uncoloured material can be visually assessed and rated on a 1-5 scale. (See also BS 1006: 1990.)

griffe (jacquard mechanism)
The knife assembly that operates to lift the hooks and harness in the process of forming a **shed**.

grin; grinning
A defect in a compound structure e.g., a **double cloth** in which one fabric can be seen through or 'grinning through' the other, as a result of bad **cover**. The term can be applied to compound

woven and knitted structures including pile fabrics. (See also **seam grin**.)

grinding (rag)
A colloquial term for **pulling**.

grinny cloth
A cloth with an unsatisfactory **cover**. It is sometimes said to be **grinning** and is known also as 'hungry cloth'.

gripper Axminster
See **Axminster carpet**.

gripper-shuttle weaving machine
A weaving machine in which the weft thread is gripped by jaws(s) fitted in a projectile, which is then propelled through the shed.

gripper-spool Axminster
See **Axminster carpet**.

gripper; gripper tape; gripper web
A narrow fabric, woven or knitted, having one or more raised stripes in the warp direction, formed of a high-friction material. It is used in the inside of waistbands for skirts and trousers to keep a blouse or shirt in place.

grist
A coarseness or fineness of yarn or other linear textile material. (See also **count of yarn** and Tables, p.396-397.)

grosgrain
A plain-weave fabric with a rib in the weft direction, the rib being more pronounced than in a taffeta, poult or faille. It is usually made with a closely set continuous-filament warp and coarse-folded continuous-filament or staple weft. One example of a finished high-quality product is 76x11; 8.3xR90/6tex; 170 g/m^2; K=21.9+10.4.
Note: Grosgrain belongs to a group of fabrics having ribs in the weft direction. Examples of this group, arranged in increasing order of prominence of the rib, are **taffeta**, **poult**, **faille**, and grosgrain.

ground (lace)
The basic structure of net, known as foundation net, by which the **objects** are joined.

grown-on
Descriptive of a small garment part which, though usually cut separately and then attached, is cut in one with the adjoining larger garment part. Normally applies to collars, waistbands, facings and sleeves which are subsumed into larger garment parts.

guanaco fibre (hair)
Fibre from the guanaco (*Lama huanaco*).

guard hairs
Fibres which project beyond the under-coat of some mammals. They are usually coarser than under-coat fibres.

guide
A component for controlling the path of a running textile material.
Note: Guides may be static, reciprocating, freely rotating, or positively driven.

guide bar (warp knitting machines)
A bar running the full width of a machine and equipped with guides through which threads are passed so that the lateral motions imparted to the guide bars by the pattern control device are transmitted to the threads.
Note: In warp knitting the term bar is also used either to define the number of guide bars in the machine or the number of guide bars used to make a particular construction, e.g., 2 bar tricot machine, 24 bar raschel lace machine or 3 bar velour fabric, 42 bar lace edging.

guide bars (lace machines)
Bars running the full width of a machine and equipped with guides through which threads may be passed so that lateral motions imparted to the guide bars by cam, jacquard, or other pattern-control device are transmitted to the threads.
Note 1: On **lace furnishing** and **string warp lace machines**, the guide bars are few in number and generally consist of a substantial bar, in which guides made from flat metal stampings with an oval hole near the top are directly fixed. The number of guides corresponds to the gauge of the machine.
Note 2: On **bobbinet (plain net) machines**, there are normally only two guide bars, of which one is a substantial bar to which are fixed leads containing pigtail guides. The number of guides corresponds to the gauge of the machine.
Note 3: On **Leavers** and **bar warp lace machines**, the guide bars are known as 'steel bars'. There may be up to 200 bars in the Leavers machine and up to 40 in the bar warp machine. Steel bars consist of thin steel strips with holes for threads punched in the upper part and slots to accommodate the bar brackets in the lower part. Friction bits are applied to the surface to provide thread space between one bar and the next. The thread holes are usually spaced one-quarter of the gauge of the machine, sometimes one-half gauge, and, rarely, full gauge.
Note 4: On **raschel machines** the ground bars are fully guided, i.e., one guide for each needle. They are substantial bars to which leads of flattened metal stampings with round holes are attached. Pattern bars are only part guided with spike guides where required; these are screwed on to a blank bar.

guipure lace embroidery
A lace construction produced by a pattern of thread embroidery on to a ground fabric, the ground fabric being thereafter removed, by chemical or other means, to leave an openwork lace.

gum waste
Waste comprising all broken silk threads which have been discarded during reeling, or at the inspection of the skeins, and which have not undergone any further processing.

gum; sericin
A gelatinous protein, comprising 20% to 30% by mass of raw silk, cementing the two fibroin filaments (**brins**) in a silk **bave**.

gusset
A piece of material inserted into a garment to provide strength, shape, enlargement or freedom of movement.

gut thread
A thread incorporated in a woven, knitted, or braided structure, primarily for the purpose of limiting its extension.

haberdashery
See **trimmings**.

habit
A woollen fabric used for ladies' riding habits, made from good quality wool, generally dyed to dark shades and given a napped finish. The term is applied widely to a range of costume fabrics.
Note: The word habit also has a series of meanings associated with clothing:
(i) dress worn by ladies when on horseback;
(ii) a sleeved tunic worn by those in holy orders - originally of coarse fabric in naturally occurring undyed colours such as white, blacks, browns and greys;
(iii) the dress characteristic of a particular rank, degree, profession or function;
(iv) bodily apparel or attire.

habutae; habutai
A Japanese word meaning a smooth lightweight fabric woven from **net silk** yarn in a plain weave. Originally used for lining kimonos.

hackling (flax)
A process in which **stricks** of scutched flax are combed from end to end to remove short fibre, naps (or neps) and non-fibrous material, and to sub-divide and parallelise the fibre strands.

hair
Animal fibre other than sheep's wool or silk.
Note: It is recognised that this definition implies a distinction between sheep's wool and the covering of other animals, notwithstanding the similarity in their fibre characteristics. Thus the crimped form and the scaly surface are not confined to sheep's wool. It seems desirable in the textile industry, however, to avoid ambiguity by confining the term wool to the covering of sheep and to have available a general term for other fibres of animal origin. Normally the less widely used fibres are known by name, e.g., alpaca, mohair, etc., but collectively they should be classed as hair. A difficulty arises when it is desired to distinguish between the fibres of the undercoat and the remainder of the fleece; for instance, between the soft short camel hair used for blankets and the coarse long camel hair used for belting. The term wool is sometimes used for the shorter fibre, qualified by the name of the animal, e.g., cashmere wool.

haircloth
1. A fabric in which the weft consists of single fibres of horsehair, obtained from tails and manes and woven on a special loom which is capable of inserting **picks** of the discontinuous fibres. The fabric width is governed by the length of available horsehair and normally varies between 47cm and 76cm. The woven structure, which usually has a cotton warp, varies according to the end-use, e.g., interlinings, furnishing fabrics, sieve and press cloths, and the horse hair may be dyed.
2. A fabric produced from yarns in which horsehair is mixed with cotton, polyester, linen or other fibres to make it strong and inflexible. Used in upholstery and also as a chest canvas in tailoring.

haircord
A plain-woven cotton fabric, characterized by fine rib lines in the warp direction created by alternate coarse and fine ends, or by having two (or more) ends weaving as one alternately with a single end. A possible construction is 28 ends x 26 picks per cm; 2 ends 18 tex and 1 end 21 tex cotton warp x 21 tex soft spun cotton weft.

haircord carpet
A **cord carpet** in which the pile is 100% hair.

hairline
An effect obtained by either colour and weave or printing, producing fine hair-like lines either lengthways (warp hairline) or widthways (weft hairline) in a fabric.

hairweight
The mass per unit length of a fibre (usually cotton) expressed in

$$\frac{g}{cm} \times 10^{-8}.$$

It is usually denoted by H. (See also **standard hairweight**.)
Note: This is numerically equivalent to millitex.

hairy (fabric effect)
A fibrous appearance in a fabric made from continuous-filament yarn, which may be localised or general and is caused by damage to individual filaments.

halching
1. The operation of looping the external yarn end around a **cop** or **bobbin** to facilitate retrieval.
2. For use of this term in the silk industry, see *Note 1* under **leasing**.

half cardigan rib; royal rib
A 1x1 rib fabric in which the wales on one side consist wholly of knitted loops and the wales on the opposite side consist of a **held loop** and a **tuck loop**.

half heald
See **leno weaving**.

half-marl yarn
See **worsted yarns, colour terms**.

half-Milano rib, weft-knitted
A weft-knitted rib-based fabric, consisting of one row of 1x1 rib (A) and one row of plain knitting (B) made on either set of needles. The appearance and characteristics of the fabric are related to the ratio of the course lengths of A and B. (See also **Milano rib, weft-knitted** and **double jersey, weft-knitted**.)

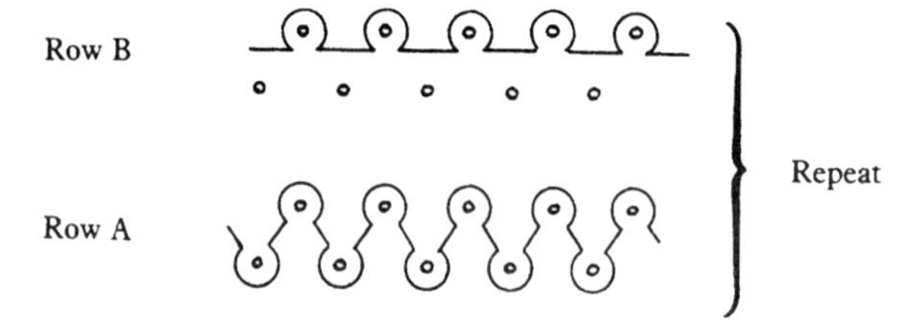

Half-Milano rib

half-point (knitting)
See **point (straight-bar and flat-knitting machines)** under **point (knitting)**.

half-point transfer (knitting)
See **spread loop**.

hand
See **handle**.

hand loom
A hand operated machine for producing cloth by **weaving**. In some instances, the shedding is performed by foot operation. (See also **weaving machine**.)

hand wheel (sewing machine); balance wheel
A wheel at the drive end of the machine **arm**, enabling the movement of the **take-up lever** and needle to be controlled by hand during sewing or, more commonly, when rethreading.
Note: Technological improvements have reduced the need for operators to touch the hand wheel during sewing, but the wheel is retained to assist when making adjustments and rethreading.

hand-hold
See **roving courses**.

hand-knotted carpet
A carpet made by knotting tufts around the warp yarns during hand weaving.
Note: The two most common types of knot are the Persian (or sehna) and the Turkish (or ghiordes).

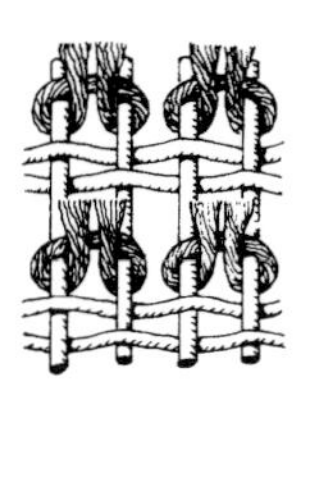
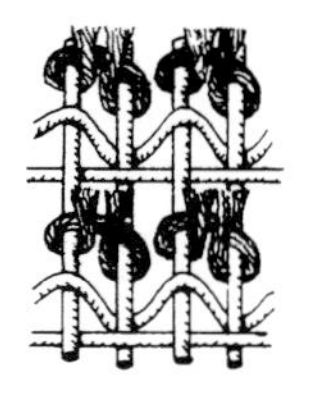
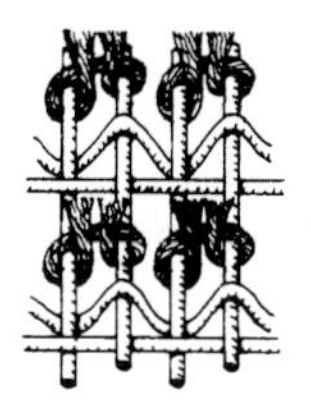

Ghiordes or Turkish knot

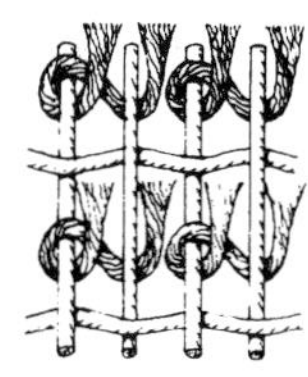
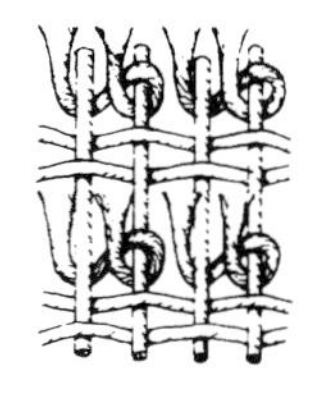
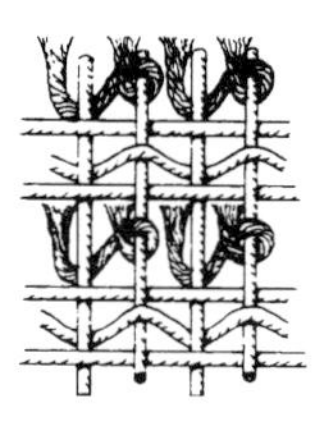

Sehna or Persian knot

handknitting yarn
A yarn which has been manufactured, finished and packaged for use in handknitting.
Note: The most common form of package is a ball of 20, 25, 50 or 100 grammes. **Hanks** and occasionally **cones** can also be used. Handknitting yarns are produced in a number of generic forms, e.g., fingering, double knitting and 'knits-as-four-ply'. These forms have historic structural definitions, but due to the wide diversity of product range existing today, a recommended knitting tension should be specified for a handknitting yarn. Handknitting tension can not easily be measured and is indirectly expressed in terms of the number of stitches and rows that will produce a given area of the knitted fabric structure or pattern using a specified needle size.

handle; hand
The quality of a fabric or yarn assessed by the reaction obtained from the sense of touch.
Note: It is concerned with the judgement of roughness, smoothness, harshness, pliability, thickness, etc.

hank; skein; reel
1. An unsupported coil (e.g., approximately 135cm in circumference) composed of wraps of yarn or sliver and wound on a **reeling machine** with a cross-wound pattern that allows the resulting assembly of strands to be leased (see **leasing**).
Note 1: Hanks may be bundled for ease of transport and storage. After processes such as bleaching, dyeing, mercerizing and drying, yarn from a hank, mounted on a swift is wound (see

hank; skein; reel
backwinding (spinning)) on to a suitable **package**.
Note 2: The hank is the traditional unit of length in the indirect system of yarn count, e.g., the cotton **count** of a yarn is calculated as the number of hanks of 840yds per pound (see Table, p.397).
2. A synonym for **count** as applied to **sliver**, **slubbing** or **roving**. (See also **lea**.)

hank sizing; skein sizing
The application of size solution to yarn in hanks.

hank swift
See **swift**.

hanking machine
See **reeling machine**.

harateen
A cloth used for furnishing in the 18th and early 19th century, when it was considered interchangeable with **moreen**. It was made with a worsted warp and a thicker worsted weft to form horizontal ribs, then finished by watering and stamping.

hard laid rope
See under **rope**.

hard-twist pile
See **curled pile** under **pile (carpet)**.

hardening
Treatment of manufactured regenerated-protein filaments so as to render them completely insoluble in cold water and cold dilute saline solutions.

hardening (felt manufacture)
A process in the pressed **felt** industry and in hat manufacture in which a mass of loose fibres, after being roughly shaped by carding and forming, is subjected to a high-speed vibratory motion in the presence of steam while under considerable mechanical pressure.

hare fibre (hair)
Fibre from the hare (*Lepus europaeus* and *Lepus timidus*).

harness (weaving)
Healds and heald shafts and/or jacquard cords (see **jacquard harness (weaving)** under **jacquard (mechanism)**) used for forming a **shed**.

harness cord
See **jacquard harness (weaving)** under **jacquard (mechanism)**.

harvard
A shirting cloth with a 2/2 twill weave, usually with a coloured warp and white weft. These cloths are often ornamented by stripes of white or coloured threads or by simple weave effects or by both. Typical construction: 64x56; 30x49tex; K=35.1+39.2.

haul-off roller; haul-off roll
The first driven roller around which an extruded yarn passes after leaving the spinneret, and whose surface speed determines the **spin-stretch ratio**.

hauteur
See under **fibre length**.

hawser laid rope
See under **rope**.

head (jute)
One of a number of bunches of raw jute forming a bale. The heads are each given a twist and folded over before being made into the bale.

head (sewing machine)
1. That part of the machine **arm** containing the needle bar, the presser bar and other upper thread and upper feed devices above the **throat plate**.
2. The part of a sewing machine that interacts directly with the materials being sewn to form stitches.

headband
A woven **narrow fabric** used in bookbinding, and having a coloured piped or 'beaded' edge woven integrally.

heading (narrow fabric)
See **fringe** 2.

heald; heddle
A looped cord, shaped wire, or flat steel stripe with an eye in the centre through which a warp yarn is threaded so that its movement may be controlled during weaving.

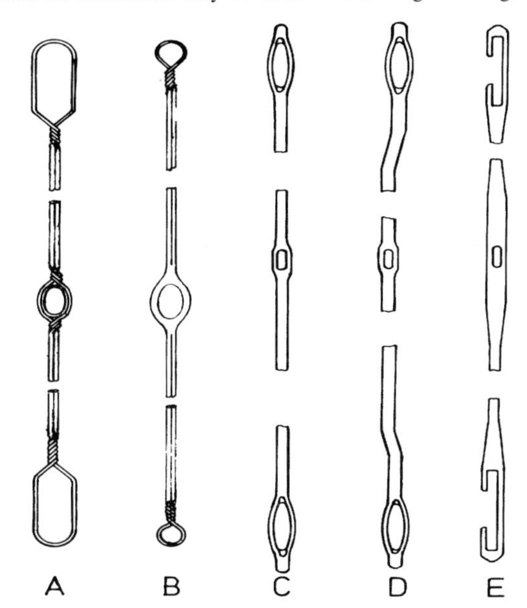

Examples of healds
A Twin-wire heald with oval end loops
B Twin-wire heald with inset mail and round end loops
C Flat steel heald, straight form
D Flat steel heald, cranked form
E Heald for riderless heald frame

heald; heddle *(continued)*

cord mail heald
A heald consisting of textile cords in which the central eye is a heald mail (see **heald mail** 1).

doup heald
See **leno weaving**.

flat steel heald
A heald made from flat steel strips in which the central and guide eyes are punched or stamped out. There are two main types, single and double (duplex).

knitted cord heald; noosed heald
A heald consisting entirely of knitted textile cords, in which the central eye (or noose) is formed by looping.

wire heald
A heald formed from shaped wire. There are two main types: (i) twin-wire healds, in which the central eye is formed by the separation (for the length of the eye) of the twin wires forming the heald; (ii) inset-wire healds, in which a wire **heald mail** (known as an inset eye) is soldered in position between the separated twin wires to form the central eye. *Note:* Both types of wire heald have wire loops at each end so that they may be attached to the **heald frame**.

heald frame
A rectangular frame, which is used to hold wire healds or healds (see **heald**) made of flat metal strips in position. Loops or holes at each end of the healds enable them to be placed upon the bars across the frame and to slide on these bars.

heald mail
1. An oval metal stamping (usually of steel or brass) containing a central eye and two holes for fixing cords.
2. A wire heald mail (known as an inset eye). This is an oval metal stamping containing a central eye. The mail is fixed (i.e., inset) by solder in a formed loop in the centre of the wire heald.

heald shaft; gear
1. A **heald frame** complete with healds.
2. An upper and lower wooden stave, to which are attached by a knitting process cord healds with noose or mail.

heat setting
See **setting**.

heddle
See **heald**.

held loop (knitting)
A loop which, having been pulled through the loop of the previous course, is retained by the needle during the knitting of one or more additional courses.

helical selvedge
See under **selvedge, woven**.

hemmer
A type of **folder** in which the raw edge is turned under or over for sewing.

hemp, true
A fine, light-coloured, lustrous, and strong bast fibre, obtained from the hemp plant, *Cannabis sativa.*
Note: The colour and cleanliness vary considerably according to the method of preparation of the fibre, the lower grades being dark cream and containing much non-fibrous matter. The fibre is obtained by **retting**. Its principal use is in twine and cordage, but some of the finer grades are used in weaving. The fibre ranges in length from 1 to 2.5m. The term hemp is often incorrectly used in a generic sense for fibres from different plants, e.g., manila hemp (abaca) from *Musa textilis*; sisal hemp (sisal) from *Agave sisalana*; sunn hemp (sunn fibre) from *Crotalaria juncea.*

henequen
The fibre obtained from the leaf of *Agava fourcroydes.*
Note: This closely resembles **sisal**.

henrietta
A fine, soft, lustrous fabric for dresses made from a fine silk warp and fine botany weft in a 1x2 twill.

herringbone
1. A combination of twill weaves in which the direction of the twill is reversed (usually by drafting) to produce stripes resembling herring bones.
2. A cloth in which this weave is used.

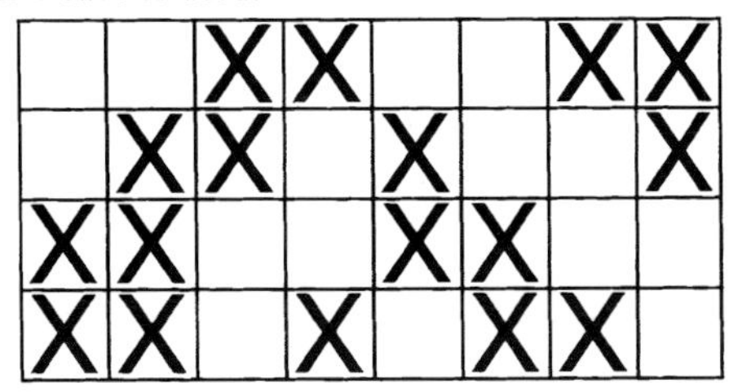

Herringbone weave

hessen
See **barras**.

hessian; burlap
A plain cloth made from single yarns of approximately the same linear density in warp and weft, usually made from bast fibres, particularly jute.

high volume instrumentation; HVI
An assembly of integrated semi-automatic electronic instruments for rapid determination of the fineness, length, impurity content and strength of samples of raw cotton.

high-bulk yarn
See **bulked yarn**, *Note* 1.

high-charged system
See **charged system**.

high-speed spinning (melt spinning)
A **melt spinning** process in which filaments are drawn from the spinneret and collected at high speeds. (See also **fully oriented yarn** and **partially oriented yarn**.)
Note: Speeds above about 3000 m. min^{-1} are currently classed as high.

high-temperature dyeing; HT dyeing
Dyeing at high pressure (above atmospheric) in order to dye at a temperature above the normal boiling point.

highly oriented yarn
See **fully drawn yarn**.

hockle (cordage)
See **cockle**.

Hoffman press
See **pressing (finishing)**.

hog wool; hoggett wool; teg wool
The first clip from a sheep not shorn as a lamb.

hole count
See *Note* under **lace quality**.

holland
1. Originally, a fine, plain-woven linen fabric, made in many European countries, but especially in Holland.
2. A plain, medium-weight cloth of cotton or linen with a beetled or glazed finish, used chiefly for window blinds, interlinings, and furniture covering.

hollow fibre; hollow filament
A tube-like manufactured fibre or filament.

hollow spindle spinning; wrap spinning
A system of yarn formation in which the feedstock (**sliver** or **roving**), is drafted and the drafted twistless strand is wrapped with a yarn as it passes through a rotating hollow spindle. The binder or wrapping yarn is mounted on the hollow spindle and is unwound and wrapped around the core by rotation of the spindle. The technique may used for producing a range of **wrap-spun yarns** or **fancy yarns**, by using different yarn and fibre feedstocks fed to the hollow spindle at different speeds. (See also **spinning**.)

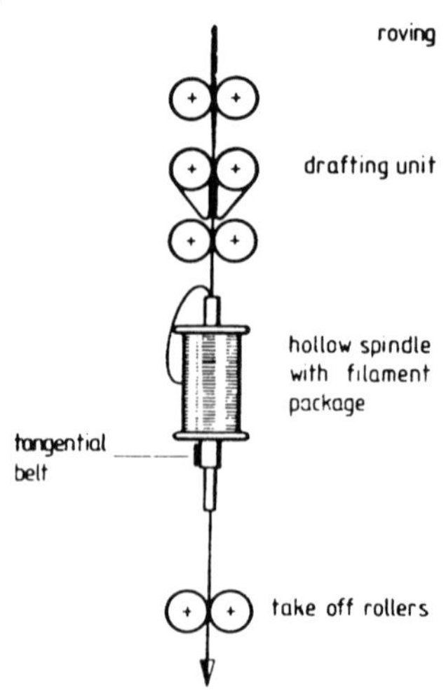

Hollow spindle spinning

homespun
Descriptive of coarse handwoven 2x2 twill-weave fabrics of tweed character. The yarns are handspun from domestic wools.

homopolymer
A polymer in which the repeating units are all the same. (See also **copolymer**.)

honan
A Chinese word meaning a silk fabric handwoven from water-reeled **net** Tussah silk, usually of medium weight.

honeycomb
A fabric in which the warp and weft threads form ridges and hollows, which give a cellular appearance (see **cellular fabric**). Three types of weave produce this effect:
(i) ordinary honeycomb, which gives a marked cellular effect on the face and back of the cloth;
(ii) Brighton honeycomb, which develops the effect more prominently on the face but in a less regular manner and with large and small cells; and
(iii) **Grecian**.

Ordinary honeycomb (actual size)

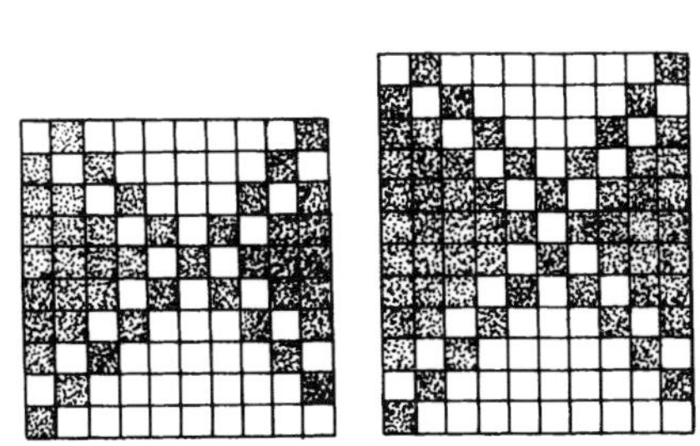

Ordinary honeycomb weaves

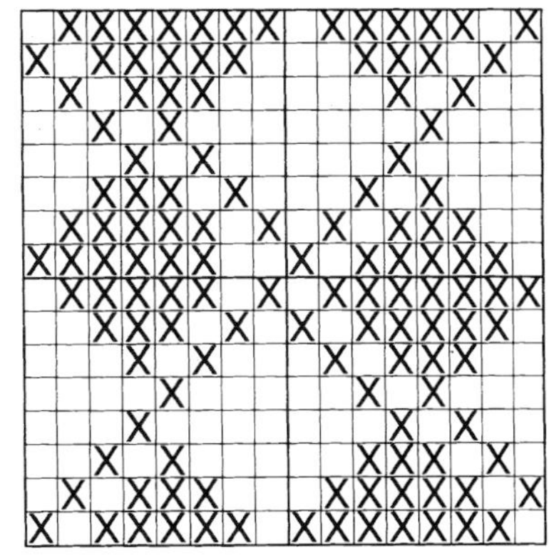

Brighton honeycomb weave

honeydew
The result of infestation of growing cotton by aphids or whitefly. It takes the form of more or less randomly distributed droplets of highly concentrated sugars, causing cotton stickiness.

hood; body
Acid-milled cone-shaped felt used in hat manufacture.

hook and loop fastener
See under **touch and close fastener**.

hopsack; mat; matt; basket
A modification of plain weave in which two or more ends and picks weave as one, or a fabric made in such a weave. The basic hopsack weaves may be modified in various ways, e.g., by introducing additional interlacing to give firmer cloth (stitched hopsack), or by arranging small square blocks of figures to form diagonal lines in the fabric (twilled hopsack).

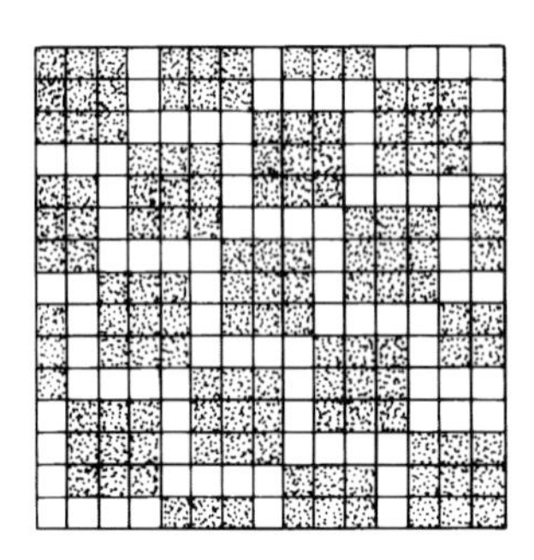

Twilled hopsack weave

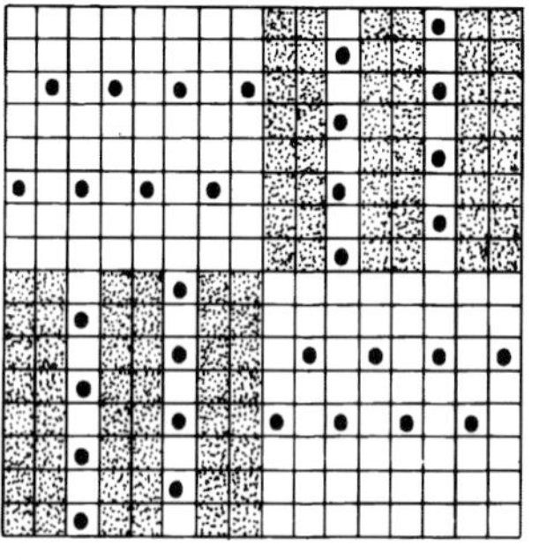

Stitched hopsack weave

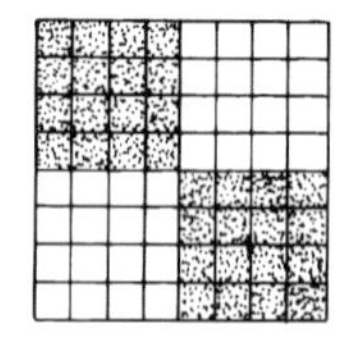

Hopsack weave

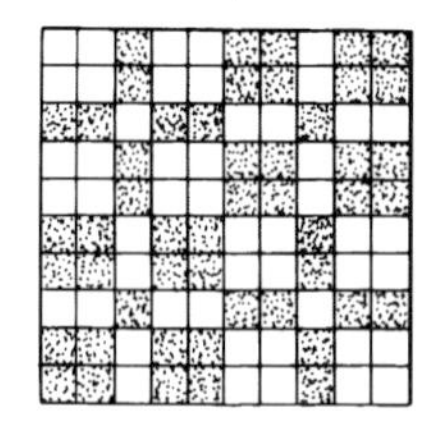

Fancy hopsack weave

horizontal-mill section warping
See **section warping**.

horse fibre (hair)
Fibre from the horse (*Equus caballus*).

hose (narrow fabrics)
A tubular woven fabric for conveying fluid under pressure.
Note: Hose is manufactured in both unlined and lined forms. When unlined, the weave is plain and the material is generally flax or hemp with a weaving density so arranged that when the fibres swell on wetting, the fabric becomes tight enough to reduce percolation under pressure to a negligible amount. For lined hose, fibres other than flax or hemp may be used in a plain or twill weave. Light-weight hose woven from synthetic yarns may incorporate an independent tubular palstic lining which is introduced during the weaving process or afterwards.

hosiery
1. Knitted coverings for the feet and legs.
2. Formerly in the United Kingdom the term was used in the generic sense of all types of knitted fabrics and goods made up therefrom.

hosiery knitting machine
A **knitting machine** for the production of **hosiery**. Most are small-diameter latch-needle **circular knitting machines**.

hot drawing (synthetic filaments and films)
See **drawing, hot (synthetic filaments and films)**.
hot flue
A machine in which hot air is used to dry fabric carried on rollers along a serpentine path.

hot mercerization
See **mercerization**.

hot-fluid jet textured yarn
See **textured yarn**, *Note 1* (iii).

Hottenroth number
A measure of the degree of **ripening** of **viscose**. (See also **salt figure**.)
Note: A Hottenroth number is the volume in millilitres of 10% ammonium chloride solution needed to induce incipient coagulation in a somewhat diluted viscose (solution) under standard conditions.

houndstooth check; dogstooth check
A small colour-and-weave effect in a fabric produced by a combination of 2/2 twill weave and a 4 and 4 order of colouring in warp and weft. (See also **shepherd's-check effect**.)

Houndstooth check

HOY
See **fully drawn yarn**.

HT dyeing
See **high-temperature dyeing**.

huckaback
1. (Weave) A weave used principally for towels and glass-cloths in which a rough-surface effect is produced on a plain-ground texture by short floats, warp on one side and weft on the other.
2. (Fabric) A fabric made in huckaback weave.

huckabuck *(continued)*

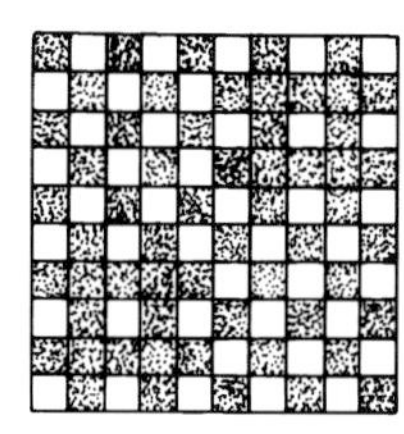

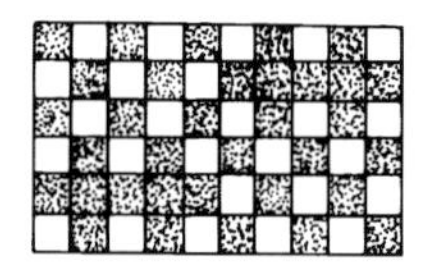

Huckaback weaves

Hudson's Bay Point blanket; Hudson's Bay blanket
A well-milled and raised heavy-weight blanket, made from coarse long-staple wool, first supplied to the Hudson's Bay Company in 1780. Made with a solid ground and coloured end border, or striped all over, or solid colour with darker end border. Near one corner a number of short stripes or 'points' are woven into the edge of the blanket at right-angles to the selvedge. A 'point' stripe is about 11cm long, and a 'half-point' about 6cm long. Originally each point indicated a barter-value of one Made-beaver, a prime quality skin from an adult beaver. Today the 'point' system is used as an indication of blanket size.

hue
That attribute of colour whereby it is recognised as being predominantly red, green, blue, yellow, violet, brown, etc.

humidity
See **absolute humidity** and **relative humidity**.

hungback
A light weight **warp-backed** (or **weft-backed**) overcoating fabric, usually having a 2/2 twill ground weave, made from woollen yarns, with extra 'hung' threads (cotton, silk, worsted or manufactured fibres) forming checks or stripes on the back of the fabric. The 'hung' threads are finer than the ground threads and in the case of 2x2 twill ground weave will have a 3x1 (or 1x2) twill weave.

hungry cloth
See **grinny cloth**.

husks (silk)
See **bisu**.

HVI
See **high volume instrumentation**.

hydro extraction
Removal of liquid from textile material.

hydroentangled fabric; spunlaced fabric
A **nonwoven fabric** made from a fibre **web** or **batt**, in which entanglement by high-pressure water jets provides the bond.

hydrostatic head test
See **water penetration resistance**.

hygral expansion
The reversible changes in length and width of fabrics containing hygroscopic fibres as a result

of changes in regain.

ikat
A process originating from Indonesia and Malaysia in which a warp is dyed to a pattern through the use of a resist agent. The warp is made first in rope form and predetermined sections are bound tightly to protect that section from the dye. When the warp is opened out after dyeing a pattern is shown on the warp which comes through into the woven fabric (warp ikat). A variation can be achieved by dyeing the weft yarn in a similar way (weft ikat) and in some cases both warp and weft may be treated (double ikat). Careful control and planning can give interesting and unusual figured effects in the cloth. The pattern shapes tend to have soft edges due to some movement of the threads in weaving, and a certain amount of bleeding and capillary action in the dye process.

ikat fabric
Fabric made from ikat dyed materials.

imitation gauze
See **perforated weave**.

immature cotton
See **maturity (cotton)**.

immature fibre (cotton)
See **maturity (cotton)**.

impact-textured yarn
See **textured yarn**, *Note 1* (iv).

imperial sateen
A heavily wefted fabric based on an eight-end sateen weave with one or more extra risers added. The weft face may be smooth or raised (see **beaverteen** and **fustian**).

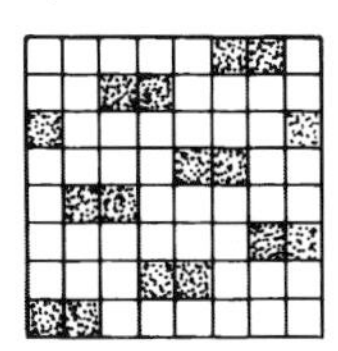

Imperial sateen weave

impervious backing (carpet)
See under **substrate (carpet)**.

independent beams (lace)
A leavers-lace construction made with beam and bobbin threads only, in which each beam thread in a **set-out** is supplied from a separate beam, and its movement is controlled by an individual steel bar (see **guide bars**). This permits independent control of the movements of all beam threads in the repeat.

India tape
A cotton **tape** typically R42/2tex x 17tex: 27.5 ends x 12 double picks per cm.

indirect warping
The transference of yarn from a package creel on to a swift from which it is subsequently wound on to a beam (see **section warping**).

industrial textiles
1. Textile materials and products intended for end-uses other than non-protective clothing, household, furnishing and floorcovering, where the fabric or fibrous component is selected principally (but not exclusively) for its performance and properties as opposed to its aesthetic or decorative characteristics.
2. A category of **technical textiles** used either as part of an industrial process, or incorporated into final products.

ingrain (filament yarn)
Descriptive of a yarn composed of filaments of different colours, the ingrain effect being produced by the random exposure of the differently coloured filaments at the yarn surface.

ingrain (yarn)
Yarn spun from a mixture of fibres of different colours, where the mixing of coloured fibres is carried out at an early stage. (See also **worsted yarns, colour terms**.)

ingrain carpet
A reversible coarse carpeting woven on a jacquard loom accommodating up to six coloured weft threads. It may be two-ply (Kidderminster carpet) or three-ply (Scotch carpet).

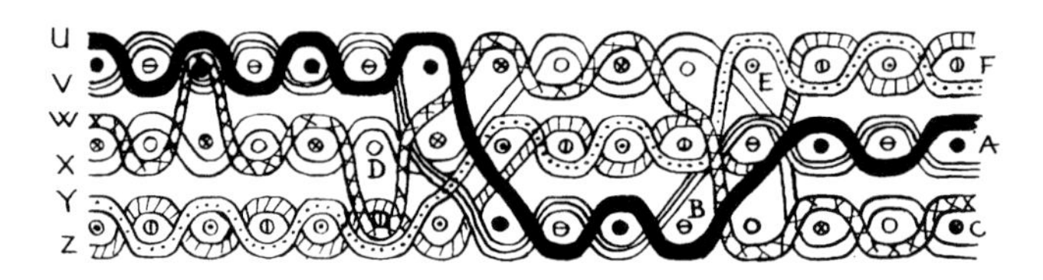

Ingrain carpet

initial modulus
The ratio of stress to corresponding strain below the proportional (Hookean) limit.

initial recovery
The decrease in strain in a specimen resulting from the removal of force, before creep recovery takes place.
Note 1: This is sometimes referred to as instantaneous recovery.
Note 2: Recovery is usually determined at constant temperature. Thermal expansion is excluded.
Note 3: For tests on plastics, the initial recovery is generally included as part of creep recovery.

inkle
An old term for **narrow fabric**.

inkle, beggar's
A **narrow fabric** constructed as **linsey-wolsey** in the 18th and early 19th centuries.

inkle loom
A simple form of narrow hand loom. It consists of a frame with pegs which hold a narrow continuous warp (a characteristic feature of an inkle loom), a simple device for making two sheds, and an adjustable peg to control warp tension.

insertion (lace)
Lace, having two straight edges, used for inserting between the edges of two pieces of material.

insertion braid
A **piping** made by a **braiding machine** which produces a flat and a tubular braid in combination.

insertions
See **banded laces.**

inspection
Activity such as measuring, examining, testing, gauging one or more characteristics of a product or service and comparing these with specified requirements to determine conformity.

insulating tape
A **narrow fabric** used in the manufacture of electrical equipment.

insulation tape
A woven fabric, impregnated with a non-conducting adhesive, which is then cut into strips.

intarsia
Weft-knitted plain, rib, or purl fabrics containing designs in two or more colours (or textures) within the same course in which each area of colour is knitted from a separate yarn, which is contained entirely within that area.

integral knitting (weft knitting)
The combined knitting of more than one component during the knitting sequence of a garment, for example, a body plus trimming and a pocket.

interlaced yarn
See **intermingled yarn**, *Note 1.*

interlining
Any one of a wide variety of fabrics used between the inner and outer layers of a garment to improve shape retention, strength, warmth or bulk. An interlining may be of woven, knitted, or nonwoven material and may be produced with or without a fusible adhesive coating.

interlock, weft-knitted
A double-faced rib-based structure consisting of two 1x1 rib fabrics joined by interlocking **sinker** loops. It is made on machines equipped with two sets of opposed needles (see **gating**) and capable of knitting in the following sequence:

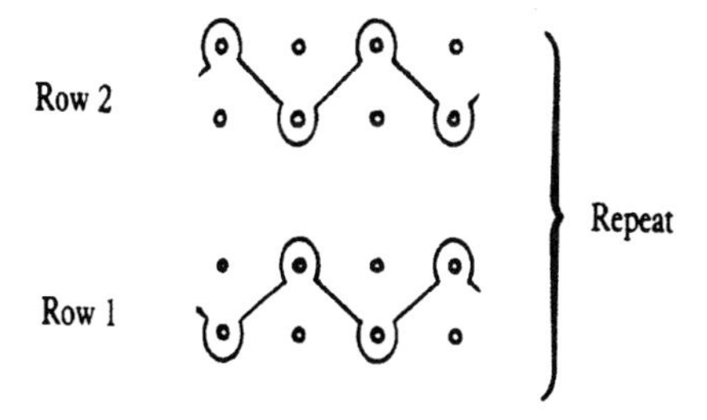

Interlock

(See also **double jersey, weft-knitted.**)
Note: Interlock was originally knitted from cotton and used for underwear, but today it is knitted from various materials for a variety of purposes, including outerwear.

interlock gating; interlock gaiting (knitting)
See **gating (knitting)**.

intermingled yarn
A multifilament yarn in which cohesion is imparted to the filament bundle by entwining the filaments instead of or in addition to twisting. The effect is usually achieved by passing the yarn under light tension through the turbulent zone of an air-jet.
Note 1: Some manufacturers describe such a product as an 'interlaced yarn'.
Note 2: Intermingling should be distinguished from air-texturing (see **textured yarn**, *Note 1* (viii)) in which a much higher level of entanglement is achieved with the objective of producing texture or bulk.

intermingling jet
An air-operated device used as an ancillary to some processes of yarn extrusion, of drawing and of texturing to induce intermingling of filaments and hence yarn coherence. (See also **intermingled yarn** and **co-mingled yarn**.)

inturned welt (knitting)
See under **welt (knitting)**.

inverted pleat
See under **pleats**.

invoice mass
The mass, however determined, of a consignment of a textile material, which is declared on the invoice. It is usually numerically equivalent to the commercial mass.

Irish lawn
A **lawn** fabric produced from fine linen yarns.

Irish linen
A woven fabric produced in Ireland from 100% flax yarn.

Irish linen yarn
100% flax yarn spun in Ireland.

Irish poplin
A **poplin** made from a dyed pure silk **organzine** warp and a three-fold **genapped** fine worsted weft. It is a silk-face and silk-back fabric, the weft being completely covered by the warp. The fabric is usually handwoven and is frequently given a **moiré** finish.

islands-in-the-sea bicomponent fibre
See under **bicomponent fibre**.

isotactic polymer
See **polymer, isotactic** under **polymer**.

Issitt's shaker
See **shoddy shaker**.

istle; ixtle (fibre)
Generic Mexican term for various species of **agave**.

Italian
A cotton fabric of five-end sateen weave with a lustrous finish, used chiefly as a lining material. Examples were 30x49; 16x17tex; K=12.0+20.2 or 33x47; R15/2x16tex; K=12.8+18.8.

J-box
An upright J-shaped vessel for the continuous steaming or wet processing of textiles.
Note: The material enters the top of the long limb of the J, is stored for a time, and is withdrawn through the short limb.

J-scray
A J-shaped trough or vessel for the processing of textiles from one process to another. The long limb of the J forms an inclined plane on which the material accumulates during the dwell period.

jack (knitting)
A term used to describe an intermediate selecting element on a knitting machine.

jack (lace machines)
A spring-steel wire part comprising a straight stem containing a loop near the top, to which a string from the jacquard of a lace furnishing machine is attached, with a point at the top at right angles to the stem. The point can enter between the warp and pattern threads so as to restrict their lateral movement.

jack bar (lace machines)
A bar on a lace furnishing machine to which **jack leads** are attached over the lace-making width of the machine. It imparts a motion to the jacks so that their points enter between the warp and pattern threads, above the **guide bars**, but below the combs, unless restrained by the jacquard. It is shogged as part of the **foundation bar**.

jack lead (lace machines)
A number of lace furnishing machine **jacks** cast to the gauge of the machine in a lead-alloy base.

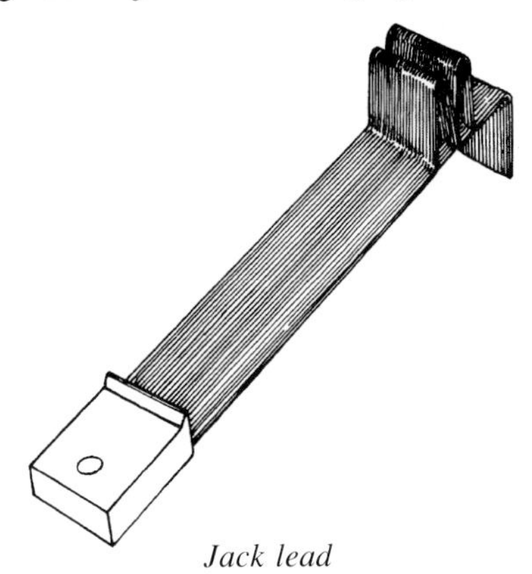

Jack lead

jaconet
Light-weight, plain-woven cloth of a lawn or muslin type with a smooth and slightly stiff finish.

jacquard (mechanism)
A patterning device and mechanism used to select individual warp threads in weaving or warp knitting, individual threads in lace making, and knitting elements in weft knitting.
Note 1: Jacquard control may be by pegged card, punched card, punched tape etc, or by electronic means.
Note 2: Named after the inventor, Joseph Marie Jacquard, 1752-1834.

jacquard card
A punched card used to control a jacquard mechanism. A series of such cards strung together control the production of the required pattern.

jacquard (mechanism) *(continued)*

Note: In many applications jacquards are now controlled by endless paper or an electronic store instead of cards.

jacquard control mechanisms (lace machines)
Mechanical or electro-mechanical devices that control the movement of a large number of patterning elements by means of punched cards or a punched continuous strip.

jacquard harness (weaving)
The series of cords and their attachments, from the hooks of a jacquard machine downwards, that control the lifting of the warp threads.
Note: The main parts include neck cords (attached to hooks by connectors and 'V'-links in the case of double-lift jacquard machines), couplings for the main cords that pass through the comber board, wire healds and **lingoes**. For high speed weaving, lingoes are being replaced by elastics or springs.

jacquard mechanism (weaving)
A shedding mechanism, attached to a loom, that gives individual control of up to several hundred warp threads and thus enables large figured designs to be produced.

jacquard selection mechanism (knitting)
Any system used to select knitting elements to produce jacquard designs.
Note 1: Originally the term referred only to jacquard punched card and harness mechanisms.
Note 2: In warp knitting the term generally refers to a device for selecting individual guides within a guide bar.

jacquard tie (lace machines)
The arrangement of the strings used to connect the jacks (see **jack (lace machines)**) to the jacquard needles on furnishing and string warp machines. Two main systems are employed to obtain the most economic use of the available jacquard capacity, each having its own particular limitations:
(i) independent or divisional tie: each upright needle controls the corresponding jack in each division across the machine width. The carriage-way pattern repeat is restricted to these divisions or sub-multiples of them;
(ii) universal tie: the majority of the upright needles control two jacks each. Each jack from one end of the machine is thus paired with the corresponding jack at the other end of the machine. A smaller number of jacks (usually one-fifth or one-seventh of the total) in the centre of the machine are controlled by individual upright needles. In making an odd number of divisions, the maximum carriage-way repeat that requires separate jack selection over its entire width is determined by the number of needles controlling individual jacks. This restriction does not apply to an even number of divisions.

jacquard tie (weaving)
The order in which the harness cords are attached to the neck cords and their arrangement in the comber board (see **jacquard harness (weaving)**). The tie is known as:
(i) Norwich or London according to the position of the jacquard and harness in relation to the loom. In the Norwich tie, the harness hangs straight and the card cylinder is at right angles to the warp. In the London tie, the jacquard is placed so that the cylinder is parallel to the warp, the harness having a 90° twist. In either case, various arrangements are possible, for example: (a) single tie, in which each jacquard needle controls only one harness cord and only one warp thread, there being only one repeat of the pattern in the cloth width; (b) repeating tie, in which each jacquard needle controls several harness cords and the pattern is repeated several times across the cloth width;
(ii) centre (point) tie, in which each needle controls one warp thread in one half of the pattern repeat and one warp thread in the other, the result being that one half of the pattern

is a mirror image of the other;
(iii) border tie, in which some of the jacquard needles and hooks are used to produce a design close to and parallel with the cloth selvedges, the other needles and hooks being used to produce a different design in the rest of the cloth.
Note: Mixed ties (combinations of two or more of the above) are also possible.

jacquard design (knitting)
A patterned design produced by the individual selection of knitted stitches.
Note: The term may be further qualified, for example, as 'small area rib jacquard'.

jacquard fabric
A fabric in which a large number of warp threads, in excess of the capacity of a dobby, weave differently and therefore require a **jacquard mechanism**.

jacquard repp
A **repp** structure figured by allowing threads to float as a result of jacquard selection.

jaffer
A plain-weave cotton fabric with warp and weft in different colours producing a **shot effect**.

jappe
A fine plain-weave fabric woven from continuous-filament yarns (originally silk) and of approximately **square** construction.

jaspé; jasper
A fabric that has a shaded appearance created by a warp thread colour pattern. It may be woven plain or figured, and is for bedspreads or curtains.

jaspé carpet
Carpet having a flame-like regular pattern. It was traditionally produced by using pile yarn dip-dyed in hanks to two tones of the same colour or to two different colours, in a fixed proportion of the lengths of the hanks in any one lot.

jaspé yarn (filament)
A yarn that has the appearance of, but is not, two differently coloured yarns folded together. It is made by texturing together two continuous-filament yarns of different chemical composition, e.g., nylon and polyester, and then dyeing only one of the two components.

jaw break
Failure of a specimen, during a tensile test, at the edge of the jaw or within 5mm of its edge.

jean
A 2/1 warp-faced twill fabric used chiefly for overalls or casual wear. Typical construction: 35x24; 32x21tex cotton; K=19.8+11.0.
Note: The term 'jeanette' is sometimes used to describe the lighter weights and these may be used for linings.

jerker bar (tufting)
See under **tufting machine**.

jersey fabric
A generic name applied to weft-knitted fabric. (See also **single jersey, weft-knitted** and **double jersey, weft-knitted**.)

jet craters
Annular deposits that sometimes form around the holes on the face of jets used in the extrusion of viscose. (See also **jet rings**.)

jet loom
See under **weaving machine**.

jet rings
Annular deposits formed occasionally inside the holes of metal jets or spinnerets when used in the extrusion of viscose, particularly into coagulants containing much zinc sulphate. (See also **jet craters**.)

jet spinning; air-jet spinning
A system of staple-fibre spinning which utilises an air vortex to apply the twisting couple to the yarn during its formation.
Note: The air is blown through one or more nozzles inclined to the axis of the cylindrical yarn passage (see diagram 1). This generates a vortex in the passage which applies a torque to the yarn as it passes through. Two such twisting devices may be used (see diagram 2). The majority of systems using this spinning technique produce a type of **fasciated yarn**. (See also **spinning**.)

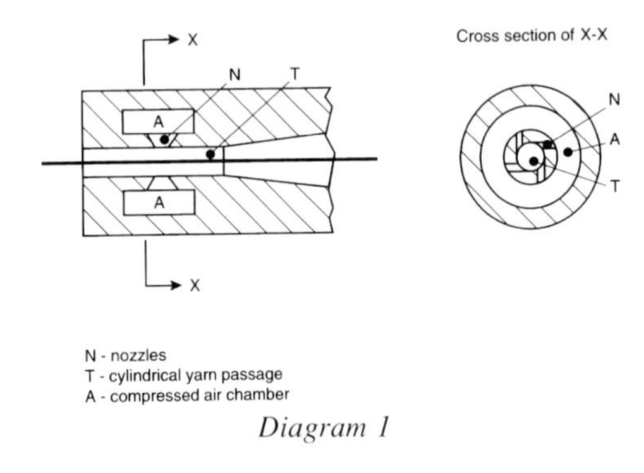

Diagram 1

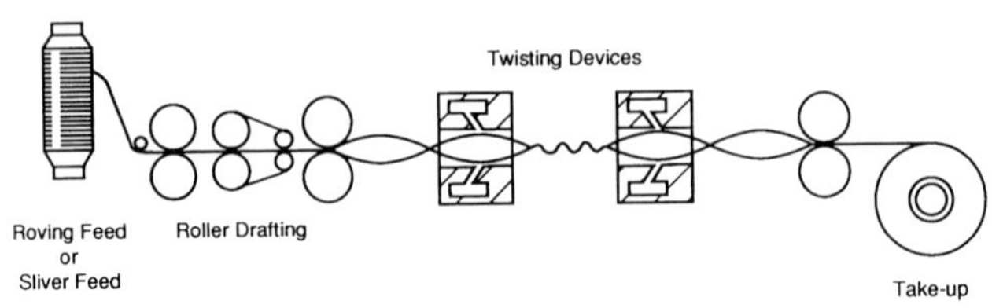

Diagram 2

jet weaving machine
See under **weaving machine**.

jet-dyeing machine
1. A machine for dyeing fabric in rope form in which the fabric is carried through a narrow throat by dye-liquor circulated at high velocity.
2. A machine for dyeing garments in which the garments are circulated by jets of liquid rather than by mechanical means.

jet-textured yarn
See **textured yarn**, *Note 1* (iii) and (viii).

jetted pocket

A **pocket** having a mouth on which the edges are finished by the application of **jettings**.

jetting; jet (clothing)
A narrow strip of matching or contrasting material sewn parallel to a pocket opening, forming a decorative neatened edge on both sides of the pocket mouth.

jig; jigger
A machine in which fabric in open width is transferred repeatedly back and forth from one roller to another and passes each time through a relatively small volume of a dyebath or other liquid. Jigs are frequently used for dyeing, scouring, bleaching and finishing.

jigging stenter; jigging tenter
A **stenter** in which a to-and-fro longitudinal motion can be given to the side frame carrying the clip chain while the fabric is moved forward. The two side frames are linked, one moving forward while the other moves back and *vice versa* to impart a swinging motion to the fabric.
Note: This device is used in finishing to reduce inter-yarn bonding to produce a softer fabric.

jockey satin
See **slipper satin**.

jute
The fibre obtained from the bast layer of the plants *Corchorus capsularis* and *Corchorus olitorius*.
Note: Commercially, jute is divided into two main classes, white jute generally being associated with *Corchorus capsularis*, and dark jute with *Corchorus olitorius*. These classes are further sub-divided into numerous grades denoting quality and other characteristics.

jute-spun
Descriptive of staple yarn that has been prepared and spun on machinery originally designed for spinning yarns from jute.

K_v, K_w values
Measures of the filterability of **viscose** expressed in terms of either volume, K_v, or weight (strictly mass), K_w.

kapok
A unicellular seed hair obtained from the seed pods of the kapok tree (*Ceiba pentandra*).
Note: The fibre is also called ceba, ceiba, Java cotton, silk cotton, silk floss etc. Indian kapok comes from *Bombax malabaricum*.

karakul
1. A type of sheep in Africa producing a coarse wool, used mainly in carpets.
2. A grade of lambskin (originally from Bokhara in Central Asia) with lustrous black hair which develops an attractive wavy pattern and is less curled than **astrakhan**. Also known as 'caracul' or 'broadtail'.

karakul cloth
See **broadtail cloth**.

KDK
See **knit-deknit yarn**.

keba
See **cocoon strippings**.
keel

The Scottish term for **cut mark**.

kemp
A coarse animal fibre with a wide lattice-type **medulla** that is shed from the skin at least once a year; it is often shorter than other fibres of the fleece, has a long tapering tip, and, when completely shed, tapers sharply towards the root end.

kenaf
The fibre obtained from the bast layer of the plant *Hibiscus cannabinus*.
Note 1: Kenaf is commonly known as 'mesta' in India.
Note 2: Being similar to jute in many of its properties, kenaf is used either as an alternative to, or in admixture with, jute.

kernmantel
A rope with a braided sheath and a core so constructed as to possess high extension and energy absorption under load, used in rock climbing and related activities to arrest the accidental fall of a climber with an appropriate deceleration.

kersey
A compact, lustrous, woollen fabric, diagonally ribbed or twilled that is heavily milled and finished with a short nap. It is similar to **melton**.

kettle
See **beck**.

kibisu
See **frisons**.

kick tape; trouser binding
A **narrow fabric**, often **Paris binding**, used on trouser leg bottoms.

Kidderminster carpet
See **ingrain carpet**.

kier boil; kiering
The process of prolonged boiling of cotton or flax materials with alkaline liquors in a large steel container known as a kier, either at or above atmospheric pressure (see **open boil** and **pressure boil**).

kilotex
See **tex system**.

kimono sleeve
See under **sleeve (clothing)**.

king bobbin (lace machines)
See **bobbin, king (lace machines)**.

kinky yarn
A twist lively or **snarly yarn**. (See also **twist liveliness**.)

kiss-roll applicator
See **lick-roll applicator**.
kneeing

An unstable condition arising in melt-spinning wherein the extrudate forms an inflexion on leaving the spinneret instead of drawing down directly from the orifice. The molten filament thus has a knee-like shape just below the orifice.

knickerbocker yarn; knicker yarn
See under **fancy yarn**.

knife (tufting)
See under **tufting machine**.

knife pleat
See under **pleats**.

knit braid
A **cord** produced on a small-diameter knitting machine with oscillating feeders.

knit-deknit yarn; KDK
See **textured yarn**, *Note 1* (vii).

knitted cord heald
See under **heald**.

knitted flat ruche
See under **ruche**.

knitted loop (weft knitting)
A basic unit of weft-knitted fabrics consisting of a loop of yarn meshed at its base with a previously formed loop.
Note: At the point of mesh with the previously formed loop, a knitted loop is usually open but may be crossed. Component parts of the knitted loop may be identified as:

Back loop

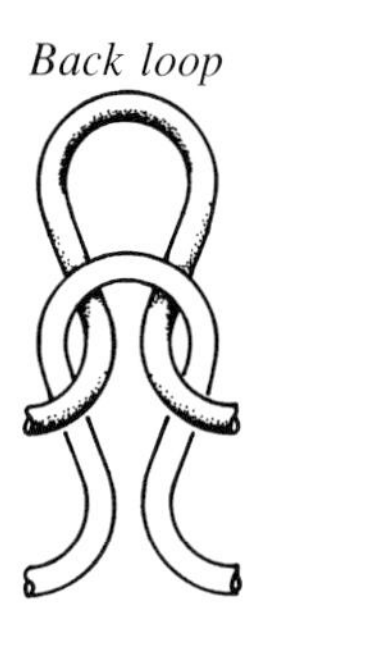

Face or front loop

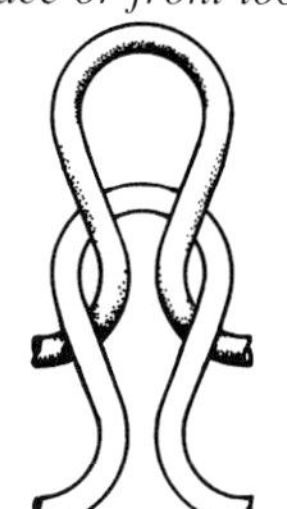

Needle loop

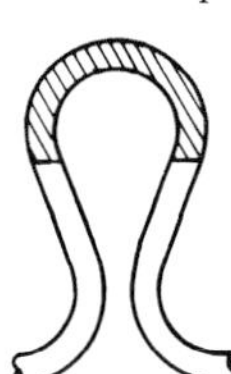

Sides or leg

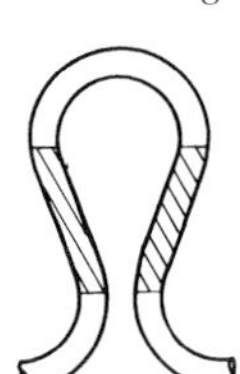

Sinker loop

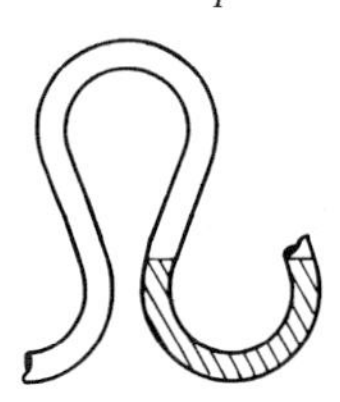

knitted loop (weft knitting) *(continued)*

back loop; reverse loop
A knitted loop meshed through the previous loop towards the back of the fabric (away from the viewer).

face loop; front loop; plain loop
A knitted loop meshed through the previous loop towards the front of the fabric (towards the viewer).

needle loop
The upper curved portion of a knitted loop.

sides; legs
The parts of the knitted loop that connect the sinker and needle loops.

sinker loop
The lower curved portion of a knitted loop.

Warp-knitted pile carpet (longitudinal section)

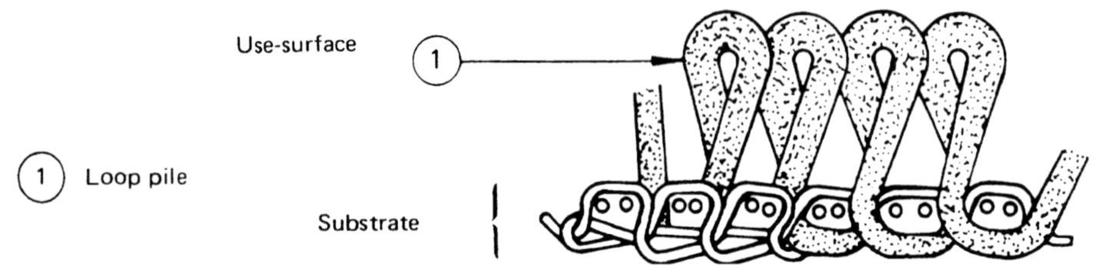

The above diagram is derived from figure 12 of BS 5557

knitted-pile carpet
A carpet made on either warp-knitting or weft-knitting machines.

knitting
The process of forming a fabric by the intermeshing of loops of yarn (see **warp knitting** and **weft knitting**).

knitting elements (knitting)
A generic term describing the loop-forming parts of a knitting machine. Also descriptive of those parts used to control and/or select the loop-forming instruments.

knitting machine
A machine for the production of fabrics, garments or yarns by **warp knitting** or **weft knitting**. The different types of warp and weft-knitting machines are classified and named, primarily, according to:
(i) the type of fabric or garment they are intended to produce;
(ii) the type of needle used;
(iii) the form, arrangement and activation of their needles or needle beds;
(iv) the type of patterning control used;
(v) whether they are hand-operated or power-operated.
Note: There are no convenient English terms to distinguish between machines in which the individual needles operate independently (German, *Strickmaschine*), and machines in which the needles are mounted so that they must be operated in unison (German, *Wirkmaschine*), although this distinction forms the basis of classifications of knitting machines. (See also **circular knitting machine, crochet-knitting machine, cylinder and dial knitting machine, double cylinder knitting machine, flat knitting machine, garment-length knitting machine, hosiery knitting machine, loop wheel knitting machine, milanese warp-knitting machine,**

multifeeder machine (circular weft-knitting machine), purl knitting machine, raschel warp-knitting machine, simplex warp-knitting machine, sinker top machine (knitting), straight-bar machine (weft knitting), tricot warp-knitting machine.)

knitwear
A generic term applied to most weft-knitted outerwear garments such as pullovers, jumpers, cardigans and sweaters.

knock-off lap (warp knitting)
A length (or lengths) of yarn received by a needle and not pulled through the loop.

knock-over (knitting)
The action of casting off the previously formed loop over the head of the needle to mesh with the newly formed loop.

knop yarn
See under **fancy yarn**.

knop-stitch (weft knitting)
A stitch, giving a raised effect, that consists of a **held loop** and more than two **tuck loops** all of which are intermeshed in the same course.

knot breaking force; knot breaking strength
In tensile testing, the breaking force of a strand having a specified knot configuration tied in the portion of the strand mounted between the clamps of the tensile testing machine.

knotting
1. The process of tying two yarn ends together.
2. The tying of the ends of a new weaver's beam to their corresponding ends on the old beam in the loom by hand or machine.
Note: Also known as 'knotting-in', 'tying-in' and 'tying back'.

knubbs
See **frisons**.

kuriwata
1. See **silk, wild**.
2. Japanese for ginned cotton.

lace
A fine openwork fabric with a ground of mesh or net on which patterns may be worked at the same time as the ground is formed or applied later, and which is made of yarn by looping, twisting, or knitting, either by hand with a needle or bobbin, or by machinery; also a similar fabric made by crocheting, tatting, darning, embroidering, weaving, or knitting.

lace; lacet (narrow fabric)
A braided or woven **narrow fabric**, flat or tubular, often cut into lengths and tagged for use as shoelaces and corset-laces.
Note: The term **lace** is used to describe narrow woven fabrics such as **carriage lace**, hood lace and uniform lace.

lace furnishing machine
See under **lace machines**.
lace machines

bar warp machine
A **warp lace machine** in which the pattern control is similar to that of a **Leavers machine**.

Barmen machine
A **braiding** machine in which threads on **king bobbins** placed on carriers are plaited with each other, and sometimes with warp threads. A jacquard controls the paths of the carriers in accordance with the requirements of the pattern.

double locker machine
A **plain net machine** in which the motion of the carriages is imparted by driving and locker bars.

Double locker plain net machine

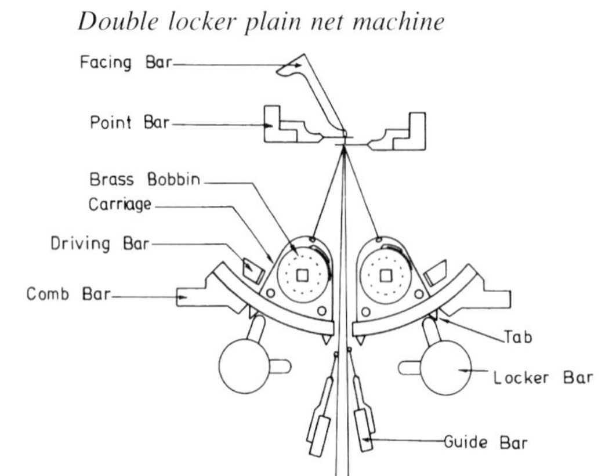

go-through machine
A **Leavers** type of machine, in which the catch bars impart motion to the carriages.

lace furnishing machine; curtain machine
A machine in which threads in brass bobbins borne in carriages, each in an allotted comb space, swing in pendulum fashion between vertical warp and pattern threads in planes at right angles to a warp sheet. The lateral movements of the warp and pattern threads are imparted by guide bars. By the interaction of a jack bar and a jacquard, spring-steel jacks modify the lateral movements of selected warp and pattern threads in accordance with the requirements of the pattern.

Leavers machine
A machine in which threads in brass bobbins borne in carriages, each in an allotted comb space, swing in pendulum fashion between vertical warp and pattern threads in planes at right angles to a warp sheet. The lateral movements of the warp and pattern threads are imparted by steel bars (see **guide bars (lace machines)**, *Note 3*) actuated by a jacquard.

mechlin machine
A **Leavers** type of machine generally without a jacquard that employs a limited number of guide bars, whose movements are controlled by cams. It is used for making a special type of net called 'mechlin'.

plain net machine; bobbinet machine
A machine in which threads in brass bobbins borne in carriages, in pairs in tandem in each comb space, swing in pendulum fashion between vertical warp threads in planes at right angles to the warp sheet and progressively traverse across the whole width of the machine and return.

rolling locker machine; roller locker machine
A **plain net machine** in which the motion of the carriages is imparted by fluted rollers.

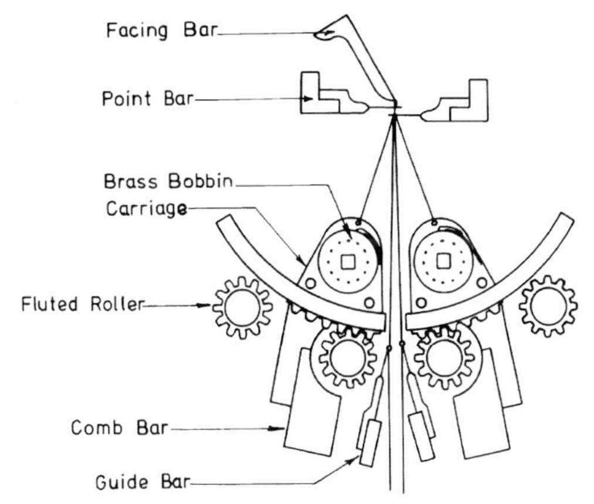

Rolling locker plain net machine

sival machine
A **Leavers** type of machine, differing from the standard version in that the frame and the catch-bar and point-bar linkages are similar to those of the **lace furnishing machine**.
Note: Its patterning principle is the same as that of the Leavers machine, and the lace produced is of the Leavers type.

string warp machine
A **warp lace machine** in which the pattern control is similar to that of the **lace furnishing machine**. The jacks work between the guide bars and the needles.

warp lace machine
A machine on which the ground threads are looped and are taken from warp beams. The pattern threads are laid in and secured by the ground threads.

lace quality
A quantitative measure of the rate of take-up (and thus the compactness warp-way) of lace on the machine. Traditional measures are: (i) Leavers and warp - the number of inches of lace per **rack**; (ii) furnishings - the number of full motions in 3 inches of lace; (iii) bobbinet - the number of meshes per inch vertically.
Note: The finished quality of lace and net differs from the quality in the machine state owing to dimensional changes introduced in **dressing**. The finished quality of plain net is traditionally expressed as the sum of the hole count per inch warp-way and the hole count per inch bobbin-way, as shown in the diagram.

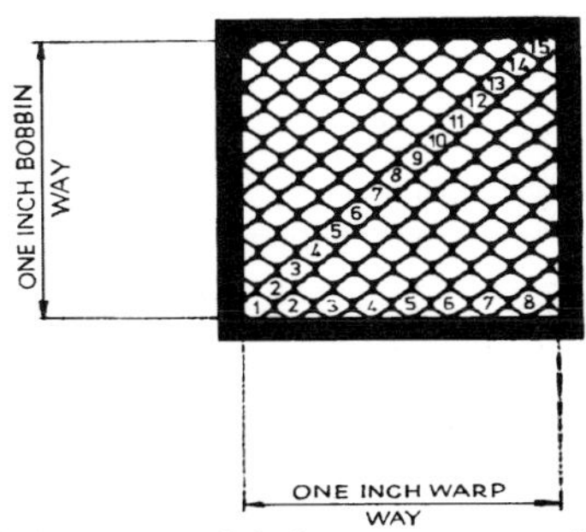

An example of a 23-hole net, i.e., 8 holes warp way and 15 holes bobbin-way

lace stitch (weft knitting)
An openwork effect (in a plain-knitted fabric) produced by transferring needle loops to an adjacent needle of the same needle bar.

Lace stitch

lacet
See **lace (narrow fabric)**.

lacing (silk)
See *Note 1* under **leasing**.

lacing cord
1. See **cable cord**.
2. See *Note 2* under **leasing**.

ladder (weft knitting)
The unmeshing of successive loops in a wale or wales, usually described as a defect.

ladder backing, weft-knitted
The reverse side of a rib jacquard fabric having fewer wales than on the face as a result of needles being inactive. Small floats are evident where the needles are absent.

ladder web
1. A four-ply woven **narrow fabric** consisting of two outer or body webs between which are woven two narrower webs in staggered relationship with each other, each being woven alternately into one end and then the outer or body web, so as to form, when opened up, supports for the slats of a Venetian blind (see diagram).
2. A crochet or warp-knitted **narrow fabric**, consisting of two **wales**, or columns of stitches spaced apart by the width of the slat of a venetian blind. Weft yarn crosses from one wale to the other to form a ladder-like structure which separates and positions the slats of the blind. Knitted web is usually made from polyester yarn and heat set after manufacture.

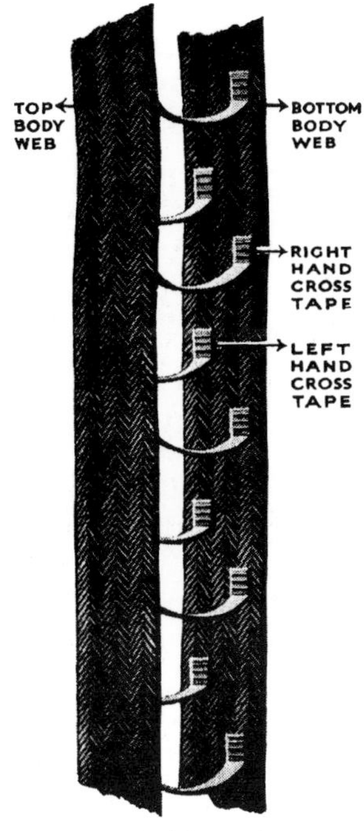

Ladder web

lag (weaving)
One unit of the pattern chain controlling the operating of a dobby, box-motion, or other mechanism.

lahore
A piece-dyed dress fabric made from cashmere in small dobby effects.

laid (cordage)
See **lay** 3.

laid rope
See under **rope**.

laid-in fabric, warp-knitted
A fabric containing one or more series of warp threads held into the ground construction by being trapped between the face loops and the underlaps of the ground construction. The laid-in yarn is connected to the ground construction by an underlap on each wale that it crosses.

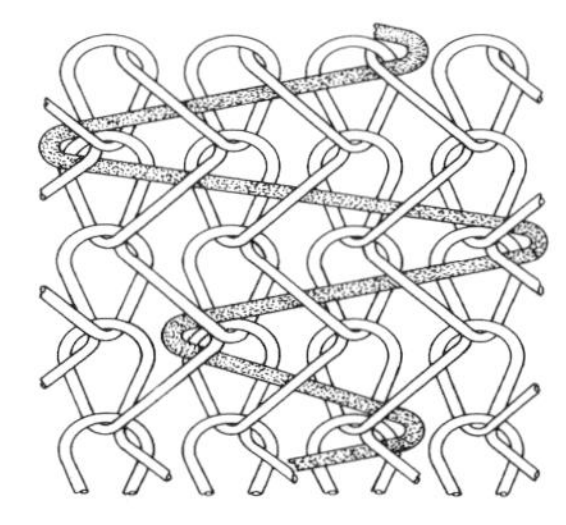

Warp-knitted laid-in fabric

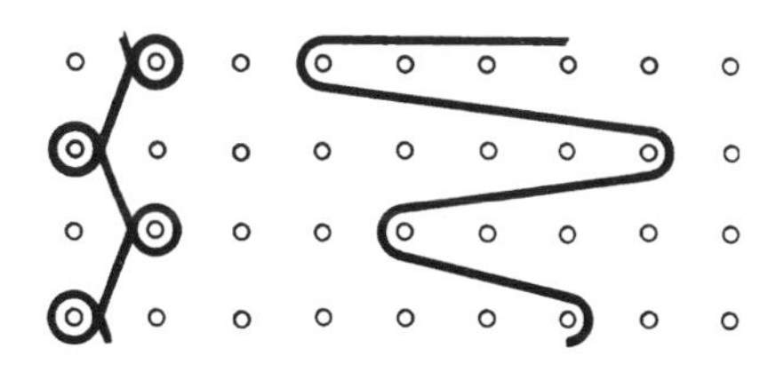

Front bar (ground) Back bar (laid-in)
Warp-knitted laid-in fabric structure

laid-in fabric, weft-knitted
A fabric containing non-knitted yarns, which are held in position by the knitted structure. These fabrics are frequently raised to produce a pile effect, e.g., fleecy fabrics.

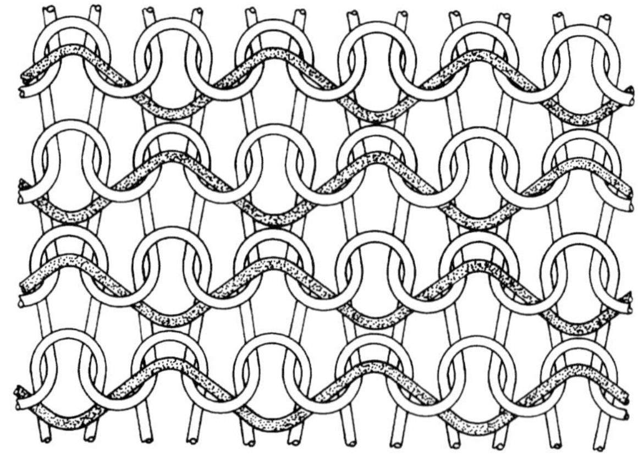

Weft-knitted laid-in fabric (plain)

laid-pile finish
See **drawn-pile finish**.

lamb's wool
Wool from the fleeces of lambs (young sheep up to the stage of weaning).
Note 1: This definition applies irrespective of the breed or type of sheep.
Note 2: It has been common practice in the trade to apply the term 'lambswool' to textile products, having a soft handle, made from 100% virgin wool of which at least one-third is lamb's wool as defined here.

lambskin cloth
A cotton fabric having a high weft **sett**, with a dense **nap** of fibre on the surface. The weave is of a weft-**sateen** character of the **fustian** type.

lamé
A general name for fabrics in which metallic threads are a conspicuous feature.
Note: Originally, lamé referred only to fabrics containing gold or silver threads.

laminated fabric
See **combined fabric**.

lampas
A multi-colour figured drapery and upholstery fabric similar to a brocade, made of silk, cotton, viscose, or combinations of yarns. Two warps, one forming the ground and one binding the wefts, in regular or irregular order, form the figure.

lampwick
A form of **wick**.

lancé
A term applied to fabrics in which the weft threads interlace with only a few warp threads at intervals across the fabric, and float over or under the rest to produce tiny dots or specks on the ground.

landing bar (lace machines)
A bar running the lace-making width of the Leavers lace machine, beneath the combs, to support the carriages. There are two landing bars, one each side of the **well**. They move in unison with the catch bars.
Note: There are no landing bars on a go-through machine, which has circle strips. This is one of the obvious features distinguishing an original Leavers machine from a go-through machine (see **lace machines**).

landings (narrow fabrics)
Those parts of a loom **batten** that guide and support the shuttles on either side of the warp.

lap
1. (General) A sheet of fibres or fabric wrapped round a core with specific applications in different sections of the industry, e.g., sheets of fibre wound on rollers or round endless aprons to facilitate transfer from one process to the next.
Note: In cotton spinning, the sheets of fibres from openers and scutchers, sliver-lap machines, and ribbon-lap machines are wrapped on cores.
2. (Flax) An arrangement of the fibre strands in scutched flax pieced out for hackling, or in pieces of hackled flax to facilitate their removal as seperate units from built-up bundles (see **scutching (flax)**).
3. (Fabric) The length of fabric between successive transverse folds when pieces are plaited down or folded.
4. (Fabric) An individual layer of fabric in roll form.
5. Fibres wrapped accidentally round any revolving machine part.
6. Silk waste after discharging and combing, but before processing into sliver or top. The staple length of the fibre decreases between the first, second and third drafts (combings).

lap (warp knitting)
Descriptive of the wrapping of the yarn around the needle. (See also **lapping movement**.)
Note: The term lap is sometimes used synonymously for **overlap**.

lap waste
A sheet of fibres accidentally wound round rollers or **aprons**, which may after removal be used as soft **waste**.

lapel
The upper part of the front edge of a coat, dress, blouse or jacket which folds back on to the main body of the garment, from the neck down to the **break**.

lapped (fabric)
See **cuttle**.

lappet
1. See **ballooning eye**.
2. A fabric in which figure is achieved by introducing extra warp threads into a base fabric that is normally plain. The figuring threads are controlled by needle bar(s) between the reed and the fell, with the amount of side-traverse given to the needle bar being controlled by a pattern wheel.

lapping movement (warp knitting)
The compound motion of the guide bars of warp-knitting machines that presents the threads to the needles so that loops can be formed. This compound motion consists in swinging motions of the guides at right angles to the needle bar, and lateral movements parallel to the needle bar. (See also **overlap (warp knitting)** and **underlap (warp knitting)**.)

lapping notation (warp knitting)
A method of portraying the successive lateral movements of a guide bar or bars. These are generally shown as figures representing the distance guide bars **shog** sideways parallel with the needle bar as well as figures representing the heights of successive pattern chain links required to bring the guide bars to the appropriate positions. For this purpose, the movements of the bars are normally drawn on **point paper** as a lapping diagram.

lash
See **warp sheet**.

lashed-in weft; lashing-in (defect)
A length of weft yarn that has been pulled inadvertently into the **shed** during weaving. This defect is most likely to occur: (i) on automatic-bobbin-change looms, as a result of a length of weft yarn extending from the selvedge to the weft-change mechanism after automatic change; and (ii) on circular-box or drop-box looms, as a result of a length of weft yarn extending from the selvedge to a stationary shuttle.

lasting
A very stout, closely woven fabric made from hard-twisted yarns. A worsted lasting, usually a seven-shaft weave, is used for protective clothing in munition works. A 'cotton lasting', which may be of sateen or weft-faced twill weave, is used chiefly for shoe tops and bag linings.

lastrile (fibre) (generic name U.S.A.)
An alternative name for some types of rubber fibre (see **rubber** 2). (See also Classification Table, p.401.)

latch needle (machine knitting)
See under **needle (machine knitting)**.

latent crimp
See **crimp, latent**.

lateral leakage
The passage of gases along or through the textile element of a coated fabric or along interstices formed by the textile element and the coating polymer of the coated fabric.

lathe
See *Note* under **sley** 1.

lawn
A fine, plain-woven cloth of linen or cotton, made in various fine, sheer qualities. Various finishes may be applied to a fabric of this type, in which case the product is known by the name of the finish used, e.g., **organdie**.

lawn finish
A light-starch finish applied to **lawn** and other fine-yarn plain fabrics to give a crisp finished material.

lay
1. See *Note* under **sley** 1.
2. Fabrics placed in identical lengths, one on top of the other, in preparation for cutting prior to making-up.
3. (Cordage) To twist two or more components about each other to form a helix about the axis of the resulting laid product which may be a **strand** or **rope**.
Note: The direction of lay in cordage is described as 'S' or 'Z' (see **twist direction**).

lay, angle of
The angle at which **strands** lie in relation to the axis of the **cordage**.

lay, length of
The pitch of one complete turn of a yarn in a **strand** or of a strand in a **rope** measured in a straight line parallel to the axis of the strand or rope.

LCSP
See **lea count-strength product**.

lea
In cotton a length of 120 yards; in worsted, 80 yards; in linen, 300 yards.
Note: In cotton and worsted these lengths are one-seventh of the standard **hank**.

lea (linen)
The count of a flax-spun yarn (see Table, p.397).

lea count-strength product; LCSP; CSP; skein break factor
The product of the lea strength (pound force), and the actual count (Ne_C) of cotton yarn.

leader fabric; leader cloth
A length of fabric used in finishing or dyeing processes to lead goods through a machine, and generally left ready for attachment to a further piece when necessary.
Note: A leader fabric is often called an **end-fent** and its use enables a piece to be finished from end to end substantially without waste.

leaf edge
The outer or fall edge of the collar (see **collar fall**).

lease
A formation of the ends of a warp that maintains an orderly arrangement during warping and preparation processes, and during weaving.
Note 1: A lease consists usually of two sheets of alternate ends, which pass alternately over and under two transverse rods or cords, and the cross formed by the sheets of ends is a characteristic of a lease. Such a formation is described as an end-and-end lease. Less frequently, other formations are used, such as groups of two ends in a desired orderly arrangement, e.g., coloured ends. A lease is often described by its function or purpose, e.g., **weaving lease**, **entering lease**.
Note 2: The orderly arrangement of warp ends, especially for warp knitting, can be maintained by gummed paper.
Note 3: The orderly arrangement of the strands in a hank is secured by a leasing band (see **leasing**).

lease, clearing
An end-and-end lease inserted in a warp at intervals, traditionally every 100yds (say 100m), for checking the orderly arrangement of the ends and correct entering in a loom. It usually consists of uncombined section leases inserted during section warping.

lease, entering
An additional lease at the beginning of a warp for use in 'gaiting-up' the loom harness and reed.

lease, false (weaving)
A weaving lease formed in a loom after entering the warp ends by raising and lowering heald shafts to form two sheets of ends and inserting the transverse lease rods. In particular, the term applies when the sheets of ends formed are dissimilar, e.g., in a loom for weaving a five-end (five-shaft) warp satin.

lease, section
A lease inserted in a warp section during warping.

lease, warping
1. A completed lease in a warp formed as the aggregate of section leases during section warping.
2. A lease inserted in each of the sheets of ends of beams of a set during back beaming.

lease, weaving
A **lease** in a warp in a weaving machine at the rear of the healds, which is maintained by two transverse or lease rods, often oval in cross-section.

lease rods
The transverse rods around which a **lease** is formed.

lease-band
See *Note 2* under **leasing**.

leasing
The operation of inserting a crossed traverse lease-cord in wraps of yarn on a reel for the purpose either of separating wraps into groups of specific numbers, e.g., 120, or of preventing tangling of wraps of yarn during processing, e.g., hank dyeing, to facilitate subsequent winding of a yarn package from hank supply after removal of the lease cords.
Note 1: In the silk industry, the latter operation is termed 'lacing' and 'halching'. The term halching is used when the lease-cord (lacing-cord) is coloured differently from the yarn.
Note 2: Synonyms for lease-cord are 'lacing cord', 'lease-band', and 'tie-band'. The diagram shows a lease-cord used in reeling in the flax industry to separate a hank of yarn into twelve 'cuts', each of 120 wraps.

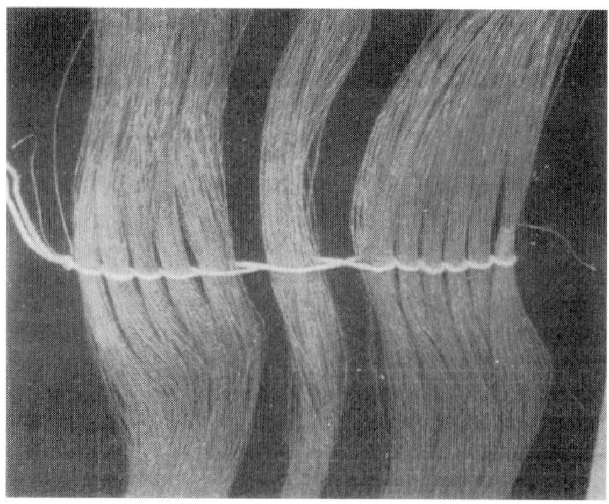

Leasing

leasing comb
A comb used in warp preparation to aid sheet separation for the formation of an end-and-end **lease**. The comb wires are provided with 'mail-eyes', and alternate ends pass through these eyes. The sheet of ends in the dent can be moved above and below the ends in the eyes.

leasing reed
See **reed, leasing**.

leather-cloth
A **coated fabric** which is embossed to give a leather-like appearance.

Leavers machine
See under **lace machines**.

legs (lace)
See **brides**.

legs (weft knitting)
See **sides** under **knitted loop (weft knitting)**.

length, fabric
Unless otherwise specified, the usable length of a **piece** between any **truth marks**, **piece-ends** or numbering when the fabric is measured laid flat on a table in the absence of tension.

leno edge
See under **selvedge, woven**.

leno fabric
A fabric in which warp threads have been made to cross one another, between the picks, during **leno weaving**. The crossing of the warp threads may be a general feature of plain leno fabrics (as in marquisette and some gauzes and muslins) or may be used in combination with other weaves (as in some **cellular fabrics**).

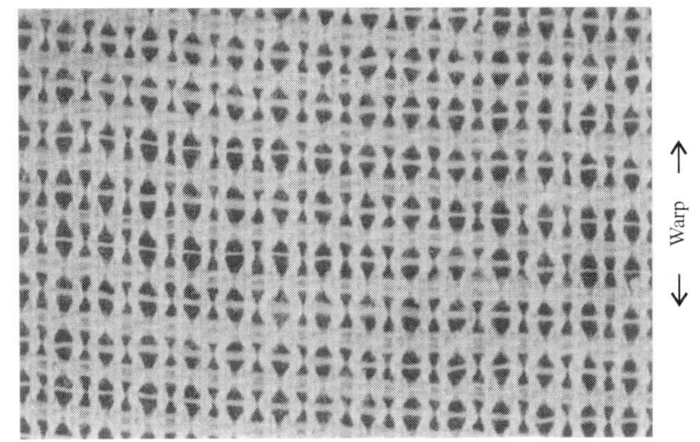

Leno cellular fabric

Leno cellular weave

leno reed
See **gauze reed**.

leno weaving
A form of weaving in which warp threads are made to cross one another between the picks.
Note 1: The simpler types of light-weight fabric produced by this method of weaving are known as 'gauze'.
Note 2: It may be necessary to use: (i) an easer motion to control the tension of the crossing ends during the formation of the crossed shed; (ii) a shaker motion to provide a partial lift to the standard heald to bring the threads approximately level and thus facilitate crossing.
Note 3: In simple leno weaving, one thread, generally called a crossing or leno end, L (see Fig. 1), is caused to lift alternately on one side and then on the opposite side of the other thread, usually referred to as the standard end, G, so as to produce 'crossed' or 'open' sheds. If the standard end is lifted a 'plain shed' (occasionally referred to as an 'ordinary shed') is formed.

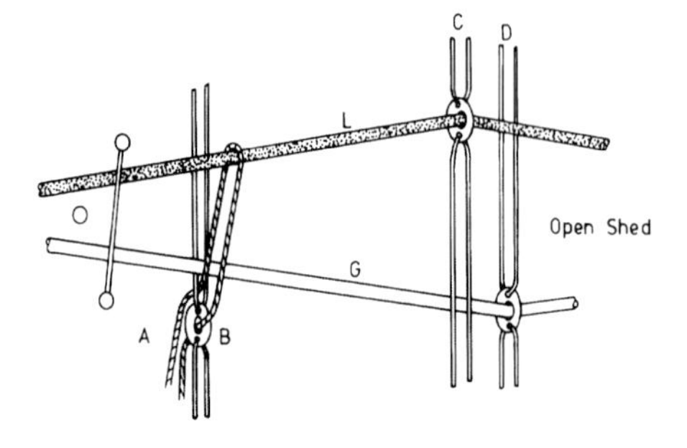

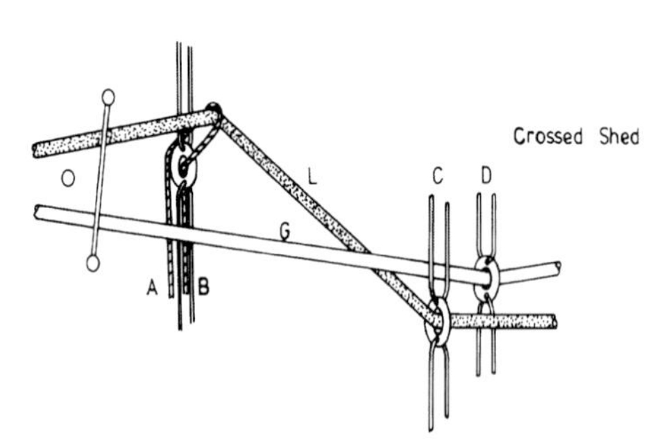

Cord doup weaving
Leno weaving, Fig. 1

Healds B and C (B working in conjunction with A on certain picks) are responsible for the operations of crossing and lifting thread L relative to thread G. A suitable name for B is front crossing heald, and for C, back crossing heald, with D referred to as the standard heald and A as the doup. With this nomenclature, the definition of the different sheds is as follows:

'crossed shed'	- operate doup and front crossing heald.
'open shed'	- operate doup and back crossing heald.
'plain shed'	- operate standard heald.

The names given to the healds and attachments operating the threads in cord doup weaving are:

Recommended Names	Also known as
A. doup	slip; half heald
B. front crossing heald	douping heald; doup; doup shaft; crossing heald; front standard
C. back crossing heald	back standard
D. standard heald	ordinary heald

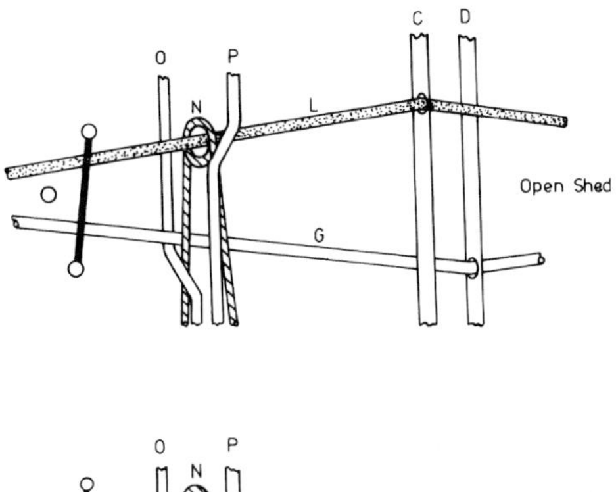

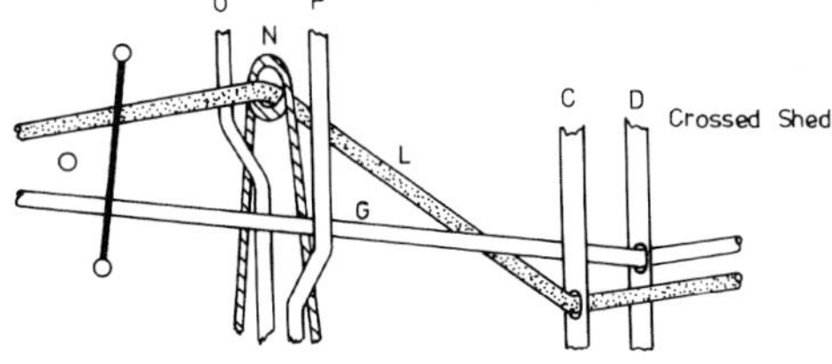

Metal doup weaving
Leno weaving, Fig. 2

In Fig. 2 the doup unit consists of legs O and P working in conjunction with a needle N and is responsible for operating the crossing and lifting of thread L relative to thread G.
The recommended names, which are generally used, are derived from their position in the loom, O being the front leg and P being the back leg, while N is referred to as the needle (the alternative but not recommended name being slip or doup). Using this nomenclature, the formation of the different sheds is as follows:

'crossed shed'	- operate front leg (including the needle).
'open shed'	- operate back leg (including the needle) and back crossing heald
'plain shed'	- operate standard heald.

Note 4: All of the diagrams illustrate bottom douping because the half heald is below the warp sheet. When the half heald is above the warp sheet, the system is known as top douping.

let-off motion

positive let-off motion
A mechanism controlling the rotation of the beam on a weaving, warp knitting, or other fabric forming machine where the beam is driven mechanically.

negative let-off motion
A mechanism controlling the rotation of the beam on a weaving, warp knitting, or other fabric forming machine where the beam is pulled round by the warp against a braking force applied to the beam.

letona
A leaf fibre obtained from the plant *Agave letonae*.

leuco dye
A reduced form of a dye from which the original dye may be regenerated by oxidation. (See also **vat dye** and **sulphur dye**.)

levelling
The migration of dye leading to a more uniform coloration of a substrate.

lick-roll applicator; kiss-roll applicator
A machine for the application of a **low wet-pick up** by a fabric passage at open-width in contact with a film of treatment liquor on the upper surface of a rotating applicator roller (termed lick-roll or kiss-roll) the lower part of which is submerged in the treatment liquor bath.

licker-in
See **taker-in**.

lift
In spinning and twisting processes, the length of that part of the take-up package which is intended to be covered by roving or yarn.

lift (weaving)
1. A term used to denote the movement of those parts of the loom mechanism associated with the formation of the **shed** and hence, in weave diagrams, to denote the representation of a warp thread over a weft thread.
2. The term is also used to describe the movement of shuttle boxes.

lifting plan
An indication of the order in which heald shafts are lifted on each pick in one weave repeat.

ligature
A **suture** used to tie a blood vessel.

ligne
A standard measure equal to 0.635mm (i.e., 1/40 inch) by which buttons and ribbons are measured.

limbric
A light-weight to medium-weight, closely woven, plain-weave, cotton cloth made from good-quality yarns. The weft is coarser and more closely spaced and has a lower twist factor than the warp, thus giving a soft cloth in which the weft predominates on both sides. Typical construction: 27x14; 12x16tex; 4x8%; 100 g/m^2; K=9.35+16.0. (See also **casement cloth**.)

limiting oxygen index; LOI; oxygen index
The minimum concentration of oxygen in a mixture of oxygen and nitrogen that will just support combustion of a material under specified test conditions.

line (cordage)
Laid, cabled or plaited **cordage** having a diameter of less than approximately 4mm. (See also **twine** 2.)

line flax
See **flax, line**.

linear density
The mass per unit length of linear textile material.
Note: The preferred unit is **tex**. (See Tables, p.396-397 for other units.)

linen
1. Descriptive of yarns spun entirely from flax fibres.

2. Descriptive of fabrics woven from linen yarns.
3. Descriptive of articles which, apart from adornments, are made of yarns spun from flax fibres.
Note: Despite some usage of this term in non-technical circles as a generic one, e.g., linen department, baby linen, household linen, it does not apply to individual articles that do not comply with the definition.

linen prover
See **counting glass**.

lingerie ribbon
Any type of fine and soft ribbon suitable for use on ladies' underwear.

lingoe
A metal weight attached to the lower end of each cord of a **jacquard harness**.

lining
Material used to cover part or all of the inside of **apparel**, the back of curtains, or the inside or back of other products. It may consist of a single layer or multiple layers of material loosely held in place along one or more edges. (See also **interlining**.)

link coner
See *Note* under **link winder**.

link winder
An automatic winding machine linked to a ring-spinning machine with automatic package doffing. The link winder forms an integral part of the spinning process working in conjunction with the ring spinner package doffing system.
Note: The term 'link coner' is also used to describe an automatic winding machine linked to a ring-spinning frame. However, it is generally agreed that the term link winder best describes the system because the winding process is not restricted to cones but can also produce cheeses.

linking; looping (knitting)
A method of joining together the edges of a piece of fabric or fabrics by a single or double chain-stitch on a linking machine, in which one or more of the pieces of fabric is run on to the points on a loop-to-point basis and is therefore stitched through adjacent needle loops. Where none of the pieces of fabric are run on to the points on a loop-to-point basis, this is referred to as **random linking**. (See also **cup seaming**.)

linking course (knitting)
See **slack course (weft knitting)**.

linking machine (knitting)
A machine, straight or circular, provided with grooved points spaced to receive loops, which are then joined together by chain-stitch.

links-links knitting
See **purl fabric, weft-knitted**.

linseed flax
Varieties of flax cultivated mainly for seed production.

linsey-wolsey
1. A coarse linen fabric.
2. A strong, coarse fabric with a linen or cotton warp and a woollen weft. The warp is entirely

linsey-wolsey *(continued)*
covered by the weft and has a nap finish.
Note: Formerly known as linsey-woolsey.

lint
1. The main **seed hair** of the cotton plant.
2. A plain-weave, highly absorbent material with one raised fleecy surface. For surgical purposes, it is sterilized.
3. See **fly**.

linters
Whole and broken **lint** fibres and **fuzz fibres**, which are removed from cotton seeds recovered from the first **ginning** by means of one or more additional ginning processes.
Note: The first ginning of cotton separates most of the lint fibres from the cotton seeds to produce the **ginned lint** which, after baling, is the ordinary raw cotton of commerce. The seed is then subjected to a second processing on a special gin to remove the **linters**, which are composed of a small proportion of whole-lint fibres, greater amounts of broken-lint fibres, and fuzz fibres, which are much coarser and shorter than the lint. The removal of lint and fuzz is not completed by this operation and the residue may be successively reginned. The products are termed 'first-cut linters', 'second-cut linters', etc., the length of the fibres in each successive cut becoming progressively shorter.

liquid ammonia treatment
A process during which textile material is immersed in or brought into contact with anhydrous liquid ammonia. The treatment confers 'flat setting', i.e., smooth drying properties and an attractive soft **handle** to cotton fabrics.

liquor: goods ratio; liquor ratio; LR; goods ratio
The ratio of the weight of liquor employed in any treatment to the weight of material treated.
Note: 'Short' and 'long' are often used to describe low and high liquor: goods ratio, respectively.

lisle thread
A highly-twisted, plied (usually 2-ply), good quality cotton hosiery yarn, spun generally in fine counts. All lisle threads are gassed and some may be mercerized (mercerized lisle). A lisle thread was formerly a plied yarn having singles of opposite twist.

list
See **selvedge, woven**.

listing
1. An undesirable uneven dyeing effect consisting of a variation in colour between selvedges and the centre of a dyed fabric, often caused in jig dyeing through difference of temperature between the selvedges and the centre of the batched-up fabric on the jig roller, or by uneven batching-up of the cloth on the rollers.
2. See **selvedge, woven**.

lively yarn
See **snarly yarn**.

llama fibre (hair)
Fibre from the fleece of the llama (*Lama glama*).

load at specified elongation
The force required to produce a specified percentage elongation.

loading
Increasing the weight of fabrics by the addition of deliquescent salts, starch, China clay, etc. *Note:* This term is not restricted to any one class of textile fabric but is used loosely in connection with the finishing of wool, cellulosic, or silk goods. (See also **filling** 1.)

locker bar (lace machines)
A bar running the net-making width of the double locker **bobbinet** machine, to which are attached two blades. There are two locker bars, one each side of the **well** of the machine, under the combs. The rocking action of these bars causes the blades to engage with the tails of the carriages and, with the **driving bars**, controls the motion of the carriages through the combs and the well.

locking course (knitting)
A sequence of knitting performed in various ways on rib or purl machines before the stitches are cast off from one bank of needles at the end of a garment or garment part. Locking courses are designed to prevent the dropped stitches from running back through the previous garment.

locknit (warp knitting)
A fabric made on a warp-knitting machine with one needle bar and two full-set guide bars which make closed lap movements in opposition to each other. The front guide bar makes a two needle underlap and the back guide bar makes a single needle underlap. The guide bar lapping movements are:

Front guide bar: 2-3, 1-0, and repeat.
Back guide bar: 1-0, 1-2, and repeat.

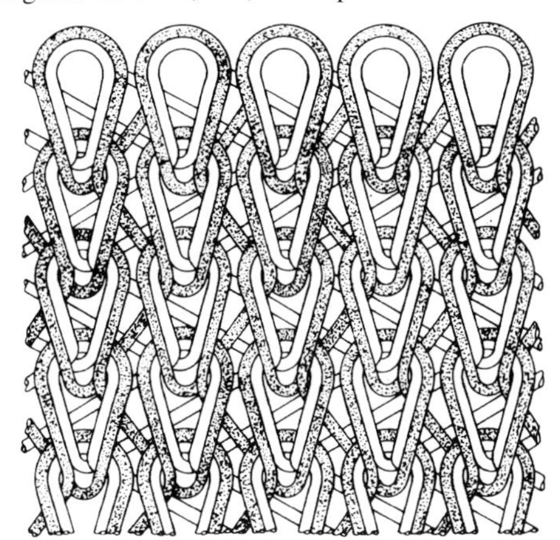

Locknit fabric (shown from the technical face)

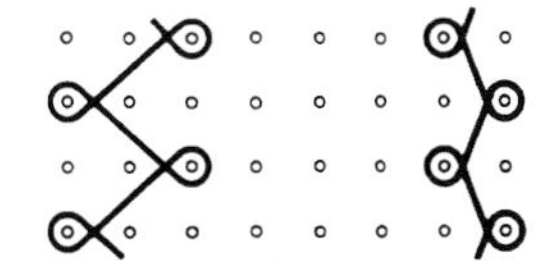

Front bar Back bar
Locknit lapping movement

locks
A term used in wool-sorting for short oddments of wool which fall from the **skirting** tables or are swept up from the boards. In some countries it can include soiled tufts and pieces from near the rumps of sheep.

loden
Coarse woollen milled water-repellent fabric used for jackets, coats, and capes.

lofty
A term applied to an assemblage of fibres to denote a relatively high degree of openness and resilience, or a large volume for a given mass.

LOI
See **limiting oxygen index**.

London shrinking
A finishing process applied to fabrics in which the fabric is thoroughly moistened and then allowed to dry naturally in the absence of tension.

loom
1. A **hand loom**.
2. A **treadle loom**.
3. A **power loom**.
4. A term used for **weaving machine**.

loom effective speed
The product of the loom speed in picks per minute and the absolute loom efficiency. (See also **loom efficiency (absolute)**.)

loom efficiency
The ratio of the number of picks inserted by a weaving machine (per unit of time, e.g., hour, shift, day or week) to the number of picks which would have been inserted if the machine had been running continuously at the specified speed.

loom efficiency (absolute)
The ratio of the number of picks inserted by a weaving machine (per unit of time, e.g., hour, shift, day or week) to the number of picks which would have been inserted if the machine had been running continuously (using actual machine speed for the calculation).

loom efficiency (overall); shed efficiency
The average of loom efficiency of all weaving machines in the shed.

loom efficiency (running)
The ratio of the number of picks inserted by a weaving machine (per unit of time, e.g., hour, shift, day or week) to the number of picks which would have been inserted if the machine had been running at the specified speed after allowing for ordinary stoppages.
Note: Ordinary stoppages are those concerning the weaver and tuner, e.g., stoppages on account of weft replenishment in non-automatic or shuttle changing looms, machine adjustments, warp adjustments, and warp replacements.

loom-state
Any woven fabric as it leaves the loom before it receives any subsequent processing. (See also **grey**.)

looming
A term covering the processes involved in preparing a weaver's beam for the loom, e.g., drawing-in, dropper-pinning, sleying, knotting, tying, etc.

loop, knitted (weft knitting)
See **knitted loop (weft knitting)**.

loop column
See **tuft column**.

loop length (pile structures)
The continuous length of yarn or fibres between two successive lowest points of bindings of the pile in a substrate.

loop pile
See under **pile (carpet)**.

loop row
See **tuft row**.

loop ruche
See under **ruche**.

loop transfer (knitting)
The process of moving loops, wholly or in part, from the needles on which they were made to other needles for the purpose of shaping or design.

loop wheel knitting machine
A circular weft-knitting machine in which knitting, takes place on a set of vertically mounted bearded needles, the yarn being manipulated with the aid of bladed wheels. (See also **circular knitting machine** and **knitting machine**.)

loop yarn
See under **fancy yarn**.

loop-raised fabric, warp-knitted
A fabric produced from continuous-filament yarns, generally polyamide, in which the long underlaps of the front or middle bar(s) are raised (see **raising**) during finishing to form a pile of unbroken filaments. A typical two bar example is: front bar 1-0/3-4; back bar 0-1/2-1.

looper (hook) (tufting)
See under **tufting machine**.

looper (sewing machine)
An eyed stitch-forming element which carries an under thread or a cover thread on some types of sewing machine. Common machine types with looper(s) include overlock and 2-thread chainstitch. (See also **spreader**.)

looper wire
See **edge wire**.

looping (knitting)
See **linking**.

loopless toe (knitting)
A fully fashioned hose toe that is so narrowed that it can be closed by seaming only.

loose reed (warp protector)
A reed so mounted in the loom sley as to yield under the pressure of the shuttle at beat-up should the shuttle fail to reach the receiving box. This displacement of the reed actuates a mechanism that stops the loom.

louring (hat manufacture)
The improvement in the lustre of felt **hoods** or **bodies** by the application of fat using a hot pad.

lousiness; exfoliation
An inherent fault in silk only apparent after degumming or dyeing. It is characterized by fine fibrils or **fibrillae** that become separated from the filament, so giving a speckled, dishevelled appearance.

low add-on finish
A finish based on a low **wet pick-up** process which enables a lower solids add-on to be used to attain the required properties.

low-charged system
See **charged system**

low orientation yarn; LOY
A yarn of low molecular orientation suitable for orientation by drawing at a high draw ratio.
Note: This term is used in contrast to **fully drawn yarn**, **partially oriented yarn** and **fully oriented yarn**.

low wet pick-up finish
Application of a chemical finish either by a topical technique (see **topical finishing**), e.g., lick roller, spray or foam applicator, or by an impregnation-removal method, e.g., vacuum extraction or use of porous bowls, to give a reduced wet pick-up significantly lower than by conventional **padding**.

LOY
See **low orientation yarn**.

LR
See **liquor: goods ratio**.

lucet; chain fork
A lyre-shaped hand tool of ancient origin, some 70cm to 150cm long, made from thin rigid material such as wood, horn, ivory, etc. It was used for making square knitted cords with low stretch and good strength characteristics, and was widely used until the advent of the industrial revolution when the manufacture of cords and laces became a machine operation.

lucetted cord
Square knitted cord produced with the aid of a **lucet**.

luminance
For a coloured object, luminance is a measure of the apparent overall reflectance. For a light source, luminance is a measure of the apparent brightness of the light.

lump
A length of unfinished fabric, usually longer than the customary piece length.
Note: This term seems to originate in its literal meaning as applied to the appearance of woven cloth when removed by hand from the loom.

lustre; luster
The display of different intensities of light, reflected both specularly and diffusely from different parts of a surface exposed to the same incident light. High lustre is associated with gross differences of this kind, and empirical measurements of lustre depend on the ratio of the intensities of reflected light for specified angles of incidence and viewing.
Note: This definition makes the difference in intensity of light the keypoint, since these form the chief subjective impression on the observer of lustre. Both specular and diffuse light must be present together, for, if diffuse light only is present, the surface is matt, not lustrous, whereas, if specular light only is present, the surface is mirror-like, and again not lustrous. The phrase 'exposed to the same incident light' has been included to rule out shadow effects, which have no part in lustre proper. The general term 'surface' is intended to apply to fibres, yarn, and fabrics, and indeed to other surfaces, e.g., that of a pearl (although there the differently reflecting

parts are very close together). In the second sentence of the definition, lustre is regarded as a positive function of the differences, the appropriate adjective of intensification being 'high'.

lyocell
A manufactured fibre of cellulose obtained by extruding cellulose dissolved in an organic solvent. (See also **modal (fibre) (generic name)** and Classification Table, p.401.)
Note 1: The organic solvent may contain a small proportion of water.
Note 2: The solution is of cellulose and not a cellulose derivative.

machine tow
See **flax tow**.

machine-finishing
See *Note* under **boarding**.

machine-washable
A term denoting that a textile article can be washed in a domestic washing machine to remove dirt and other extraneous substances using an aqueous detergent solution.

madapolam
A bleached or dyed plain cotton fabric with a soft finish in any of a wide range of qualities used for ladies wear.

made-to-measure
See **bespoke**.

made-up
See **making-up**.

Madras muslin
A **gauze** fabric with an extra weft, which is bound into the gauze texture in the figured parts and cut away elsewhere. (See also **muslin**.)

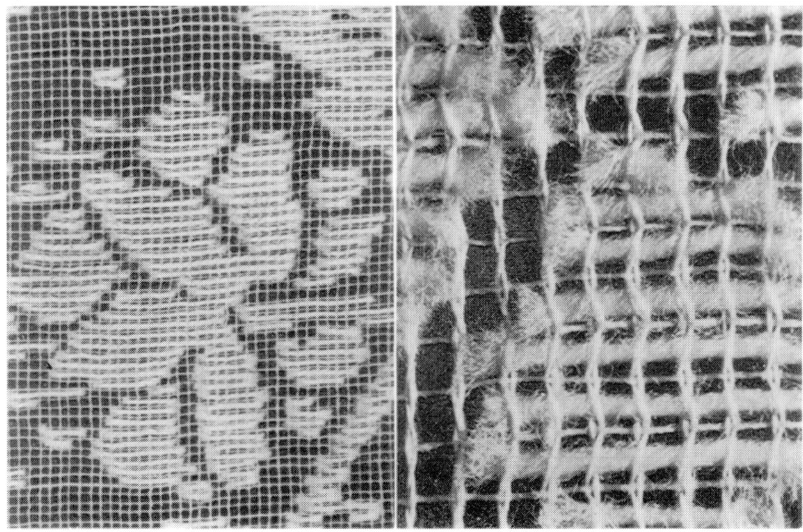

Actual size *Magnified x 5*

Madras muslin

magazine bar (weft knitting)
A **transfer bar** fitted with replaceable grooved points arranged uniformly along its upper edge. The magazine bar is designed to hold a supply of knitted pieces which are loaded on to the points one after another by **running-on** each piece on to the points on a loop-to-point basis. The garment pieces can thus be subsequently delivered from the loaded magazine bar singly or collectively as required on to the points of the transfer bar. (See also **linking**.)

magazine creel
A **creel** for mounting two or more yarn packages per end, tied top to tail, from which the yarn is withdrawn over-end, to give end-continuity from successive packages.

magyar sleeve
See under **sleeve (clothing)**.

mail
See **heald mail**.

mail heald
See **heald**.

making-up
The process of assembling garments or other products from their component parts. The finished product is said to be 'made-up'.
Note: Making-up implies sewing but it may also include cutting, other joining techniques, pressing and finishing operations. An example of the wider meaning is when a clothing factory, where all these activities are found, is referred to as a 'making-up unit'.

man-made fibre
See **fibre, manufactured** under **fibre**.

mangle
A machine consisting of two or more rollers (**bowls**) running in contact to form a **nip** (or nips) the purpose of which is to express liquid from textiles which pass through it.

manila
Fibre obtained from *Musa textilis*, also known as 'abaca'. (See also **hemp, true**.)

manufactured fibre
See **fibre, manufactured** under **fibre**.

mapleleaf braid
A woven narrow fabric similar to **oakleaf braid**, but with a mapleleaf pattern.

marcella; waffle piqué
A fancy or figured fabric of **piqué** structure. Typical construction (cotton):

Warp	Weft
28 ends per cm of 15tex face	38 picks per cm in a ratio of
14 ends per cm of 21tex stitching	10 face picks of 12tex to 2 wadding picks of 30tex

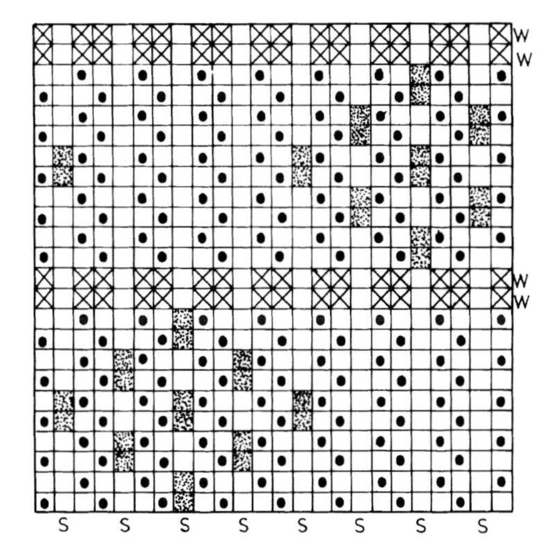

Marcella weave

marionette (narrow fabric weaving)
A mechanism for controlling the movement of shuttles in a multi-tier weaving machine.

marker
The representation or drawing of the arrangement of identified garment **pattern** pieces relevant to the cutting of a batch of material. The marker is placed on the material and provides guidance for cutting.
Note: Markers may be on fabric, paper, card or held in computer data files.

marker planning
The process of constructing a **marker**, with the primary intention of minimising wastage of the material to be cut.

marking thread
A distinctively coloured sewing thread with a high fastness to washing, bleaching and dry cleaning solvents supplied specially to commercial launderers for stitching their identification symbols on to textile items being cleaned.

marl; mottle
To run together, and draft into one, two slubbings or rovings of different colour or lustre. (See also **worsted yarns, colour terms**.)

marl effect yarn (continuous-filament)
Two single, continuous-filament yarns, of different solid colours or dyeing properties (subsequently dyed) doubled together. (See also **ingrain (filament yarn)**.)

marl yarn
A yarn consisting of two or more single ends of different colours twisted together.
Note: This definition is not applicable to worsted yarns (see **worsted yarns, colour terms**).

marocain
A crêpe fabric with a pronounced weftways rib formed by the use of a fine close-sett warp and a highly twisted weft picked two Z- and two S-twisted yarns.

marocain *(continued)*

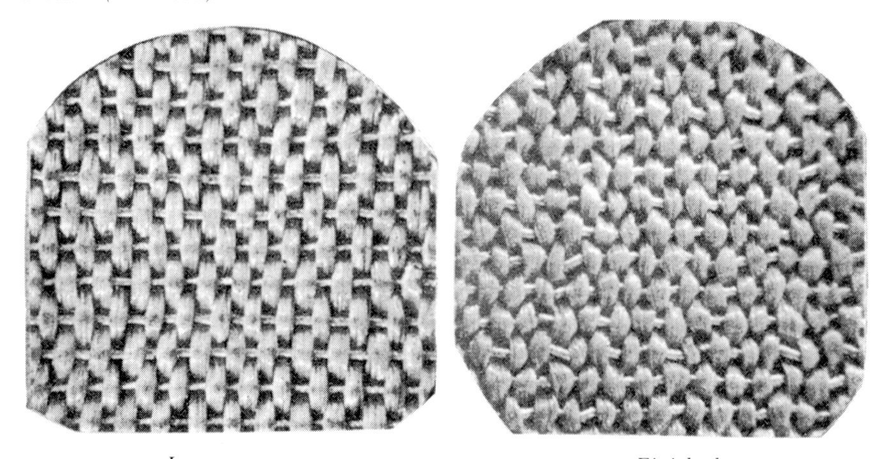

Loom-state *Finished*

Acetate-viscose marocain (magnification x 8)

marquisette
A light, open-textured, fine-quality gauze, in which slipping is reduced by crossing the warp threads by means of the leno principle (see **leno weaving**).

marquisette, warp-knitted
A square-hole net produced from two or three guide bars each using a full-set threading, the front bar making a chaining movement and the second and third bars laying-in so that they connect the **chains** or **pillars** generally every third course. Typical lapping movements are as follows:

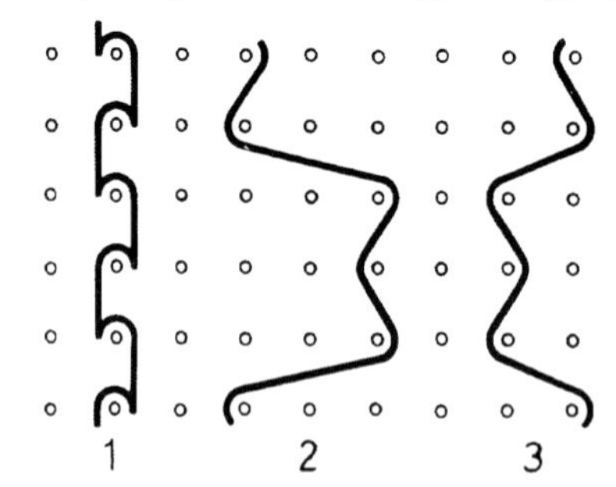

Warp-knitted marquisette structure

married yarn (defect)
See **spinners' double**.

Martindale test
See **abrasion test**.

mass coloration
A method of colouring manufactured fibres by incorporation of the **colorant** in the spinning composition before extrusion into filaments.

mass pigmentation
A form of **mass coloration** in which a pigment is used.

mat
See **hopsack**.

mat (textile floorcovering)
See **rug**.

matching
1. A process by which the amount of each colouring matter present in a material is adjusted so that the final colour resembles that of a given sample as closely as possible.
2. A comparison of dyed samples of textiles of nominally the same colour.
Note: 1 and 2 can be done by eye or by using instruments that measure chromaticity co-ordinates.

matchings
Wool that has been sorted.

matelassé
A double or compound fabric with a quilted appearance. It is commonly made with two warps and two wefts, the threads generally being arranged two face, one back in both warp and weft, but other proportions are often used. The quilted effect may be accentuated by the use of wadding threads and a tightly bound ground weave. The designs are formed by floating threads or patches of fancy weaves.

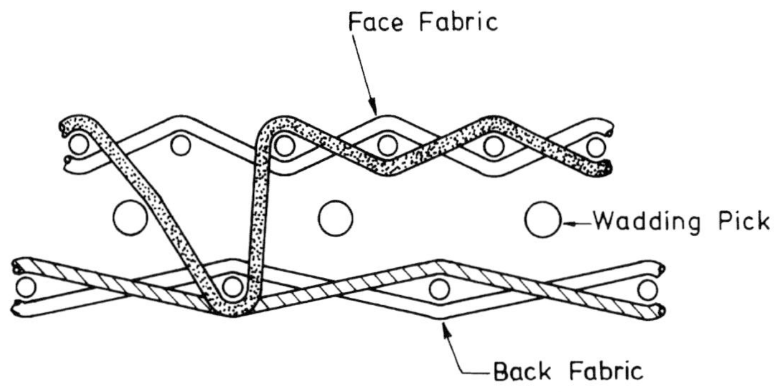

Matelassé: section through weft

matrix-fibril bicomponent fibre
See under **bicomponent fibre**.

matt
1. Descriptive of a surface with little or no lustre.
2. (Weaving) See **hopsack**.

maturity (cotton)
A cotton fibre characteristic which expresses the relative degree of thickening of the fibre wall. It is usually estimated by one or more of several indirect tests which are often used to discover the proportion of fibres having a maturity greater than some selected level. The following terms are used in relation to cotton maturity:
(i) mature fibre: fibre where a high degree of wall thickening has taken place during cotton growth;
(ii) immature fibre: fibre where little wall thickening has taken place during growth;
(iii) normal fibre: fibre whose wall has developed to greater than a specified amount;
(iv) thin-walled fibre: fibre which does not fall into either the normal or dead categories;
(v) dead fibre: an extreme form of immature cotton.

maturity ratio
A method of expressing, numerically, the maturity of a sample of cotton fibres. It is the ratio of the actual degree of wall thickening to a standard degree of thickening equal to 0.577 (see ISO 4912). (See also **percentage maturity**.)

Mauritius fibre; Mauritius hemp
A fibre from the leaf of the plant *Furcraea gigante*.

Mayo twill
See **Campbell twill**.

Maypole braider
See **braiding machine**.

mealy
Descriptive of the appearance of a print or dyeing which shows small irregularities, rather like oatmeal, caused, e.g., by too much printing paste on the fabric from too deep an engraving or unsatisfactory fabric preparation.

mechanical finish
A finish obtained by mechanical means, e.g., shearing, calendering.

mechlin machine
See under **lace machines**.

medical textile
A general term which describes a textile structure which has been designed and produced for use in any of a variety of medical applications, including implantable applications.

medulla
The central portion of some animal fibres consisting of a series of cavities formed by the medullary cells which collapse during the growth process. In **kemp** the medulla forms the greatest portion of the fibre and is surrounded by a comparatively thin layer of **cortex**. In fibres from the outer coat of some breeds of sheep and certain other animals (e.g., reindeer) the medulla also forms a large portion of the fibre.

mélange printing; Vigoureux printing
A printing process in which bands of thickened dye paste, with intervening blank areas, are applied across slubbings of wool or other fibres. The slubbing is subsequently steamed, washed, and then combed to produce a very even mixture of dyed and undyed lengths of fibre.

mélange yarn
A yarn produced from colour-printed tops or slivers. It is distinguishable from a mixture yarn in that the individual fibres are of more than one colour. (See also **worsted yarns, colour terms** and **mélange printing**.)

melded fabric
A fabric consisting wholly or in part of bicomponent fibres, in which cohesion has been achieved through the selective melting of one component of the bicomponent fibres.

melt blowing
A process in which a polymer is melt-extruded through a die into a high velocity stream of hot air which converts it into fine and relatively short fibres. After quenching by a cold air stream the fibres are collected as a sheet on a moving screen.

melt fracture
An unstable melt-spinning condition in which the surface of the extrudate becomes rough and irregular.

melt spinning (manufactured fibre production)
The conversion of a molten polymer into filaments by **extrusion** and subsequent cooling of the extrudate.

melt-spun
Descriptive of manufactured filaments produced by **melt-spinning**.

melton
A heavy-weight fabric, all-wool, or with cotton warp and woollen weft, which is finished by heavy milling and cropping. The fibres in the cloth are tightly matted together by the milling process, and this gives the fabric a felted appearance. It is usually made in a 2/2 twill, especially if all-wool, but it is sometimes made in other weaves to facilitate milling and the covering of the cotton warp.

meltonette
A lightweight fabric which resembles melton cloth, and is used for women's wear.

melusine finish
A lustrous finish produced on felt **hoods** or **bodies** by brushing with selected compounds such as an acrylic resin dispersion.

mending
The insertion of yarn into a woven fabric where the warp or weft is missing and also the correction of other faults by means of needlework, e.g., stitching. This is a skilled manual needlework operation.

mending (textile floorcoverings)
See **picking** 2.

mercerization
The treatment of cellulosic textiles in yarn or fabric form with a concentrated solution of caustic alkali whereby the fibres are swollen, the strength and dye affinity of the materials are increased, and their handle is modified. The process takes its name from its discoverer, John Mercer (1844). The additional effect of enhancing the lustre by stretching the swollen materials while wet with caustic alkali and then washing off was discovered by Horace Lowe (1889). The modern process of mercerization involves both swelling in caustic alkalis and stretching to enhance the lustre, to increase colour yield and cotton yarn strength. A related process, **liquid ammonia treatment**, produces some of the effects of mercerization. In chain mercerizing, shrinkage in fabric width is allowed, followed by re-stretching and washing on a clip-stenter. In chainless mercerizing, the fabric is effectively prevented from shrinking by transporting over rotating drums.

hot mercerization
The treatment of cellulosic fabric with a hot concentrated solution of caustic alkali to facilitate uniform penetration prior to cooling and stretching etc., so as to improve the degree of mercerization.

post mercerization (linen)
Crease resistant linen fabrics may be produced by treatment with urea formaldehyde resin followed by a mercerizing treatment to confer durability and suppleness.

mercerization *(continued)*

slack mercerization

Mercerizing of a fabric in absence of tension, or under reduced tension.

Note: After washing-off, the fabric remains in the shrunken condition, and consequently a high degree of yarn crimp is obtained and the fabric becomes more extensible. There are two reasons for operating this process: to produce a stretch fabric or as part of the process for crease resisting linen. Dye absorption is increased but lustre is not.

merchant converter; converter

An individual or organisation that obtains textile material from a supplier, procures its processing and then re-sells the finished product. (See also **textile agent** and **textile merchant**.)

merino

1. Wool from merino sheep.

Note: The merino breed of sheep originated in Spain and the wool is noted for its fineness and whiteness. It was confined to Spain until the late 1700s when merino sheep were exported to, and bred in, many other countries. Well known types of merino are: Australian, Rambouillet, Vermount, South African, Saxony. The word merino is now almost synonymous with 'fine wool', the wool being finer than 25 microns (μm).

2. A pre-20th Century term applied in France and Germany to worsted fabrics produced from yarns using merino or other fine wools.

3. A **plainback** worsted fabric developed in England in the 1820s. It was made from fine yarns spun from merino or other fine wools of 23 to 28 tex for the warp and 17 to 22 tex for the weft. The stimulus for the development of this fabric was the availability of fine machine-spun worsted yarn.

4. Woollen fabrics made in England from yarns produced from wool reclaimed from soft woollen and worsted dress goods.

5. A French shawl made from two-fold warp yarn using merino wool. The weft yarn is made from other wool or silk.

6. A fine cotton fabric used as a dress material in the Philippines. It is made from yarns of 13 tex for the warp, and 15 to 10 tex for the weft with 32 ends x 32 picks per cm.

mesh (knitting)

See **micromesh** and **spread loop**.

mesta

See **kenaf**.

metachrome process

A single-bath method of dyeing in which the fibre is treated in a dyebath containing a suitable **chrome dye** together with a chromate, whereby a dye-chromium complex is formed within the fibre.

metal (fibre) (generic name)

A manufactured fibre made from any metal. (See also Classification Table, p.401.)

metallic (fibre) (generic name U.S.A.)

A manufactured fibre composed of metal, plastic-coated metal, metal-coated plastic, or a core completely covered by metal. (See also Classification Table, p.401.)

metallic yarn

See **metallized yarn**.

metallized fabric
A textile fabric on which metal has been deposited, e.g, chemically or by electric arc or by lamination using an adhesive.

metallized yarn; metallic yarn
A yarn which has free metal as a component. (See also **tinsel yarn**.)
Note: There are several types, the best known of which are:
(i) yarns in which separate metal fibres or filaments are included;
(ii) metal of narrow strip section, usually lustrous (see **flat metal yarn**);
(iii) metal of narrow strip section, coated or laminated with film such as viscose, cellulose ethanoate (acetate), butanoate (butyrate), or polyester; the film may be coloured;
(iv) yarns on which metal is attached to or deposited on the fibres, e.g., chemically, by electric arc, or by adhesive;
(v) multi-end yarns in which at least one single yarn is metallic; (vi) a **gimp** in which the helical covering consists of a strip of (ii) above.

metameric
Descriptive of objects that exhibit **metamerism**.

metameric match
A match that is judged to be satisfactory under a particular illuminant but not under illuminants of different spectral compositions.

metamerism
A phenomenon whereby the nature of the colour difference between two similarly coloured objects changes with change in the spectral distribution (characteristics) of the illuminant.
Note 1: Metamerism is most frequently seen when two coloured objects match in daylight, but differ markedly in colour when viewed in tungsten-filament light. This arises because the visible absorption spectra of the two objects differ significantly, although the **tristimulus values** in daylight are identical.
Note 2: This term is often used loosely to describe the behaviour of a single coloured object that shows a marked change of colour as the illuminant changes, but this effect is properly described as lack of **colour constancy**.

metier
The bank of cells or compartments used in the dry-spinning of cellulose ethanoate (cellulose acetate).

microfabric
A fabric composed of **microfibres** or microfilaments.

microfibre; microfiber
1. A fibre or filament of linear density below approximately 1 decitex.
Note 1: For glass and other mineral fibres see definition 2 below.
Note 2: Some commercial fibres as coarse as 1.3 decitex are classified as microfibres by their producer.
Note 3: Fibres finer than approximately 0.2 decitex are sometimes referred to as 'ultra-fine fibres' or 'ultra-fine microfibres'.
(See also **bicomponent fibre**, *Note 3*.)
2. (Glass and other manufactured mineral fibres) A fibre or filament of thickness less than approximately 3 microns (μm).

microfibril
See **fibril**.

micromesh (weft knitting)
A ladder resistant construction used in women's fine gauge hosiery containing tuck stitches which spiral around the leg thus reducing the light reflectance and producing an attractive appearance. The tuck stitches occur at every alternate course and, in the case of 3x1 micromesh, they occur at every fourth wale.

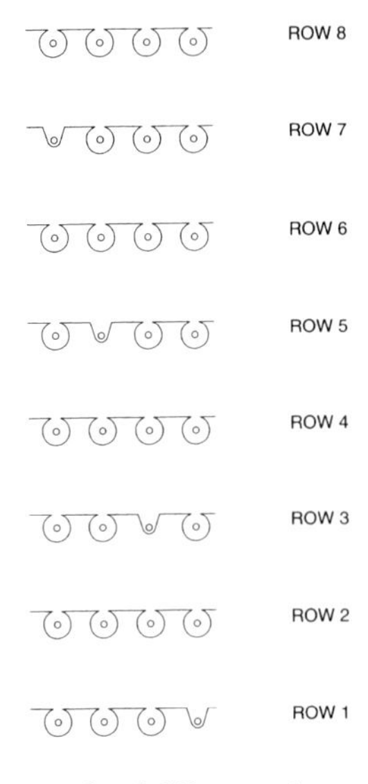

3 x 1 Micromesh

micronaire value
A measurement of cotton fibre quality which is an indication of the fibre specific surface.
Note: The micronaire value is a function of both **fibre fineness** and **maturity**. Low values indicate fine and/or immature fibres; high values indicate coarse and/or mature fibres. Micronaire value is determined in practice by measuring the resistance to air flow of a plug of fibres having a specified mass confined in a chamber of specified volume. The original micronaire instruments were calibrated against the linear densities, in micrograms per inch, of a series of American Upland cottons and this scale of values survives in modern instruments, though it is now recognised as an arbitrary scale.

microstretching
The use of inter-meshing, driven, corrugated rollers for the incremental extension of the weft yarns of woven fabrics. When applied immediately before the cross-linking of lightweight cotton fabrics, weft tensile strength loss is minimised.

migration
The movement of an added substance, e.g., a dye, pigment, cross-linking agent or alkali, from one part of a textile material to another. (See also **fibre migration**.)

milanaise
Narrow braid or corded fabric in which the cord effect is produced by **leno weaving**. One end is made to cross another coarse end in an alternate crossed and open shed sequence.

milanese (weaving)
A low-quality cotton fabric of about 17x14 tex with approximately 30 ends x 38 picks/cm.

milanese fabric, warp-knitted
A warp-knitted fabric usually made with a full set of warp threads (i.e., containing twice as many threads as there are wales in the fabric). The threads are continuously divided into two equal warp sheets, one set of which traverses to the right continuously and the other set to the left, so that any particular thread traverses the full width of the fabric and, on reaching the selvedge, is transferred to the other set.
Note: The manner of traversing the threads may be either **silk lap** or **cotton lap**.

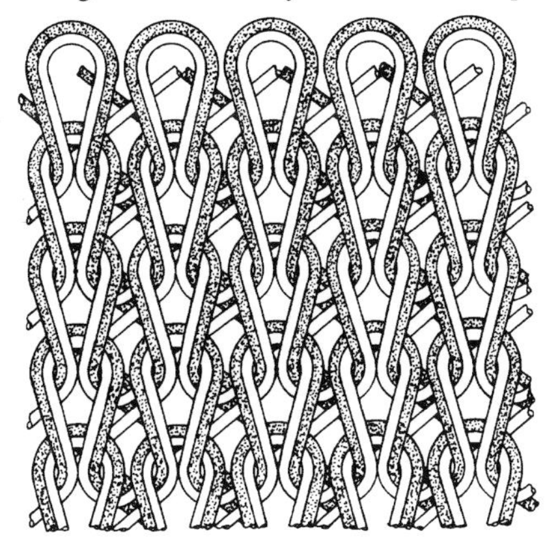

Milanese fabric, with silk lap (technical face)

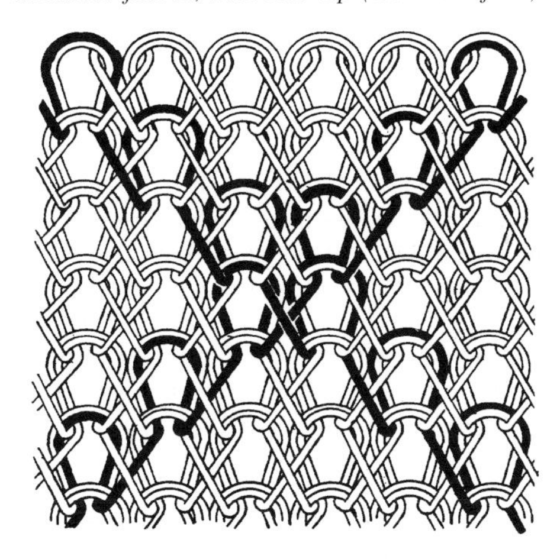

Milanese fabric, with cotton lap (technical back)

milanese warp-knitting machine
A straight-bar or circular warp-knitting machine specially constructed to knit milanese fabrics by means of atlas-type lapping traverses with two sets of warp threads which move continuously in opposition without changing direction. (See also **atlas fabric** and **knitting machine**.)

Milano rib, weft-knitted
A weft-knitted rib-based fabric. Each complete repeat of the structure consist of three components knitted in the sequences shown to give one row of 1x1 rib (A) and one row of plain tubular knitting (B), the two component parts of B usually being similar. The appearance and characteristics of the fabrics are related to the ratio of the course lengths of A and B. (See also **half-Milano rib, weft-knitted** and **double jersey, weft-knitted**.)

Milano rib, weft-knitted *(continued)*

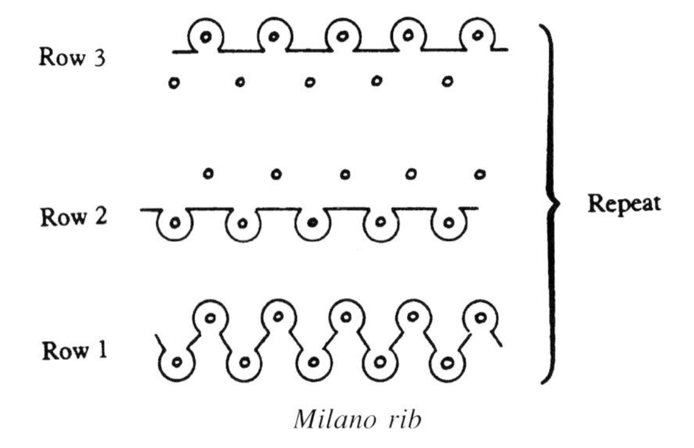

Milano rib

mildew
A growth of certain species of fungi.
Note: On textile materials, this may lead to discoloration, tendering, and variation in dyeing properties.

military braid
A plain, flat **braid** usually made from continuous-filament yarn.

mill rig; mill row
A rope mark or running mark (see **rope marks**) formed during rotary milling of pieces.

milling (fabric finishing); fulling
Consolidation or compacting of fabrics, that usually contain wool or other animal fibres.
Note: The treatment, which is usually given in a rotary milling machine or in milling stocks, produces relative motion between the previously wetted fibres of a fabric. Depending on the type of fibre, the structure of the fabric and on variations in the conditions of milling, a wide range of effects can be obtained varying from a slight alteration in handle to a dense matting with considerable reduction in area.

milling acid dye
An acid dye of good fastness to acid or alkaline milling.

millitex
See **tex system**.

mineral (fibre)
An inorganic fibre, natural or manufactured.
Note 1: Asbestos is a naturally occurring inorganic fibre and the term 'mineral fibre' has sometimes been used to mean asbestos exclusively.
Note 2: Metallic fibres are not normally described as mineral fibres (see **metallic (fibre)**). (See also Classification Table, p.401.)

mineral wool
A **web** or **batt** of manufactured, inorganic fibres used for thermal or acoustic insulation.

mini-grain
A two coloured **ingrain yarn** in which one colour predominates.

minimum-care
See **easy-care**.

mink fibre (hair)
Fibre from the mink (*Mustela (Lutreola) vison*).

mispick; wrong picking
An incorrect sequence of weft insertions.

miss (weft knitting)
A knitting cycle where a needle or needles do not take a yarn and thus produce a **float loop** either intentionally or as the result of a knitting fault.

miss-lap (warp knitting)
The effect produced when a guide bar is not shogged (see **shog**) sideways for either an **overlap** or an **underlap**.

missed loop (weft knitting)
See **float loop (weft knitting)**.

missing pick (fabric defect)
The unintentional omission of one complete length of weft thread across the full width of a fabric. The fault may appear as a narrow crack.

mistral
A plain-weave dress material, made with nub yarns.

mixed end; wrong end (fabric defect)
A thread that differs in material, linear density, filament, twist, lustre, colour, or shade from adjacent normal threads.
Note: In woven or warp-knitted fabrics the defect appears as a vertical line running warpway; in weft-knitted fabrics, as a horizontal stripe running across the fabric and repeated at regular intervals.

mixed weft (fabric defect)
An unintentional mixing of two or more lots of weft yarns.
Note: This may lead to the formation of **weft bars** (see under **bar (woven fabric)**).

mixture yarn
See **worsted yarns, colour terms**.

mock
1. Imitation.
2. To be opposed. Two **ends** that **lift** opposite to each other are described as 'mocking', e.g., in **plain weave**, adjacent ends mock in plain order.

mock cake
A package of yarn produced by winding on to a collapsible mandrel or former, which is subsequently removed. It is built up from the inside to the outside in contrast to a **cake**.

mock fashioning mark (knitting)
A loop formation which imitates the mark produced by fashioning (see **shaping (weft knitting)**).

mock gauze
See **perforated weave**.

mock grandrelle
A single yarn with a **grandrelle** effect.

mock leno
A woven fabric which imitates the appearance of **leno weaving** (See also **perforated weave** and **distorted thread effects**).

mock space weaving machine (narrow fabrics)
A **multi-tier weaving machine** in which all pieces being woven in any one row are so spaced as to lie immediately above or below the **landings** of those in a vertically adjacent row.

mock voile
A plain-weave cotton fabric, woven from hard-twisted single yarns (instead of two-fold), and woven with one thread per dent.

mock-chenille yarn
See under **fancy yarn**.

modacrylic (fibre) (generic name)
A manufactured fibre composed of synthetic linear macromolecules having in the chain at least 35% and less than 85% (by mass) of recurring cyanoethene (acrylonitrile) groups. (See also Classification Table, p.401.)
Note: In the U.S.A. the definition excludes fibre qualifying under **anidex** and **rubber** 2.

modal (fibre) (generic name)
A manufactured fibre of cellulose obtained by processes giving a high breaking strength and a high wet modulus. The breaking strength B_C in the conditioned state and the force B_M required to produce an elongation of 5% in its wet state are

$$B_C \geq 1{,}3\sqrt{Tt} + 2Tt$$

$$B_M \geq 0{,}5\sqrt{Tt}$$

where Tt is the linear density in decitex. B_C and B_M are expressed in centinewtons. (See also Classification Table, p.401.)
Note 1: This definition was introduced in ISO 2076:1989 and replaces the definition previously used.
Note 2: **lyocell fibres** with a high breaking strength and a high wet modulus should not be referred to as modal.

model
1. A designated style in a range of garments.
2. A garment made in limited numbers.
3. The designer's original **sample** garment or prototype. (See also **sealed sample**.)

mohair braid
Any type of **braid** made from mohair yarns.

mohair fibre
Fibre from the angora goat (*Capra hircus aegagrus*).

moiré fabric
A ribbed or corded fabric that has been subjected to heat and heavy pressure by rollers after weaving so as to present a rippled appearance. The effect arises from differences in reflection

of the flattened and the unaffected parts. This type of fabric is also correctly described as 'watered'.

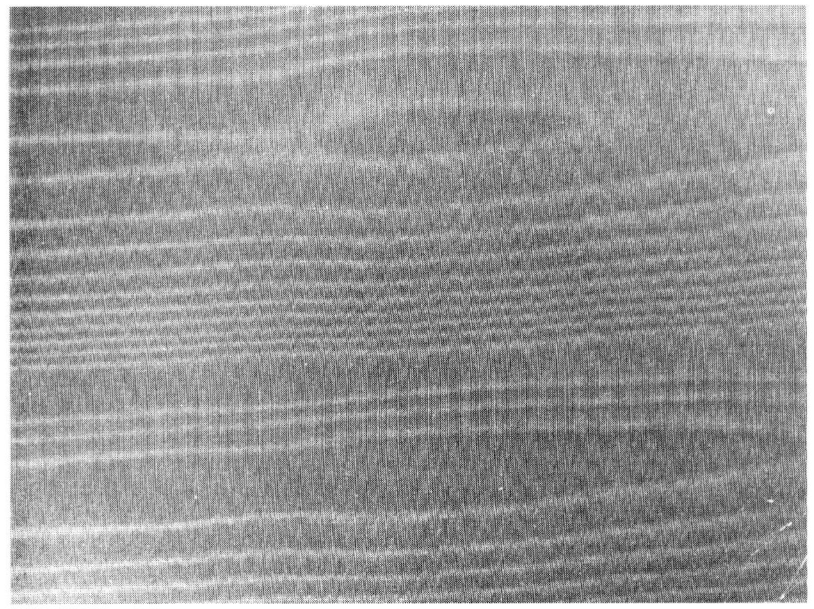

Moiré fabric

moiré fault
An undesirable shaded effect produced when the weave pattern of one fabric is accidentally impressed on to the face of another fabric, usually under heat and pressure during processing, e.g., beam dyeing.

moiré finish
A watered or rippled appearance on the surface of a fabric. The effect is obtained by passing two layers of a rib or cord fabric between heavy heated rollers or by passing the fabric between suitably engraved calender rollers. Originally developed for silk fabrics, but good results can be obtained on cotton and manufactured fibre fabrics. There are many styles of moiré finish.

moirette
A plain-woven cotton fabric, with a predominance of warp or weft, creating lines across or down the fabric respectively, e.g., 44x25; R20/2xR30/2tex; K=19.7+13.7. Polished cotton weft is used (see **polishing**). The finish is the same as for **moreens**.

moisture and extractable component
The mass of water in any form plus extractable matter in a material, determined using prescribed methods and expressed as a percentage of the mass of the extracted and dried material.

moisture content
The ratio of the mass of moisture in a material to the total moist mass. The ratio is usually expressed as a percentage and is calculated as follows:

$$\frac{\text{Total moist mass} - \text{Dry mass}}{\text{Total moist mass}} \times 100$$

moisture regain
See **regain**.

moisture regain in the standard atmosphere
The mass of water in any form which a material contains when, after preconditioning, it comes into equilibrium with the standard atmosphere, determined using prescribed methods and expressed as a percentage of the mass of the dried material.

moity wool
A term, used mainly in the U.K., for wool containing vegetable matter (straw, hay, twigs, etc.), picked up by sheep during grazing.

molar mass; molecular weight (polymer)
The average of the sum of the atomic weights of the atoms present in the chains of individual macromolecules in a polymer. This will in general depend upon the basis on which it is measured or calculated, and this should be stated, e.g., number average, weight average or viscosity average.

molecular weight (polymer)
See **molar mass**.

moleskin fabric
A thick cotton fabric, originally uncut **corduroy** having a very high weft **sett**, piece dyed and given a smooth raised finish to simulate the fur of a mole. Only one warp is used, but the picks are arranged two face and one back. Typical construction for workwear: 16x134; R74/3x37tex; K=13.8+81.5. Lighter weight constructions are used in the fashion trade. (See also **fustian**.)

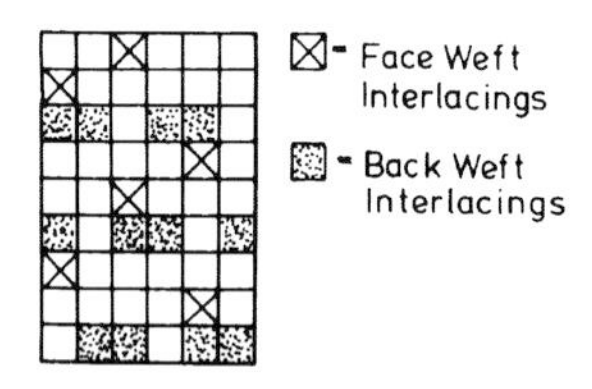

Moleskin weave

molleton
A heavy reversible cloth with a nap on both sides. It was originally made of wool.

Molleton: section through warp

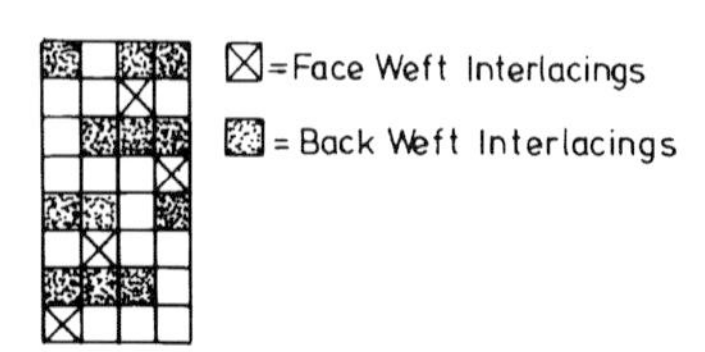

Molleton weave

molten-metal dyeing process
A method of continuous dyeing in which material is impregnated with an aqueous liquid containing dye and chemicals and then passed through a bath of liquid low-melting alloy usually below 100°C.

momme
A Japanese measurement of mass equivalent to 3.76g (approximately). It is used to indicate the mass per unit area of silk fabric, this being expressed as the weight in momme of a length of degummed fabric 22.8m in length and 3.8cm in width.
Note: The length measurements used are standard units of length in the Japanese silk kimono industry.

monofilament yarn
See **continuous-filament yarn**.

monomer
A small, simple, chemical compound from which a polymer may be formed.
Note: In most cases a given polymer can be made from a variety of alternative monomers. In some cases two or more different monomers are involved in the production of a polymer.

monovoltine silkworm
A variety of silkworm producing one generation per year.

montagnac fabric
A curly woollen fabric with an astrakhan-like pile, produced by cutting some of the weft floats and leaving others uncut. It is subsequently brushed to form a very warm and durable fabric.

moquette
An upholstery fabric in which pile warp ends are lifted over wires that may or may not have knives. On withdrawal, the result is a cut or uncut pile. Cut moquette may also be made on the face-to-face principle (see **velvet, woven** (iii)).

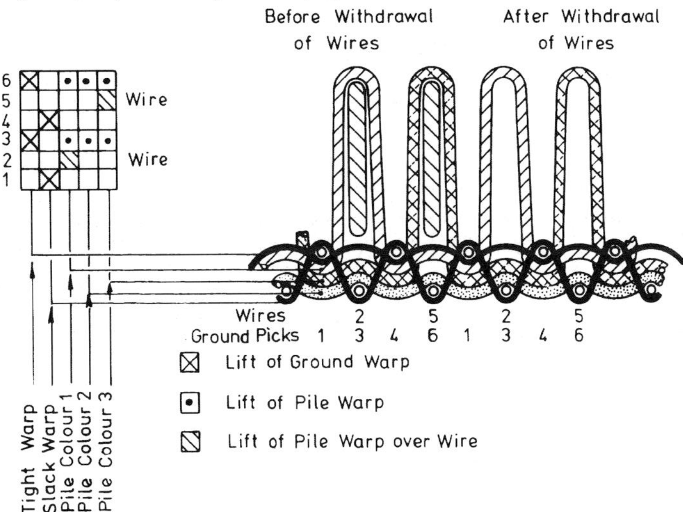

Uncut moquette: a weave and cross-section along the weft illustrating the method of production

mordant
A substance, usually a metallic compound, applied to a substrate to form a complex with a dye, which is retained by the substrate more firmly than the dye itself.

mordant dye
A dye that is fixed with a **mordant**.

moreen; morine
Originally a worsted fabric similar to **harateen**. Now a **repp** cloth, woven with a coarse cotton warp and a fine cotton weft, the cloth having a **moiré** finish.

mosquito net
See **bobbin net**.

moss cord
See **crêpe cord (narrow fabric)**.

moss-crêpe fabric
A fabric made with a moss-crêpe weave and S and Z-twist moss-crêpe yarns in warp and weft. This fabric has a characteristic spongy handle. Various combinations of (i) moss-crêpe weave with other yarn and (ii) moss-crêpe yarns with other weaves are possible. All the resulting fabrics have some but not all of the characteristics of true moss crêpes.

moss-crêpe weave
A crêpe weave with a repeat in the warp and weft directions relatively large compared with that of many other crêpe weaves.

Moss-crêpe weave

moss-crêpe yarn
A two-ply yarn made by doubling a normal-twist yarn with a high-twist yarn. A traditional moss-crêpe yarn was 11.2 tex dull viscose rayon, doubled with 8.4 tex viscose rayon, with approximately 400 turns/metre, all twists being in the same direction ('S' or 'Z').

motes (cotton)
There are two broad categories:
(i) fuzzy motes: The largest of this type of mote consists of whole aborted or immature seeds covered with **fuzz fibres** and sometimes also with very short lint fibres, the development of which has ceased at a very early stage. Small fuzzy motes originate as either undeveloped or fully

grown seeds, which are broken in ginning and disintegrate still further in the opening, cleaning, and carding processes.
(ii) bearded motes: A piece of seed coat with fairly long lint fibres attached.
Note 1: Both classes of mote become entangled with the lint cotton.
Note 2: Fuzzy and bearded motes carrying only a small piece of barely visible seed-coat are frequently termed 'seed-coat neps'. (See also **trash**.)

mothproofing
Treatment of textile materials, e.g., wool and other keratin fibres, with an insecticide to reduce damage caused by the larvae of moths feeding on the textile fibre.
Note: This is one of the more frequently used insect-resist treatments and may also provide some measure of resistance to attack from larvae of other species, e.g., beetles.

motifs (lace)
The decorative figures of a pattern. These may be cut out and applied to a garment for ornamentation.

motion (lace machines)
The passage of the bobbin threads through the sheet of warp and pattern threads. (See also **lace machines**.)
Note 1: Leavers motions are counted for each passage of the bobbin threads whether from the front to the back of the machine or *vice versa*.
Note 2: A full furnishing motion is a passage of the bobbin threads twice through the warp threads.
Note 3: A bobbinet motion is the passage of one tier of bobbin threads in one direction through the warp threads, i.e., for each complete to-and-fro passage of the carriages, four motions are counted.

motion (warp lace machines)
One complete revolution of the cam shaft.
Note: This will normally make one complete loop on each needle.

motion mark
See **pick bar** under **bar (woven fabric)**.

motion way (lace)
The direction in which the lace is made, parallel to the dressing selvedges (see **pattern repeat**).

mottle
See **marl**.

mousseline
A general term for very fine, semi-opaque fabrics, finer than muslins, made of silk, wool or cotton.

move number; step number
The number of picks by which the interlacing of a warp thread in a weave moves upwards relative to the warp thread on its immediate left. Move numbers (M) can therefore be used to describe weaves; for example, eleven-end sateen, diagram A, can be written as

$$\frac{1}{10}\ \text{M4},$$

move number; step number *(continued)*
while seven-end sateen, diagram B, can be written

$$\frac{1 \quad 1 \quad 1}{2 \quad 1 \quad 1} \text{M2.}$$

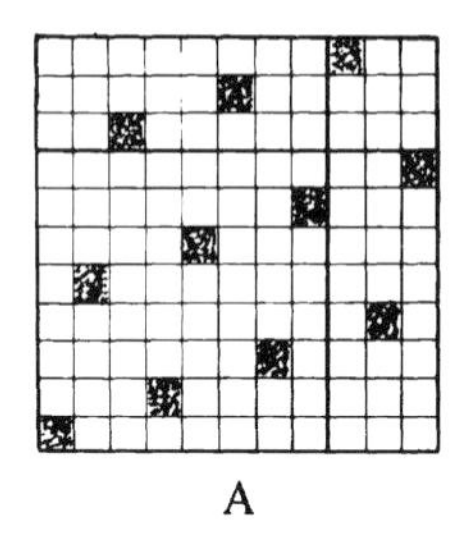

A

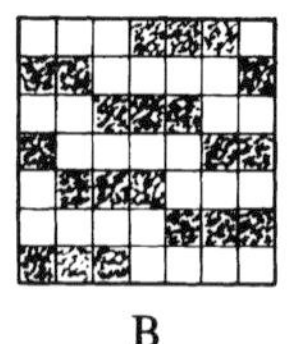

B

muff
An unsupported **cross-wound package** in the shape of a lady's muff.
Note: Muffs are normally wound directly on to a collapsible mandrel, but have been produced by winding bulked yarns in an extended form, lacing and allowing the package to contract on removal from the swift. Each muff has a piece of stockinette threaded through the centre and brought round each end to enclose it. Bulked yarns are conveniently dyed in this form.

muga; munga
See **silk, wild**.

mule draw
The maximum extent of the movement of the spindle tip during the outward (drafting and twisting) passage of the mule carriage.

mule spinning
An intermittent method of spinning, whereby the actions for the formation of a yarn, i.e., drafting and twisting, are undertaken in one operation and the winding on to the package in another.
Note: Drafting and twisting occur as the spindles and the delivery rollers separate. Winding on to the package (**cop**) takes place when the spindles and delivery rollers come together. Drafting is of the order of 1.2 to 1.6 on the woollen system. (See also **spinning**.)

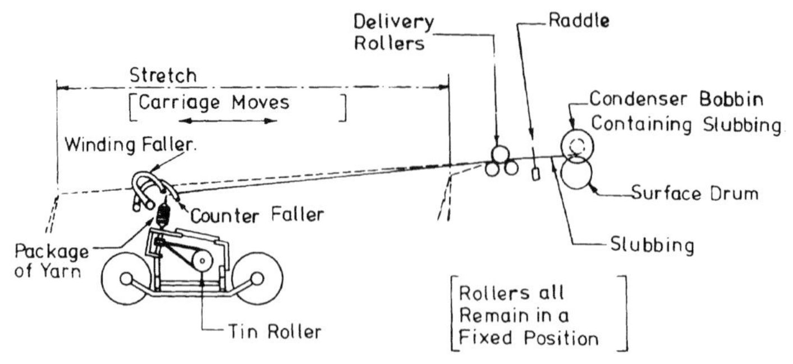

Woollen mule

mull
A plain cotton fabric of relatively open texture (traditionally with warp and weft cover factors of between 8 and 10) made from fine yarn and used for dress and other purposes. The fabric is

soft-finished and usually bleached. The construction typically lies in the range: 6-10tex warp and weft; with 25-31 ends and picks per cm.

multi-axial fabric, warp-knitted
A warp-knitted construction in which three or more substantially straight threads are inserted at different angles to each other, for example, vertically, horizontally and diagonally. (See also **triaxial weaving**.)

multi-filament yarn
See **continuous-filament yarn**.

multi-piece weaving machine; multi-space weaving machine
A **narrow fabric** weaving machine equipped for the simultaneous weaving of two or more **pieces**.
multi-shuttle weaving machine
A **multi-piece weaving machine** in which weft insertion is by shuttle.

multi-space weaving machine
See **multi-piece weaving machine**.

multi-tier weaving machine (narrow fabrics)
A **multi-shuttle weaving machine** in which the **batten** is provided with more than one horizontal row of shuttles. In such weaving machines where two or more wefts per piece are introduced sequentially, the batten is raised or lowered as required, so as to align one row of shuttles only at each pick with the row of pieces being woven.

multifeeder machine (circular weft-knitting machine)
A circular weft-knitting machine with a multiplicity of feeders (see **feed**), normally distributed evenly around the circumference of the cylinder and/or dial of the machine. Adjacent feeders supply yarn to successive courses in the fabric. (See also **knitting machine**.)

multilobal
Descriptive of a fibre or filament whose cross-section resembles a polygon but has concave sides and rounded vertices (lobes).
Note: The prefixes tri (3), penta (5), hexa (6), octa (8), etc., are used with the suffix lobal to indicate the number of lobes.

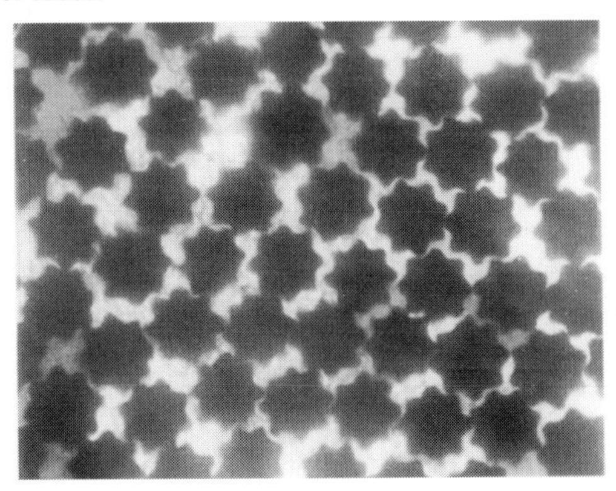

Octalobal

multilobal *(continued)*

Trilobal

multiphase weaving machine
See **weaving machine**.

multivoltine silkworm
See **polyvoltine silkworm**.

munga
See **silk, wild**.

mungo
The fibrous material made in the woollen trade by **pulling** new or old hard-woven or milled cloth or felt in rag form. (See also **shoddy**.)

mushroom fastener
A fastener consisting of 2 tapes, both being pile fabrics. One tape is woven by the face to face method (see **velvet, woven** (iii)), the monofilament pile being severed with the aid of heat thus providing molten mushroom shaped pile ends. The other tape with a loop pile is usually slit from wide knitted fabric. On offering one tape to the other, a secure closure is obtained which can be opened by peeling from either end. (See also **touch and close fastener** and **hook and loop fastener**.)

muskrat fibre (hair)
Fibre from the muskrat (*Ondatra zibathicus*).

muslin
A generic term for a light-weight, open cloth of plain weave or simple leno weave traditionally with a cover factor of 4.5 - 9.0 in the warp and 4.5 - 8.0 in the weft. Normally, muslins did not exceed 68 g/m^2.(See also **Madras muslin**.)
Note: Some of these cloths are used in the **grey** (butter muslin and cheese cloth), whereas others (dress muslins) are bleached and dyed.

nainsook
A fine, light, plain-woven cotton cloth with a soft finish.

name selvedge
See **selvedge, woven**.

nap
1. A fibrous surface produced on a fabric or felt by **raising** in which part of the fibre is lifted from the basic structure.
Note: Nap and **pile** are often used synonymously, but the practice of using the two terms for different concepts is to be encouraged as providing a means of differentiation and avoidance of confusion.
2. A variant of **nep**, used in the flax-processing industry.
3. In raw cotton, matted clumps of fibres which are entangled more loosely than those in neps (see **nep**).

napier
1. A double-faced fabric for dresswear or overcoats that has a wool face and a backing mainly of hair.
2. A floor covering of hemp and jute.

napping
1. An alternative term for **raising**.
2. Cloth which has been raised to obtain a dense full **nap** is treated in a machine which rubs the raised fibres into small pills, balls or curls. The process is sometimes called 'friezing' (see **friezé**).

narrow fabric
Any fabric made by interlacing fibres or yarns which (in the U.K.) does not exceed 45cm in width. In the U.S.A., and for the purpose of the Tariff code in the EC, the upper limit of width is 30cm. A characteristic of a narrow fabric is that its edges are an essential feature.

narrow fabric selvedge (needleloom)
See under **selvedge, woven**.

natural fibre
See **fibre, natural** under **fibre**.

natural flax
See **flax, green**.

navel
In **rotor spinning**, a device, aligned on the axis of the rotor, through which the yarn is withdrawn, different types of which can influence yarn quality and processibility.

neck
In the process of drawing synthetic filaments or films, the relatively short length over which a reduction in cross-sectional area occurs as a result of stretching beyond a critical value.
Note: Commercial drawing processes for manufactured fibres and films do not necessarily involve the formation of a neck.

neck cord
See **jacquard harness (weaving)** under **jacquard (mechanism)**.

necking
The abrupt reduction in cross-section that may occur when an undrawn filament is stretched.
Note: The term is most commonly associated with **undrawn yarn** but also occurs when conducting tensile tests on knitted fabrics (see **wasting**) and thermoplastic materials, particularly in film form.

needle (machine knitting)
A hooked element used for intermeshing **needle loops**; there is normally one needle for each needle wale.

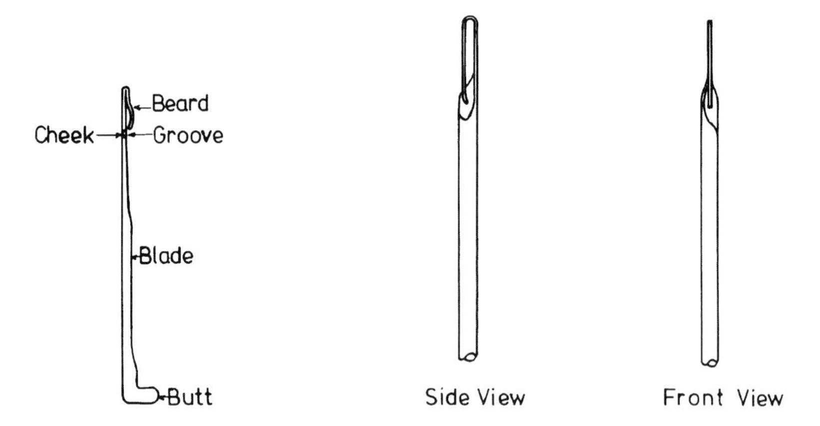

Bearded needle or spring needle *Carbine needle*

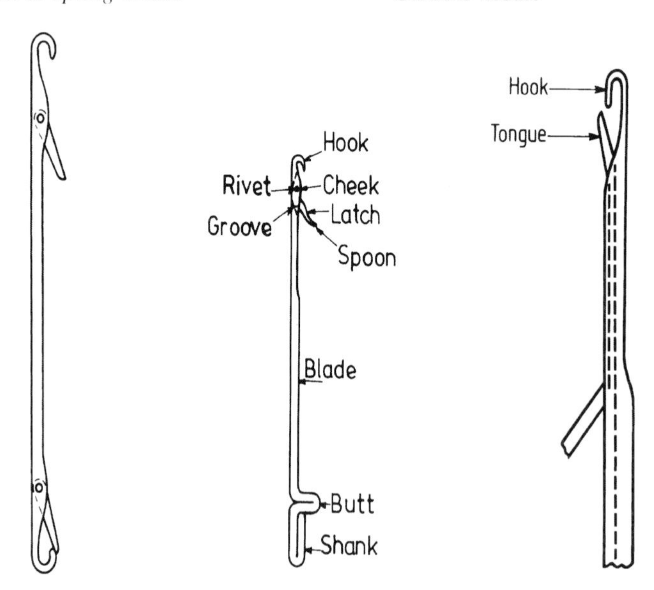

Double-ended needle *Latch needle* *Two-piece or compound needle*

bearded needle; spring needle
A needle having an extended terminal hook or beard which can be flexed to close the hook. The beard returns to its original position when the force is removed.

carbine needle
A needle similar in shape to a bearded needle but with the beard shielded by a shoulder on the stem.
Note: The needle may only be lapped in one direction for the yarn to pass under the beard. A **presser** is not necessary as the needle is self-acting, the shoulder passing the loop on to the beard. The needle has a limited use, mainly in crochet-type machines.

compound needle; two-piece needle
A needle in which the hook and hook closing parts are separately controlled.

double-ended needle
A needle having an operative hook at each end.

latch needle
A needle having a terminal hook closed by a pivoting latch. Its knitting action is self-acting since the fabric loop pivots the latch and allows the loop to be cleared from the open hook or knocked over the closed hook depending on the direction of needle movement.

needle (needlefelt); felting needle
A barbed needle mounted in a needleloom, (see **needleloom (nonwoven)**) to cause fibre reorientation and entanglement during **needling**.

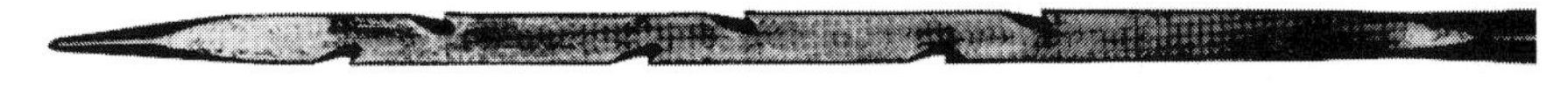

Blade of a barbed needle

needle (sewing)
A shaft-like element used for sewing, normally pointed at one end with an eye for thread or yarn.
Note 1: Needles for machine sewing are manufactured, normally from steel, with a great variety of point shapes, different diameters and surface finishes according to the intended end-use. Though usually straight, they may be curved (see **needle, curved** below).
Note 2: A wide variety of hand-sewing needles is similarly available. Again, the design depends on the materials to be sewn and other operational and end-use considerations.
Note 3: Generally speaking, machine sewing needles have the eye at the pointed end; hand-sewing needles have the eye at the opposite end.

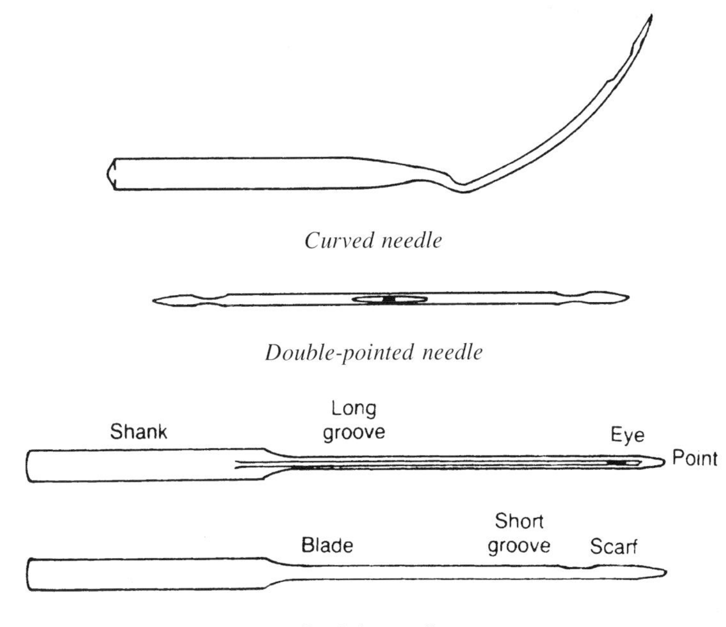

Curved needle

Double-pointed needle

Straight needles

(Source: H. Carr and B. Latham. *The Technology of Clothing Manufacture*, 1994)

needle (sewing) *(continued)*

needle, curved

A needle with a curved blade used in some overedging and in blindstitch machines. In hand sewing, curved needles are sometimes used when the back of the material being sewn is not easily accessible.

needle, double-pointed

A needle with points at both ends and an eye midway between.

Note: This needle is used to form simulated hand stitching, Stitch Type 209.

needle, straight

A needle with a straight blade (see diagram).

needle (tufting)

See under **tufting machine**.

needle bar (tufting)

See under **tufting machine**.

needle damage

See **stitching damage**.

needle feed

See **feed mechanisms (sewing)**.

needle gauge (sewing machine)

The distance between the points of any two adjacent needles on a multiple needle machine.

needle loop (weft knitting)

See under **knitted loop (weft knitting)**.

needle plate (sewing machine)

See **throat plate**.

needle positioner

An automatic device on a sewing machine which ensures that the needle will stop either 'in' (needle down) or 'out' (needle up) of the fabric as required.

needle transfer (knitting)

The transfer of a double-headed latch needle from a slider in one bed or cylinder to the slider in the bed or cylinder opposite.

needle-bonded fabric

See **needlefelt**.

needle-punched fabric

See **needlefelt**.

needle-run (lace)

Lace in which the **objects** are formed or outlined by embroidering on to a net or lace base.

needlebonding

See **needling**.

needlecord

A fine-ribbed **corduroy**.

needled fabric
See **needlefelt**.

needlefelt; needle-punched fabric; needled fabric; needle-bonded fabric.
A nonwoven structure formed by the mechanical bonding of a fibre **web** or **batt** by **needling**.
Note: Originally needlefelts were produced as flat structures without pile but recent advances in needleloom design have generated options for the production of various pile textures including both cut and loop.

needlefelting
See **needling**.

needleloom (narrow fabrics)
A shuttleless loom in which the weft is drawn from a stationary supply and introduced into the shed in the form of a double-pick by a weft inserter needle. The weft is retained at the opposite selvedge by the action of knitting, or by the introduction of a locking thread from a separate supply.

needleloom (nonwoven)
A machine for producing **needlefelt**. A needle beam reciprocates vertically at rates of up to 3000 cycles per minute. **Felting needles** are mounted in a board at a density of 300-5000 per metre width and pass through a fibre **web** or **batt**, which is supported between bed and stripper plates.

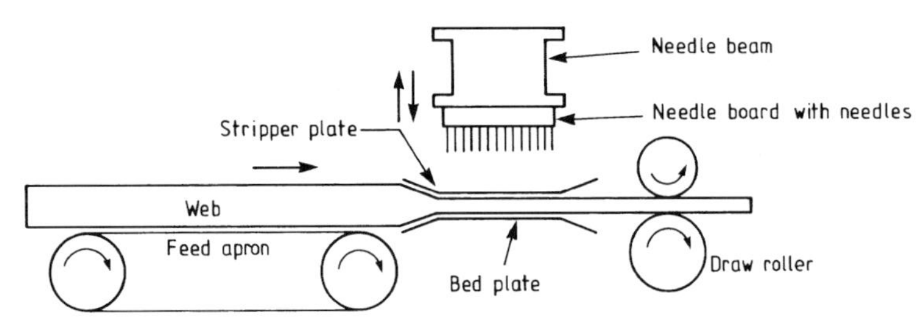

Needleloom machine

needleloom carpeting
See **fibre-bonded floorcovering**.

needleloom selvedge (narrow fabric weaving)
See **narrow fabric selvedge (needleloom)** under **selvedge, woven**.

needlepoint (lace)
A hand-made lace made by stitching according to a predetermined plan, a sewing needle and thread being used.
Note: A skeleton of outlining threads is held in position by tacking in accordance with the pattern drawn directly on to layers of linen or on to parchment attached thereto. The ground net and the pattern are stitched on to this framework without stitching through the backing, and the backing is removed, when the lace is made, by cutting the tacking between the two layers of linen.

needlepunching
See **needling**.

needling; needlepunching; needlefelting; needlebonding
The use of barbed needles (see **needle (needlefelt)**) mounted in a needleloom (see **needleloom (nonwoven)**) to entangle a fibre **web** or **batt** by mechanical reorientation of some of the fibres within its structure.

negative let-off motion
See **let-off motion**.

negative shedding
An operation in which the movement of the healds or harness is controlled in one direction only, the return movement being effected by springs or weights (see **shedding**).

negative take-up motion
See **take-up motion**.

nep
A small knot of entangled fibres.
Note 1: In the case of cotton it usually comprises dead or immature cotton hairs.
Note 2: A local variation of the term, referring to flax, is 'nap'.

nepp yarn
See **knickerbocker yarn** under **fancy yarn**.

neppy yarn
A yarn in which the incidence of nep occurs at a relatively high level and so constitutes a fault.
Note: Neppy yarns are sometimes used purposely as decoration, for example, **knickerbocker yarn** (see under **fancy yarn**).

net
An open-mesh fabric in which a firm structure is ensured by some form of twist, interlocking, or knitting of the yarn. It may be produced by gauze weaving, knitting, or knotting, or on a **lace machine**. (See also **plain net**.)

net silk; nett silk
1. Raw-silk filaments or strands that have been processed into yarns by twisting and folding or both, as opposed to **spun silk**.
2. Descriptive of fabrics produced from net silk.

neutral-dyeing acid dye
An acid dye which has substantivity for wool, silk or polyamide, when applied from a neutral bath.

New Zealand flax; New Zealand hemp
See **phormium**.

new wool
See **virgin wool**.

nib (lace machines)
1. A flattened projection at the tip of the carriage spring that enters the lips of the bobbin and holds the latter in position.
2. A projection on the carriage at the base of the breast that forms the recess into which the catch bar drops to enable it to propel the carriage.

nick
See **notch**.

ninon
1. A fabric originally made from very fine highly twisted silk yarns with two or three ends weaving as one and with two or three threads lightly twisted together to form the weft so giving

the effect of two or three picks in a shed, and known as double or triple ninons respectively. Typical construction: 130x44; 20dtex x R60dtex/3.Single ninon appears to have been relatively uncommon.
2. A voile fabric of manufactured fibre yarns, typically intended for use as curtains.

nip
The line or area of contact or proximity between two contiguous surfaces that move so as to compress and/or control the velocity of textile material passed between them.

nip padding
See **padding mangle**.

nip roller
One of a pair of rollers intended to run with their cylindrical surfaces in contact or separated only by yarn or other textile material.
Note: The two rollers are intended to have the same surface speed and one normally drives the other by frictional contact, either directly or through the textile material.

noil
The shorter fibres separated from the longer fibres in combing during the preparatory process before spinning.

noil; bourette (silk)
The fibres extracted during silk dressing or those that are too short for producing spun silk. These fibres are usually spun on the condenser system to produce what are known as silk noil yarns.

nominal gauge length (testing)
The length of a specimen under specified pre-tension, measured from nip to nip of the jaws of the holding clamps in their starting position.

non-automatic weaving machine
See **weaving machine**.

non-fibrous matter (quantitative analysis of fibres)
Oils, fats, waxes, dressings, salts and other soluble materials, all or some of which may be present in a fibre mixture, which could interfere with dissolution of fibres in quantitative analysis. This must be removed by **pre-treatment** before analysis.

non-woven (carpets)
A term used in the carpet industry to describe carpets that are not woven. (See also **nonwoven fabric**.)

nonflammable
Not capable of burning with a flame under specified test conditions.

nonionic dye
A dye that does not dissociate electrolytically in aqueous solution.

nonwoven; nonwoven fabric
Opinions vary as to the range of fabrics to be classified as nonwovens. In general, they can be defined as textile structures made directly from fibre rather than yarn. These fabrics are normally made from continuous filaments or from fibre **webs** or **batts** strengthened by bonding using various techniques: these include adhesive bonding, mechanical interlocking by needling or fluid jet entanglement, thermal bonding and stitch bonding. (See also **non-woven (carpets)**.)

nonwoven; nonwoven fabric *(continued)*
The controversial areas are:
(i) wet-laid fabrics, containing wood pulp, in which the boundary with paper is not clear;
(ii) stitch-bonded fabrics which contain some yarn for bonding purposes;
(iii) needled fabrics containing reinforcing fabric.
There are two standard definitions:
1. Nonwovens are defined under ISO 9092:1988 as: 'A manufactured sheet, **web** or **batt** of directionally or randomly orientated fibres, bonded by friction, and/or cohesion and/or adhesion, excluding paper (see *Note 1*) and products which are woven, knitted, tufted, stitch-bonded incorporating binding yarns or filaments, or felted by wet-milling, whether or not additionally needled (see *Note 2*). The fibres may be of natural or manufactured origin. They may be staple or continuous filaments or be formed *in situ* (see *Note 3*).
Note 1: To distinguish wet-laid nonwovens from wet-laid papers, a material shall be regarded as a nonwoven if (i) more than 50% by mass of its fibrous content is made up of fibres (excluding chemically digested vegetable fibres) with a length to diameter ratio greater than 300; or, if the conditions in (i) do not apply, then (ii) if the following conditions are fulfilled: (a) more than 30% by mass of its fibrous content is made up of fibres (excluding chemically digested vegetable fibres) with a length to diameter ratio greater than 300 and (b) its density is less than 0.40 g/cm^3.
Note 2: The commonly used term **needlefelt** has given rise to some confusion since it restrictively associates **needling** with **felting** or felt-like products. In fact, needling (mechanical interlocking of fibres by specially designed needles or barbs) is a major bonding method of nonwovens in its own right and is frequently the only consolidation route for nonwovens ranging from medical/hygienic disposables to spunlaid geotextiles.
Note 3: The appearance of a relatively new group of products such as split-films, extruded meshes and nets, etc., presents a further borderline case between nonwovens and related technologies (in this case, plastics). For the purpose of this International Standard, products shall be regarded as nonwovens if they meet the requirements of the core definition above and where their main structural element can be regarded as fibrous, however derived.
Note 4: Viscose is not considered to be a chemically digested vegetable fibre.
Note 5: Density may be determined using the methods specified in ISO 9073-1 and ISO 9073-2.'
2. Nonwoven fabric is defined under ASTM D 1117-80 as: 'A textile structure produced by bonding or interlocking of fibers, or both, accomplished by mechanical, chemical, thermal, or solvent means and combinations thereof. Discussion: The term does not include paper or fabrics that are woven, knitted, or tufted.'

noosed heald
See **knitted cord heald** under **heald**.

normal fibre (cotton)
See **maturity (cotton)**.

notch; nick
A small cut made on the edge of a garment part which is used as a guide during assembly. Notches may indicate the positions of **darts**, pleats, pockets, seams, style features, zips, gathers and **balance marks**.

novoloid (fibre) (generic name U.S.A.)
A manufactured fibre containing at least 85% by weight of a cross-linked novolac. (See also Classification Table, p.401.)
Note: A novolac is a phenol-formaldehyde condensate of low molecular weight made using an excess of the phenol.

number of yarn
See **count of yarn**.

nun's veiling
A light-weight, clear-finished, plain-weave fabric, usually made of worsted, silk, or cotton yarns and usually dyed black.

nurse cloth
A coloured, woven, plain, matt or twill fabric, principally in stripes of blue and white, having approximately 27x20 threads/cm with 25 tex cotton warp and weft. It is used for overalls and nurses' uniforms.

nutria fibre (hair)
Fibre from the coypu (*Myocastor coypus*).

nylon (fibre) (generic name)
See **polyamide (fibre)**.

nytril (fibre) (generic name U.S.A.)
A manufactured fibre containing at least 85% of a long-chain polymer of 1,1-dicyanoethene (vinylidene dinitrile) where the vinylidene dinitrile content is no less than every other unit in the polymer chain. (See also Classification Table, p.401.)

oakleaf braid
A jacquard woven **narrow fabric** having a conventional oakleaf and acorn design contained within a border, customarily used as a uniform cap-band.

oatmeal crêpe
A weave that creates a rough irregular surface effect by a random arrangement of binding points. It can be produced on a plain base or a satin base, or by reversing small motifs or superimposing weaves. The effect is intended to simulate the surface appearance of a crêpe produced in plain weave using highly twisted S and Z-twist yarns.

OBA
See **fluorescent brightening agent**.

objects (lace)
Ornamental devices (such as flowers) appearing regularly in various parts of a piece of lace.

OE yarn
An abbreviation for open-end spun yarn. This term is principally used for rotor-spun yarns. (See also **open-end spinning** and **rotor spinning**.)

off-grain
1. (Fabric) A general term used to describe fabrics in which the warp and weft, although straight, are not at right angles to each other.
Note: This term is applied to skewed or drawn pieces when lack of information as to the cause prevents the use of the more precise term. (See also **grain**.)
2. (Clothing) Off-grain **pattern** pieces are produced during **marker planning** by intentionally skewing the pattern in relation to the **grain** of the fabric.

oil cloth
A cotton fabric that has been treated on one side with a drying oil to make it impervious to water.

oil-combed top
A wool or hair **top** produced by adding 3% extra oil above the dry-combed top standard to give a total fatty matter content of 3.5% when calculated by the traditional Bradford Method. On the International Wool Textile Organisation's method of testing the maximum fatty matter content becomes 4.6% based on the oven-dry fat-free weight. This method specifies Soxhlet extraction with dichloromethane as the solvent.
Note 1: The standard regain of an oil-combed top is 19% based on the combined weight of oven-dry fat-free wool and the fatty matter.
Note 2: A top that has been dry-combed and subsequently oiled is also described as an oil-combed top.

oil-repellent
Descriptive of textile material on which oil does not spread.

oiled silk; oiled viscose
Silk and viscose fabrics, respectively, made impervious to water by treatment with a drying oil.

oiled wool
Unscoured or undyed knitting wool or wool dyed before spinning and containing added oil not subsequently removed.

olefin (fibre) (generic name U.S.A.)
See **polyolefin (fibre)**.

oligomer
A polymer containing a small number of repeating units.
Note: The oligomer most frequently encountered in the textile industry is the cyclic trimer of ethylene terephthalate, which is the repeating unit of the polymer used for the most important of the **polyester fibres**. This material can form undesirable deposits during the physical processing and dyeing of yarns and fabrics.

ombré
A French term meaning shaded. It is used in relation to textiles: (i) as an adjective to describe fabrics with a dyed, printed, or woven design in which the colour is graduated from light to dark and often into stripes of varying shades; and (ii) as a noun, meaning (a) a shaded design or (b) a fabric with a shaded design.

on-call cotton
Raw cotton purchased under a procedure whereby the price (points on or off futures) is agreed between buyer and seller, but the actual futures price is left to be fixed within a stipulated period. The buyer has the right to 'call' (i.e., demand fixation of the futures price) at any time within the stipulated period.

ondé
A French word meaning waved, used in relation to textile fabrics to describe a wave effect produced by calendering or weaving.

ondine
A thick, cord **bengaline** in which every cord is crinkled.

ondulé
A fabric having a wavy effect in the warp direction produced by an **ondulé reed** that moves up and down during the weaving of a series of picks.

ondulé reed
A special reed that has groups of dents spaced more widely at the top and more closely at the base and *vice versa* alternately, used to produce an **ondulé** fabric.

one-by-one purl fabric, weft-knitted
See **purl fabric, weft-knitted**.

one-by-one rib fabric, weft-knitted
See under **rib fabric, weft-knitted**.

onium dye
A cationic dye that is solubilized by a labile ammonium, sulphonium, phosphonium, or oxonium substituent which splits off during fixation to leave an insoluble colorant in the fibre.

open boil
Scouring of cellulosic textiles with alkaline liquors in open vessels at or near the boiling point (see **scouring**).

open lap (warp knitting)
A lapping movement in which the underlap is non-existent or is made in the same direction as the preceding overlap. This results in a thread entering and leaving a loop at opposite sides without crossing over itself.

open leg
A term to describe the provision for open leg stance in the angle of leg seams in, for example, jodhpurs.

open loop (warp knitting)
A loop in which the same thread enters and leaves the loop at opposite sides without crossing over itself.

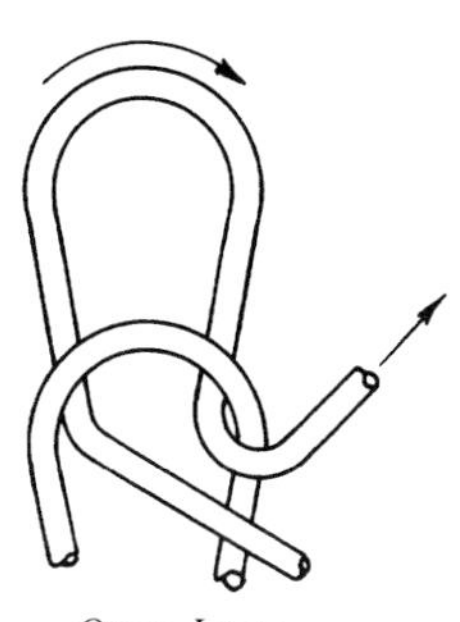

Open Loop

open shed (leno weaving)
The shed formed when the crossing end is lifted over the weft in its non-crossed position when using bottom douping (see **leno weaving**, *Note 4*).

open shedding
A method of forming a **shed** in which, between the insertion of one weft pick and the next, the only warp threads moved are those that are required to change position from the upper to the lower line of the shed, or *vice versa*.

open soaper
A machine consisting of a number of compartments, each having rollers and/or nips, which is

open soaper *(continued)*
used for continuous wet treatment of textiles in open width. By suitable arrangement of the liquids in the series of compartments, a sequence of operations, e.g., fixing, rinsing, soaping and rinsing, can be carried out.

open width washer
A machine for washing fabrics in open width continuously or in batch form.

open-end spinning; break spinning
A spinning system in which sliver feedstock is highly drafted, ideally to individual fibre state, and thus creates an open end or break in the fibre flow. The fibres are subsequently assembled on the end of a rotating yarn and twisted in. Various techniques are available for collecting and twisting the fibres into a yarn, the most noteworthy being **rotor spinning** and **friction spinning**. (See also **spinning**.)

open-width processing
The treatment of fabric at its full width in the unfolded state in contrast to **rope-form processing**. The fabric may be carried on rollers through the processing media or be held on a roller, as in **beam dyeing**.

opening
The action of separating closely packed fibres at an early stage in the processing of raw material.
Note: The separation may involve the removal of non-fibrous material (cleaning).

opening roller
See **beater**.

optical brightening agent; OBA
See **fluorescent brightening agent**.

ordinary heald
See **leno weaving**.

ordinary lay
A laid rope (see **lay** 3) in which the direction of twist in the **roping yarn** and the finished rope are the same, and in which the direction of lay of the **strand** is in the opposite direction.

organdie
A plain-weave transparent fabric of light weight and with a permanently stiff finish.

organic cotton
Cotton grown under conditions prescribed by one of various local or regional organic certification schemes. (See also **transitional cotton**.)

organzine
Silk yarn used as warp for weaving or for knitting, comprising single threads that are first twisted and then folded together two, three, or four-fold, and then twisted in the direction opposite to that of the singles twist.

orientation
1. Parallelism of fibres, usually as a result of a combing or attenuating action on fibre assemblies, causing the fibres to lie substantially parallel to the axis of the web or strand which they constitute.
2. A predominant direction of fibrils and/or linear macromolecules in the fine structure of fibres.
Note 1: In manufactured fibres, orientation is usually parallel to the fibre axis as the result of

extrusion, stretching, or drawing. In natural fibres the predominant direction is determined during growth, for example a helix around the fibre axis in cotton.
Note 2: Unoriented structures are those in which orientation is absent. Disoriented structures are those in which orientation has been reduced or eliminated as a result of a disrupting treatment.

orthocortex
See **cortex**.

osnaburg
Originally, a fabric of plain weave from coarse flax yarns in the province of Hanover, it is now made in cotton, with a coarse weft that may be condenser-spun. Stripes and checks may be introduced and it may be used in the unbleached state. Typical example: 22x14; 37x74tex; K=13.3+12.0; 2 ends in a heald and 2 healds to a dent.

ottoman
A warp-faced fabric showing a bold weftway-rib effect on the face. It was originally made with silk warp and a wool weft.

oven-dry mass
The constant mass obtained by drying textiles at specified temperatures varying according to fibre between 77°C and 110°C.

overedge bed
See under **bed (sewing machine)**.

overfeed fabric, warp-knitted
A fabric produced on a warp-knitting machine in which, generally, one warp is fed faster than would be required to form normally shaped loops. The excess yarn results in large loops and underlaps which appear as surface pile. The fabric may subsequently be brushed and raised or sueded.

overflow-jet dyeing machine
A general term for **soft-flow jet** and partial immersion jet dyeing machines. Their action is characterised by the textile material in rope form being lifted briefly from the dyebath, by a small diameter winch or reel, into an overflow reservoir and then carried along a transportation tube by means of a relatively gentle flow of dye liquor.

overhand
A way of examining textile materials by viewing horizontally at eye-level.

overlap (warp knitting)
Lateral movement of the guide bars on the beard or hook side of the needles. This movement is normally restricted to one needle space.

overlock seam
See under **seam type**.

overlocking
The joining of two or more pieces of fabric by means of a double or treble chain-stitch, which is brought round to join and cover the edges, one or more of which have been cut by knives incorporated in the machine. This operation is performed on overseaming machines.

overnit
See **double piqué, weft-knitted**.

overtufting
See under **tufting machine**.

Oxford
A plain-weave shirting of good-quality yarns that has two warp ends weaving as one. Fancy-weave effects can be incorporated, and dyed yarns are used to form stripes. Typical cotton-type construction: 35x20; 20x30tex; K=15.6+10.9.

oxidized oil staining; gilding
Staining of textiles caused by oxidation of oil acquired or applied during processing.
Note: The presence of oxidized oil may cause discoloration and affect the dyeing property of the material. In the manufacture of woollen and worsted yarns, this discoloration is sometimes referred to as 'yellowing' or 'bronzing'.

oxygen index
See **limiting oxygen index**.

ozone fading
An irreversible change in hue that occurs when dyed or printed textiles are exposed to ozone. Ozone fading is particularly prevalent under severe conditions of atmospheric pollution.

pack (manufactured fibre extrusion)
A replaceable assembly, usually comprising filter media, spreader plates, and one or more spinnerets.

package dyeing
A method of dyeing in which the liquor is circulated radially through a wound package.
Note: Wound packages include slubbing in top form and cheeses or cones of yarn. (See also **beam dyeing**.)

pad
Abbreviated form of **padding mangle** or **padding**.
Note: It is often used in conjunction with other process terms to describe sequential operations in dyeing or finishing, e.g., pad-bake, pad-batch, pad-dry, and pad-steam. It is occasionally used also to describe processes carried out on a padding mangle as opposed to batchwise treatment, e.g., pad-develop.

pad-batch
A sequence of operations involving padding and batching without intermediate drying.

pad-roll
A sequence of operations involving **padding** and wet batching in an atmosphere of steam. (See also **pad-batch**.)

padder
See **padding mangle**.

padding
1. (Finishing) Impregnation of a substrate with a liquor or a paste followed by squeezing, usually by passage through a nip, to leave a specific quantity of liquor or paste on the substrate.
Note: This term is often used as a synonym for 'slop padding'.
2. (Clothing) A method of imparting shape and/or firmness to **collars** and **lapels** in tailored garments by the introduction of successive rows of stitching.
3. (Making-up) Descriptive of the use of **wadding** to create shape in a garment, as in shoulder padding, or to create a relief or contoured surface as in quilting.

padding mangle; padder; foulard
A form of mangle for the impregnation of textiles in open width in which the textile is passed through one or more nips. The textile may be saturated before passing through the nip, as in slop padding, or the impregnating liquid may be carried as a film on the surface of one of the bowls forming the nip and transferred to the textile as it passes through the nip, as in nip padding.

padding thread
See **wadding thread**.

Paisley pattern
A decorative pattern featuring an Indian cone or pine, used on shawls and fabrics.

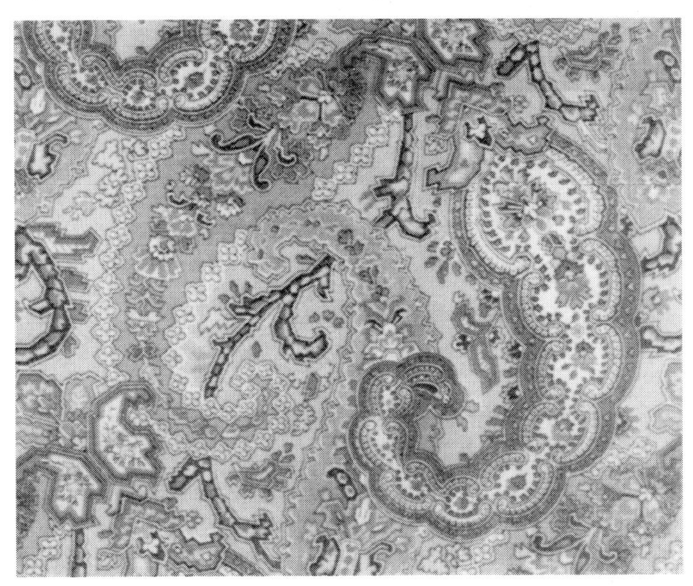

Paisley pattern

palmer finisher; palmer dryer
A single, driven, heated cylinder, usually of large diameter, over the larger part of the peripheral surface of which endless felt is passed, so that fabric being treated passes between the felt and the cylinder and is held in close contact with the latter.
Note: The objective may be to dry fabric or to produce a characteristic finished effect.

PAN (fibre)
See **polyacrylonitrile (fibre)**.

Panama (fabric)
A clear-finished, plain-weave fabric, approximately square construction, typically with a density of 170-240 g/m^2; generally used for men's tropical suitings.

Panama canvas
A canvas of matt weave that is given a **beetled** finish and used for embroidery purposes.

panel wrap
See **embroidery-plated fabric, knitted** under **plated fabric, weft-knitted**.

panne velvet
See under **velvet, woven**.

paper machine clothing; paper-making machine clothing
Fabrics used to convey paper or board during manufacture through the forming, pressing and drying sections of the machine. Woven or needled structures are generally used in specific sections of the machine. (See also **forming fabric**, **press felt** and **dryer fabric**.)

paper yarn
A yarn consisting of one or more continuous lengths of paper strip, or a yarn incorporating one or more continuous lengths of paper strip as a major component.
Note 1: Paper in normal widths is wound into rolls of substantial length, and cut or slit into strips ranging from 0.5mm (0.20 inches) wide upwards. By appropriate treatment (which may include turning-over the edges or the application of adhesives or water or both), strips are twisted sufficiently to make a round-section, tubular form of yarn. Coloured paper may be used.
Note 2: Single paper yarns may be doubled, and one or more twisted with textile yarn(s), around a core yarn.

papering
The insertion of cold or heated board elements (papers) into folds of fabric prior to pressing in a hydraulic press.

paracortex
See **cortex**.

parallel line gratings
Transparent plates containing uniformly spaced parallel lines in the cross-wise direction. It is possible to determine the number of threads per unit length (cm or inch) in a fabric by selecting an appropriate grating and placing it parallel to a set of threads. The number of lines appearing on the grating indicates the difference between the total number of lines on the grating and the total number of threads in the area covered by the grating. By placing a grating at a small angle to a set of threads, irregularities in their spacing can be detected.

parallel shed machine
See **multiphase weaving machine** under **weaving machine**.

parallel-laying
The production of a **web** or **batt** from single or superimposed card webs, laid in the direction of carding, for **nonwoven fabric** manufacture.

paramatta
A fine quality 1/2 twill fabric with worsted weft, used particularly in the making of double-texture rubber-proofed garments.
Note: The term was originally applied to a dress fabric with silk (later cotton) warp, woven in Parramatta, New South Wales.

parchmentizing
A finishing treatment, comprising a short contact with, e.g., sulphuric acid of high concentration, the aim of which is to produce a variety of effects, depending on the type of fabric and the conditions used, ranging from a linen-like handle to a transparent organdie effect. The treatment is applied mainly to cotton. Reagents other than sulphuric acid will also produce the effect.

Paris binding
A **binding**, with firm handle, of twill or herringbone twill weave (usually 3/1) originally made

with silk warp and polished cotton weft, but usually made with mercerized cotton or continuous-filament warp and weft. Often used to reinforce clothing subject to high abrasion.

partially oriented yarn; POY
A continuous-filament yarn made by extruding a synthetic polymer so that a substantial degree of molecular orientation is present in the resulting filaments, but further substantial molecular orientation is possible.
Note 1: The resulting yarn will usually require to be drawn in subsequent processing in order to orient more fully the molecular structure and optimise tensile properties.
Note 2: Yarns of this type made by **high-speed spinning** are commonly used as a feedstock for producing draw-textured yarns (see **draw-texturing**) or for **draw-warping**.

patch pocket
A pocket formed by the application of a piece of material to the surface of a garment.

pattern
A representation of the shapes and sizes of all the component parts to be cut from material necessary to make a garment or other made-up product, including seam and hem allowances. These individual components are referred to as 'pattern pieces'.
Note: Patterns may be constructed from card or other sheet material, or may be held as digital information in a computer database, and are used primarily as a means of controlling the cutting of fabric.

pattern attachment (tufting)
See under **tufting machine**.

pattern blanket; designers' blanket
The product of a system for designing fabrics, especially suitings, whereby a warp consisting of a number of different block stripes is woven in a given pattern and weft in different colours or picking patterns is introduced at intervals. The resulting 'blanket' exhibits a number of combinations of colours in a single construction, some of which constitute a random range. Pattern blankets are used to provide samples for commercial selection.

pattern repeat (lace)
1. (Leavers and raschel lace) The distance motion-way of one complete repeat of the design.
2. (Furnishing lace) The distance motion-way and carriage-way over which motifs and design repeat. This is determined motion-way by the number of motions of the jacquard required to complete one repeat of the design, and carriage-way by the cutting of the cards; the maximum possible repeat is determined by the jacquard tie-up.
Note 1: In the finished lace, the repeat carriage-way is at right angles to, and the repeat motion-way is in line with, the warp and bobbin threads.
Note 2: In certain products, e.g., lace panels, patterned borders are used to surround an area in which there are pattern repeats.

pattern wheel (knitting)
1. A tooth-edged wheel or disk applied to a circular knitting machine for the selection of needles or other loop-forming elements.
2. A wheel composed of sectors of different radii, the circumference of which determines the lateral positions of, for example, the guide bar of a warp-knitting machine or the point bar of a straight-bar knitting machine.

PBI (fibre) (generic name U.S.A.)
A manufactured fibre in which the fibre-forming substance is a long-chain aromatic polymer having reoccurring imidazole groups as an integral part of the polymer chain. PBI is an

PBI (fibre) (generic name U.S.A.) *(continued)*
abbreviation for polybenzimidazole. (See also Classification Table p.401.)

pearls; purls (lace)
Small loops either at the edge of a piece of narrow lace or used as decoration on **brides**.

peau de soie
A French term, meaning literally 'skin of silk' applied originally to a fine silk fabric in a modified satin weave that had a ribbed or grained appearance and was sometimes reversible. The term nowadays includes fabrics made from manufactured fibre yarns. It is recommended that in such contexts the name of the fibre should be indicated.

pebble
A term often used for the characteristic appearance of a **crêpe fabric**.

pegging plan
Synonymous with **lifting plan** for dobbies where pegs in lags control selection.

Pekin
A design in which wide stripes of equal width are woven in different colours or weaves.

pelerine (knitting)
The effect produced by transferring sinker loops (see **eyelet (weft knitting)** and **point (pelerine)** under **point (knitting)**).

percale
A closely woven plain-weave fabric, usually of Egyptian cotton, of lighter weight than **chintz**. A percale may be glazed or unglazed.

percentage cover
cover factor as a percentage of the maximum possible for a particular weave structure.

percentage maturity; percentage mature fibres
A method of expressing, numerically, the maturity of a sample of cotton fibres. It is the percentage of fibres in the sample whose walls have developed to greater than a specified degree. (See also **maturity ratio**.)

percentage moisture content
See **moisture content**.

perch
1. A manually or mechanically operated machine consisting of a system of rollers over which fabric is drawn at open width for the purpose of inspection.
2. To inspect fabric in a vertical (hanging) position or at an angle inclined upwards away from the source of light.
Note: The inclined position on a manual perch is obtained by holding the fabric forward when required. On a mechanical perch the angle is fixed by a low front roller. The purpose of perching is to inspect the product at different stages of manufacture and processing.

perforated weave
An open mesh character of **mock-leno** fabric created primarily by the weave. Two examples and the corresponding weaves are given below. Light-weight open texture structures are sometimes referred to as 'imitation gauze' or 'mock gauze'.
Note: Arrows on the weave diagrams show where the spaces will develop, because at these

places the interlacings completely reverse. Elsewhere, the interlacings are such that the threads crowd together. The effect may be emphasised by leaving one or more dents empty and varying the rate of take-up.

Perforated mock-leno fabrics

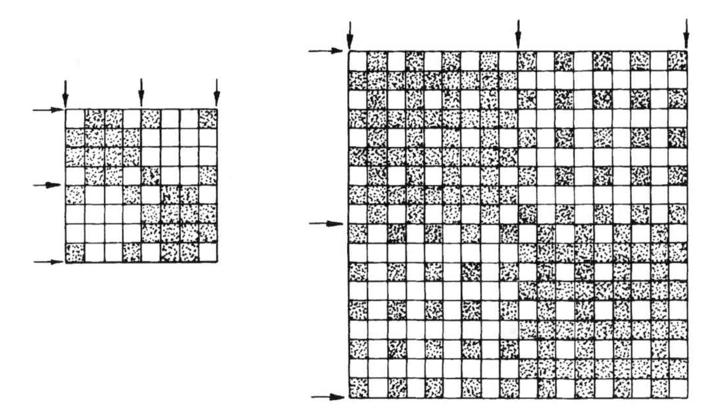

Perforated mock-leno weaves

permanent deformation; permanent set
The net long-term change in dimensions of a specimen after deformation and relaxation and completion of creep recovery.

permanent press
See **durable press**.

permanent set
1. See **setting**.
2. See **permenant deformation**.

permanent shading (carpet)
See **shading (carpet)**.

perry
See **reach**.

petersham ribbon (millinery)
Ribbon in plain weave, originally used on ladies' hats, which usually has a continuous-filament

petersham ribbon (millinery) *(continued)*
warp, typically with 10-12 picks per cm of cotton or spun viscose, giving it a pronounced rib. The edge is formed by the turn of the weft.

petersham ribbon (skirt)
A **narrow fabric** having a pronounced rib, usually with 9-12 picks per cm and having lateral stiffness produced either by the high density of the weave or by a finishing process. In former times, some were woven with pockets for whale bone or plastic strips to give added lateral stiffness. In contrast to **petersham ribbon (millinery)**, it has an edge of contrasting weave.

pfleidering
The process of shredding pressed alkali-cellulose in a machine named a Plfeiderer, after its inventor.

pH
A value taken to represent the acidity or alkalinity of an aqueous solution, and defined as the logarithm of the reciprocal of the hydrogen-ion concentration of a solution

$$pH = \log^{10} \frac{1}{\left[H^{+}\right]}$$

It is expressed numerically on a scale from 0 to 14, neutral solutions having a pH of 7, acidic between 0 and 7, and alkaline between 7 and 14.

phormium
A fibre obtained from the leaves of the New Zealand plant *Phormium tenax*. It is sometimes called New Zealand flax or hemp, although now grown in other countries.

photodegradation
Degradation caused by the absorption of light or other radiation and by consequent chemical reactions. Ultra-violet radiation is an especially potent cause.

pick; shot
1. A single operation of the weft-inserting mechanism in weaving.
2. A single weft thread in a fabric as woven.
Note: A single picking operation in weaving may insert more than one pick (i.e., weft thread) in the fabric.
3. To pass the weft through the warp **shed** in weaving.

pick, dead; crammed pick
A pick on which the take-up motion is put out of action by a retarding or cramming motion.

pick bar
See under **bar (woven fabric)**.

pick-and-pick
1. Descriptive of a woven fabric in which alternate picks are of different colours or yarns.
Note: If the weft is inserted by shuttles, the fabric must be produced on a **pick-at-will** loom.
2. Descriptive of a shuttle loom that picks alternately from opposite sides.

pick-at-will
Descriptive of a shuttle loom with mechanisms which can insert picks from either side in any sequence.

pick-found
Descriptive of a fabric that, ideally, contains no missed or broken picks.

pick-up
See **wet pick-up**.

picker
The part of the picking mechanism of a loom that actually strikes the **shuttle**.

picker mechanism (weft knitting)
A mechanism which raises or lowers needles into or out of action during the knitting of heel and toe pouches on circular hosiery machines.

picker point (warp knitting)
See **point (milanese machines)** under **point (knitting)**.

picking
1. The operation of passing the weft through the warp **shed** during weaving.
2. The rectification of the face and back of a carpet after manufacture including inserting missing tufts, replacing faulty ones, and repairing broken yarns in the backing. Also known as 'mending'.
3. A process preferably carried out before the final stage of fabric finishing, which involves removing by hand any contamination (such as kemp, wrong fibre, coloured hair, etc.) which has not been removed by previous processing. This process is carried out particularly during the finishing of suitings, face-finished fabrics, and cream or off-white fabrics.
4. See **scutching (cotton)**.

picking-out
See **unweaving**.

picklock wool
A term used in wool-sorting, mainly in the U.K., for second-best sorts from fleeces.

picot
A decorative feature at one or both edges of a **narrow fabric** consisting of a series of loops produced either by: (i) extending the weft beyond the **selvedge**; or (ii) deflecting, usually by means of a wire, a cord which weaves as an **end** concealed within a tubular selvedge.

picotage
A speckled effect on the surface of a pile fabric owing to differential light reflection from deformed tips of tufts.

piece; bolt
A length of fabric of customarily accepted unit length.
Note: A frequent contract practice is for the purchaser to specify a minimum piece length below which no pieces will be accepted. Alternatively, a 'cut-through' allowance is specified, which the seller has to make in the case of all pieces less than the specified figure. The reason for such practices is the greater liability to waste in cutting out from short-length pieces than from standard-length pieces. The term 'piece' is applied at all stages of fabric manufacture, and although often qualified, e.g., grey piece, or loomstate piece, the qualification is generally understood in commercial practice.

piece (flax)
The small handful that is the unit of scutched flax.

piece(s) (narrow fabrics)
A term used when describing the number of narrow fabrics being produced simultaneously in a machine.

piece dyeing
Dyeing in fabric form.

piece glass
See **counting glass**.

piece-end
See **end-fent**.

piece-goods
1. Fabric sold by or from the **piece**.
2. (Weft knitting) Fabric knitted in a continuous piece with no indication of a garment sequence. It is later converted into a garment.

pieces
Small bunches of wool fibres taken during sorting from various fleeces.

piecing; piecening
The joining of fibre assemblies, usually by overlapping two ends.
Note: This is usually performed at machines used in yarn manufacture up to and including the spinning process, and may be done manually or automatically. Sometimes the ends of the fibre assemblies are tapered. (See also **splicing (yarn joining)**.)

pierced cocoons
Cocoons from which the moths have been allowed to emerge so that they may reproduce.

pigment
A substance in particulate form that is substantially insoluble in a medium, but which can be mechanically dispersed in this medium to modify its colour and/or light-scattering properties.

pigment padding
The application of an aqueous dispersion of a pigment to a fabric by **padding**.
Note: It is commonly used to describe the first stage of a process for the application of vat dyes to fabrics, followed by fixation of the vat dye through its leuco form. It is also used in the application of resin-bonded pigments.

pigtail
A yarn-guide in the form of a short open-ended helix, sometimes used as a **ballooning eye**.

pile
A surface effect on a fabric formed by tufts or loops of yarn that stand up from the body of the fabric.
Note: **nap** and pile are often used synonymously, but the practice of using the two terms for different concepts is to be encouraged as providing a means of differentiation and avoidance of confusion.

pile (carpet)
That part of a carpet consisting of textile yarns or fibres, cut or looped, projecting from the substrate and acting as the use-surface.

carved pile
The pile of a carpet that is subjected after manufacture to a shearing operation with the object of creating different levels of pile, often on the periphery of certain elements of design formed by the pile.

curled pile; hard-twist pile; frisé pile; Wilton, plain, hard-twist
The pile of a carpet, in which curl has been induced by over-twist or by other means.

cut pile
The pile of a carpet consisting of legs of tufts or individual fibres.

cut-loop pile
The pile of a carpet, formed during manufacture by loops and tufts of different lengths or of the same length.

loop pile; uncut pile
The pile of a carpet consisting of loops.

sculptured pile
A pile in which a pattern is created by having areas of different lengths of pile and/or by omitting pile in certain areas. (See also **Wilton, carved**.)

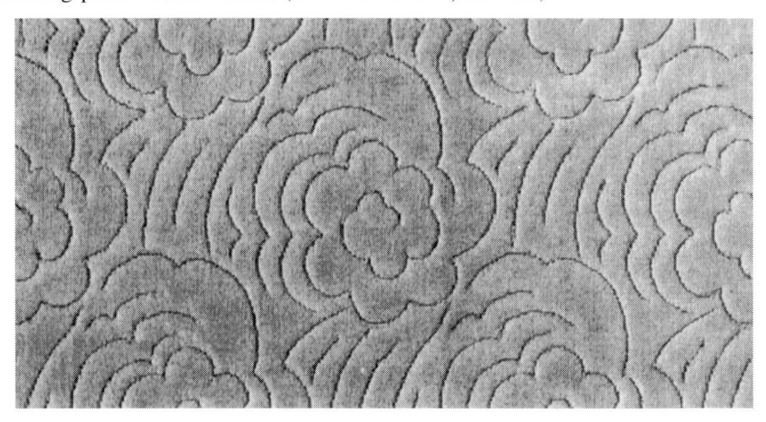

Sculptured pile

shag pile
A carpet texture characterized by long pile tufts laid over in random directions in such a manner that the sides of the yarn form the use surface.
Note: Modern shags are usually made from plied heat-set yarns and are either cut-pile or cut-and-loop styles.

textured pile
A pile in which the surface character is varied, e.g., by having areas of different characteristics or by combinations of different yarn or pile types, e.g., soft and hard twist.

tip-sheared pile
The pile of a carpet, originally consisting of loops of different lengths, which has been subjected after manufacture to a shearing process to cut the tips of the longer pile loops.

pile, effective (carpet)
That part of the pile which is above the **substrate**.

pile density
Pile mass per unit area relative to pile height. It is normally expressed as pile mass in g/m^2 divided by pile height in mm.

pile fabric
A fabric with a pile surface, which may be of cut and/or uncut loops (see **pile**).

pile length, effective (carpet); pile height, effective (carpet)
The length of fibre or of one leg of a tuft from the place where it emerges from the **substrate** to its furthest extremity, or half the length of a loop measured between the two points where it emerges from the substrate.

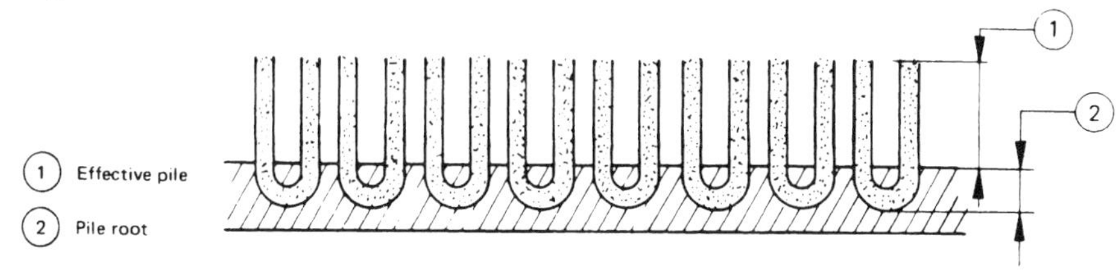

Pile root (longitudinal section)
The above diagram is derived from figure 27 of BS 5557

pile mass; pile weight
The mass of pile yarn or fibre in a unit area of pile textile fabric or floorcovering. It may be expressed as total pile weight, where the mass of yarn or fibre in the base fabric is included, or surface pile weight, where only the pile mass above the base fabric is taken into account.

pile root (carpet)
That part of the tuft and/or loop, excluding dead yarns (see **dead yarn (carpet)**), which is in the substrate of a carpet.

pile ruche
See under **ruche**.

pile thickness, effective (carpet)
The difference in the thickness of a carpet, measured under a defined pressure, before and after the pile above the backing has been shorn away.

pill
See **pilling**.

pillar (lace)
Two or more threads from warp, beam, or spool, encircled and bound by one bobbin thread.

pillar stitch, warp-knitted
A stitch which produces a vertical chain of consecutively knitted loops on the same needle from the same yarn. The chains may be connected together by other yarns or they may be entirely separate from each other.

pilling
The entangling of fibres during washing, dry cleaning, testing or in wear to form balls or pills which stand proud of the surface of a fabric and which are of such density that light will not pass through them (so that they cast a shadow).

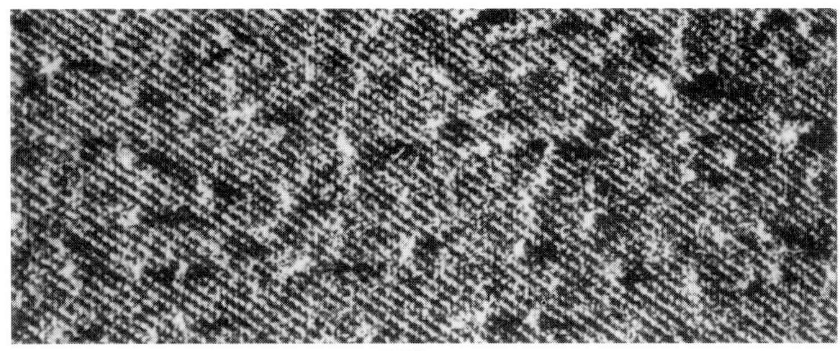

Pilling

pilling test
A test to assess the propensity of pills to form and/or be retained on the surface of a fabric when it is subjected to specified conditions. (See also **pilling**.)

pillow lace
See **bobbin lace**.

pilot
A woollen fabric, generally made in navy blue and used for seamen's coats. It is usually in 2/2 twill weave, heavily milled, with a raised brush finish.

pin drafter
See **gill box**.

pin drafting
See **gilling**.

pin holes (knitting)
Very small holes usually caused by loop distortion.

pin-twisting
The generation of false-twist (see **false-twisting**) by a device in which a yarn is wrapped around a small pin or peg of wear-resistant material mounted across a rotating tube through which the yarn passes. One rotation of the tube generates one turn of twist. (See also **friction-twisting**.)
Note: Twist pins are usually made of sapphire or of a ceramic material and in some designs more than one pin has been used.

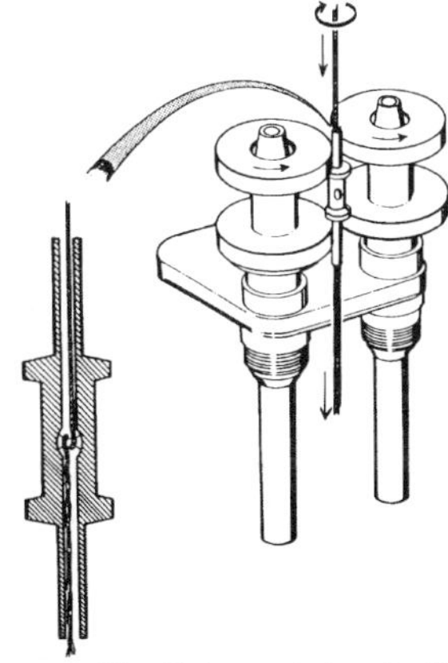

Pin (false-twist) spindle

pinching
See under **gathering**.

pineapple cone
A low-angled yarn package in which the traverse length decreases as the diameter increases. It is used mainly for continuous-filament yarns in order to produce a more stable package.

pineapple fibre
A fibre from the leaf of the plant *Ananas comosus*, capable of being processed into fine fabrics.

pintuck
A narrow fold secured by **top stitching** and generally used decoratively.

piping
A braided, woven or knitted **narrow fabric** comprising a flat flange or sewing foot to one edge of which is attached a **cord** that remains visible when the flange is sewn into a seam. Piping may also be fabricated from **bias binding**. (See also **insertion braid**.)

piping cord
A cord, usually of twisted cotton, used as the **core** of a piped seam.

piqué, warp-knitted
A fabric, normally made with two guide bars, that shows pronounced cord effects in the warp direction. The portions between the cords are made by omitting one or more threads from the guide bar that is making the smaller underlap.

piqué, weft-knitted
See **single piqué**, **double piqué** and **texipiqué**.

piqué, woven
A plain-weave fabric with sunken lines between rounded cords and having a plain-weave face. When the cords are in straight lines across the fabric the structure is known as **welt** and when they create waved lines it is known also as waffle piqué or **marcella**.

piquette (weft knitting)
A non-jacquard double-jersey fabric made on an interlock basis consisting of a selection of knitted and float loops in the following sequence. (See also **double jersey, weft-knitted**.)

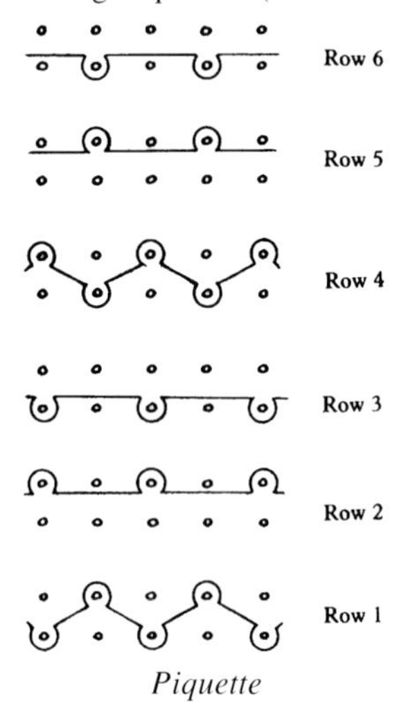

Piquette

pirn; quill; weft bobbin
1. A support, slightly tapered, with or without a conical base, on which yarn is spun or wound for use as weft.
2. The weft package wound on the support defined above.
3. A relatively long but narrow package of yarn taken up on a cylindrical former during the draw-twisting of continuous-filament yarns.

pitch
1. A term often used as a synonym of **gauge**.
2. A term often used as a synonym of **sett**.
3. Traditionally the number of knots per square inch in the pile of hand-made carpets.
4. The distance between jack centres on a dobby, or needle centres on a jacquard.

placket; plaquet
An opening in a garment or an extra piece of material applied to that opening for reinforcement or as a style feature. It may incorporate fastenings.

plaid
1. See **tartan**.
2. The outer article of Highland costume usually having a chequered or tartan pattern.

plain back
A twill-face, **plain-weave** back structure made from single worsted yarns.

plain fabric, weft-knitted
A fabric consisting wholly of knitted loops which are all meshed in the same direction.
Note: The fabric may also be described as single jersey, plain web or stockinette. The appearance may be described with reference to the surface of the structure:
(i) face; technical face: the surface of a plain weft-knitted fabric that consists wholly of face loops;
(ii) back; technical back: the surface of a plain weft-knitted fabric that consists wholly of back loops;
(iii) effect side: the surface of the fabric intended to be used outermost on a garment or other construction; and
(iv) reverse side: the surface opposite to the effect side.

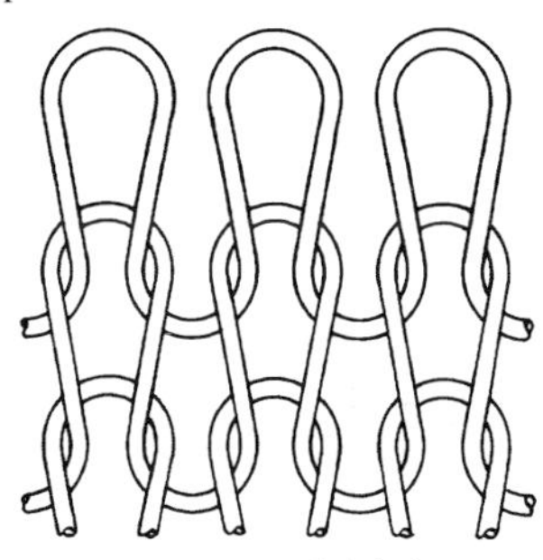

Plain knitted fabric

plain loop
See **face loop** under **knitted loop (weft knitting)**.

plain net
A twist lace fabric made with equal numbers of warp and bobbin threads. The warp threads run lengthwise in the fabric. The bobbin threads twist round the warp threads and traverse

plain net *(continued)*
diagonally in the fabric. Equal numbers of bobbin threads are always traversing in opposite directions. A fine plain net made of silk is sometimes described as **tulle**.

plain net machine
See under **lace machines**.

plain seam
See under **seam type**.

plain shed (leno weaving)
The shed formed when the standard end is lifted when using bottom douping (see **leno weaving**, *Note 3*).

plain-weave
See **weave, plain**.

plait
1. The intersection of the strands of a **braid**.
2. See **braid**.

plait pitch
The longitudinal distance along the axis of a **braid** from a point on a plait to the corresponding point on the next plait along.

plaiting
1. See **braiding**.
2. The arrangement of fabric in width-way folds. (See also **cuttle**.)

plaits
The products of the **braiding** process, e.g., mending plaits.

planking; bumping (hat manufacture)
Acid milling of settled **forms** by the combined action of hot sulphuric acid and mechanical treatment to produce a cone-shaped felt known as a **hood** or **body**.

planting (carpet)
The inclusion of additional colours in any frame of solid colour of yarn.

plasma treatment
An electrical discharge treatment carried out under vacuum pressure which in textiles is used to modify the surfaces of fibres.

plated fabric, weft-knitted
A fabric knitted from two yarns of different properties, both of which are used in the same loop whilst positioned one behind the other. The special feature of the fabric is that each loop exhibits the characteristics of one yarn on the face side and the characteristics of the other yarn on the reverse side. In plain weft-knitted fabrics, where the loops are all formed in the same direction, the characteristics of one yarn are visible on the surface composed of the face loops whilst the characteristics of the other are only visible on the reverse surface composed of the back loops. Types of plated fabric are:

> **cross-plated fabric, knitted**
> A fabric constructed by the feeding of two yarns to one set of needles so that one plates on the face and the other on the reverse side of the stitches and reversing the plating positions of the two yarns at predetermined course intervals.

embroidery-plated fabric, knitted
A fabric constructed by the superimposing of one or more additional ends of yarn over a restricted but variable width of weft-knitted fabric, so that the loops formed by the additional yarns appear on the face of the fabric. 'Panel wrap' and 'wrap stripe' are examples of embroidery plating produced in this way.

float-plated fabric, knitted
A fabric constructed by the knitting of a plated fabric in which the face yarn is floated at certain needles to allow the other yarn to appear on the face.

reverse-plated fabric, knitted
A plated knit construction in which the relationship of the two yarns is reversed in certain stitches within the same course to give a patterned or fancy effect.

sandwich-plated fabric, knitted
A plated construction usually knitted on one set of needles in which the ground yarn is positioned (sandwiched) within the structure so that it is not visible from either the face or the reverse side of the fabric. (See also **polar fleece**.)

Plain plated knitted fabric

pleated fabric, warp-knitted
A fabric produced from two or three guide bars in which the front warp is stopped while the front bar **miss-laps**. The fabric produced by the back bar (or back and middle bars) while the front bar miss-laps is raised out of the plane of the fabric in the form of a pleat extending across the complete width. All bars are full-set threaded.

pleats
Regular folds in material, most commonly sharp, having one of a variety of configurations. The folds usually run in the lengthways direction of a garment giving a decorative effect and allowing for expansion of the garment in the cross-section.
Note 1: True pleats comprise two opposing folds flattened to form three layers of material. The top layer is visible in the garment; the two lower layers are the underlays. Other formations known as pleats (such as **accordion pleats** and **sunray pleats**, see below) are simply crease patterns in which the entire upper surface of the material is visible.
Note 2: If the pleat is not sharp, it is specified to be unpressed.

pleats *(continued)*

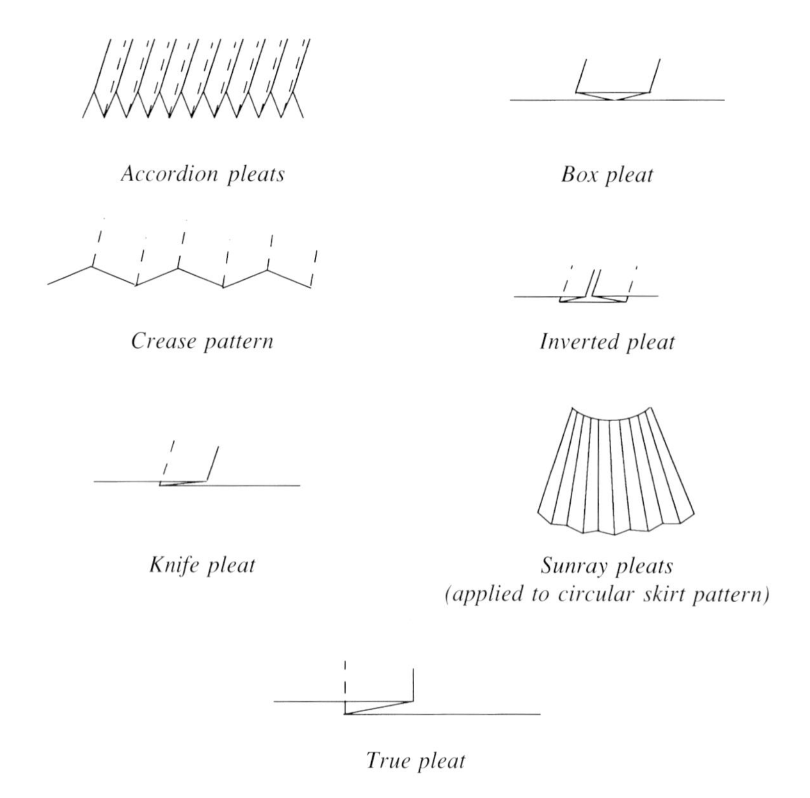

accordion pleats
Narrow, regularly spaced straight creases, usually parallel, and capable of being closed up against each other.

box pleat
Two opposed pleats meeting on the underside.

inverted pleat
Two opposed pleats meeting on the face of the material.

knife pleat
A single pleat with sharp folds.

sunray pleats
A radial crease pattern used in circular, half-circle or quarter-circle skirts.

plied yarn
An alternative to the term **folded yarn**.

plissé
A French term meaning pleated, applied to fabrics with a puckered or crinkled effect. (See also **seersucker.**)

pluckings
Short, clean fibre produced at the end of a scutching machine where operatives dress and square **pieces** of flax ready for selection. In grading, pluckings are classed as **tow**.

plush, double, warp-knitted
A pile fabric produced on a double needle bar **raschel warp-knitting machine**, by knitting separate ground fabrics on each needle bar and connecting them by pile threads which knit on both needle bars so that two fabrics are produced face to face. The fabrics are then separated to produce two cut pile fabrics.

plush, warp-knitted
A fabric in which one series of threads form pile loops standing at approximately 90° to the fabric plane, being connected to the ground construction by knitting-in or laying-in. The pile loop may be cut or uncut. (See also **plush, double, warp-knitted**.)

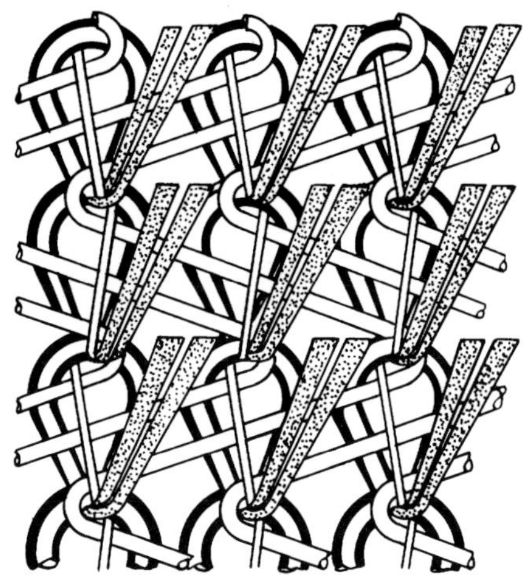

Warp-knitted plush

plush, weft-knitted
A knitted fabric in which two yarns are fed to the needles in a plating relationship where all or some of the sinker loops of one yarn are elongated to form plush loops on the technical back. It is sometimes known as 'knitted terry'.

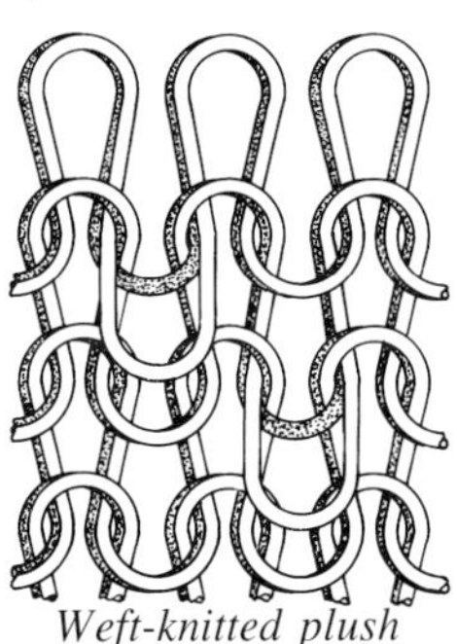

Weft-knitted plush

plush, woven
A pile fabric, with a longer and less dense pile than that of velvet (see **velvet, woven**).

ply (fabric)
1. One component or layer of a **compound fabric**.
2. (Making-up) A single layer or thickness of material. The number of plies in an assembly is the number of material layers.
Note 1: Plies of material of identical length are placed one on top of the other in preparation for

ply (fabric) *(continued)*
cutting.
Note 2: Seams are constructed from two or more plies of material.

ply (tyres)
See under **tyre textiles**.

ply (yarn)
See **fold (yarn)**.

pneumatic structure (technical textiles)
A fabric structure constructed of small tubes containing high-pressure air at, for example, ten times atmospheric pressure. The tubes may support fabric forming an enclosure. (See also **air-supported structure**.)

pocket
A bag, pouch or similar container formed by the insertion or application of material to a garment. Pockets may have style features such as **flaps**, **jettings** or **welts**. (See also **patch pocket**.)
Note: Garments may have style features which have the appearance of pockets but have no functional use.

poile; poil
A raw silk yarn made of eight to ten singles twisted together. Usually silk of inferior quality is employed. Used as the core of tinsel yarn and some pile and flat silk fabrics.

point (knitting)

point (filling-in-point)
An instrument used on a straight-bar knitting machine for picking up a loop from the previous course and placing it on a needle left empty within the knitting width by loop transference.

point (linking or point-seaming machines)
One of a set of collectively controlled grooved instruments on which fabric is placed for linking or point-seaming. (See also **gauge (linking)**.)

point (milanese machines)
An instrument used on the English-type milanese machine to help in performing the functions of conventional guides. According to the work they do, they are classed as (i) still points, (ii) traverse points, or (iii) picker points.

point (pelerine)
An instrument for the collection and transference of sinker loops.

point (raschel machines)
An instrument used in the manufacture of plush and pile fabrics on raschel machines.

point (rib transfer)
1. One of a set of grooved instruments, placed in a straight bar, that is used to facilitate running-on and loop-transferring operations in the knitting of fully fashioned goods.
2. A component part of a quill ring, on which ribs are placed in the process of transference to an open-top circular-knitting machine.

point (straight-bar and flat-knitting machines)
An instrument used for transferring loops from one needle to another. A half-point is used to spread a loop across two needles.

point (lace machines)
An alternative term for **gauge**.

point bar (lace machines)
A bar to which **point leads** are attached across the lace-making width of the machine. There are two point bars, one at the front and one at the back of the machine. Each bar moves alternately so that its points enter between the warp and pattern threads and the bobbin threads just above the carriages, take up the crossed threads and then hold the lace just below the facing bar while the points of the other bar perform a similar action. Their action completes the formation of the lace and resembles the beating-up action of the reed in weaving.
Note: In the bobbinet and furnishing machines, a shogging motion is given to the points and this completes the other motions of the machine in forming the lace.

point bonding
A method of making **thermally-bonded nonwoven fabric**, in which heat and pressure are applied to specific areas by the use of embossed calender rollers so as to cause local bonding.

point lead (lace machines)
A number of points (see **gauge (lace machines)**) cast to the gauge of the machine in a lead-alloy or other base.

point paper
See **design paper**.

point-paper design (woven fabrics)
The representation on **design paper** (point paper) of the order of interlacing the threads in a fabric.

points (lace machines)
A series of thin steel pins, usually tapered, one to each comb space, used to take up the crossed threads of the lace (see **point bar (lace machines)**).

polar fleece, weft-knitted
A plush-plated fabric where the plush yarn is caused to sandwich plate on the surface of both the face and the reverse of the fabric. (See also **plated fabric, weft-knitted**.)

polishing
The treatment of tanned skins, or of fabrics, particularly pile fabrics, to increase lustre by mechanical means, without compressing the material.

polishing (yarn)
Operation(s) for conferring on yarns a relatively high degree of smoothness of surface.
Note 1: It is usually done by sequential yarn processes as follows: (i) single-end yarn sizing; (ii) drying; (iii) frictional treatment by passage of yarn over contacts with stationary and/or revolving burnishers and/or brushes at predetermined temperatures. Drying and burnishing may be concurrent.
Note 2: The application of a size coating to a yarn promotes smoothness of surface, and in fibrous yarns helps to lay protruding fibres in one direction.
Note 3: A typical objective of polishing is to reduce friction between yarn and machine, e.g., during the passage of sewing yarns through eyes of needles in power-sewing machines.

polka gauze
A cotton gauze fabric ornamented by small spots introduced in **swivel weaving**.

polka rib
See **full cardigan rib**.

polyacrylonitrile (fibre); PAN
See **acrylic (fibre)**.

polyamide
A synthetic linear polymer in which the linkage of the simple chemical compound or compounds used in its production takes place through the formation of amide groups, e.g.,

$$[\text{—NH—R—CO—NH—R—CO—}]_n \text{ or}$$
$$[\text{—NH—R}_1\text{—NH—CO—R}_2\text{—CO—}]_n,$$

where R, R_1, and R_2 are generally, but not necessarily, linear divalent hydrocarbon chains $(\text{—CH}_2\text{—})_m$.
Polyamides are distinguished from one another by quoting the number of carbon atoms in the reactant molecule or molecules. For polyamide derived from an amino-acid or lactam, this is a single number. In the case of a polyamide made from a diamine and a dicarboxylic acid, the number of carbon atoms in the former is given first followed by a punctuation mark and then the number of carbon atoms in the latter.

$$\text{hexanolactam } (\varepsilon - \text{caprolactam}) \rightarrow \left[-\text{NH}-\left(\text{CH}_2\right)_5-\text{CO}-\right]_n \text{(nylon 6)}$$

$$\text{1,6} - \text{diaminohexane} + \text{hexanedioic acid (adipic acid)} \rightarrow$$
$$\left[-\text{NH}-\left(\text{CH}_2\right)_6-\text{NH}-\text{CO}-\left(\text{CH}_2\right)_4-\text{CO}-\right]_n \text{(nylon 6,6)}$$

$$\text{1,6} - \text{diaminohexane} + \text{decanedioic acid} \rightarrow$$
$$\left[-\text{NH}-\left(\text{CH}_2\right)_6-\text{NH}-\text{CO}-\left(\text{CH}_2\right)_8-\text{CO}-\right]_n \text{(nylon 6,10)}$$

(See also Classification Table, p.401.)

polyamide (fibre) (generic name); nylon (fibre) (generic name)
A manufactured fibre composed of synthetic linear macromolecules having in the chain recurring amide groups, at least 85% of which are attached to aliphatic or cyclo-aliphatic groups. (See also Classification Table, p.401.)
Note 1: This limited definition was introduced by ISO in 1977 as a consequence of the creation of a separate class for **aramid** fibres.
Note 2: The term polyamide is not used generically in the U.S.A.

polyamide, natural (fibre)
Natural fibres consisting of polymers containing the repeating group —CO—NH—. Examples are silk, wool, and other animal hairs.

polycaproamide (fibre)
See **polyamide (fibre)**.

polycarbamide (fibre); polyurea (fibre)
A manufactured fibre composed of synthetic linear macromolecules having in the chain recurring aliphatic groups joined by ureylene groups which together comprise at least 85% (by mass) of the chain.
Note: This name was formerly an ISO **generic name**. (See also Classification Table p.401.)

polyester
A polymer whose repeating units contain ester linkages in the main chains of the macromolecules.
Note: Cross-linkable polyesters are resin-forming and linear polyesters are fibre-forming (see **polyester (fibre)**).

polyester (fibre) (generic name)
A manufactured fibre composed of synthetic linear macromolecules having in the chain at least 85% (by mass) of an ester of a diol and benzene-1,4-dicarboxylic acid (terephthalic acid). (See also Classification Table, p.401.)
Note 1: This term is more restrictive than the chemical definition of **polyester**.
Note 2: In the U.S.A., the **generic name** is more broadly defined to encompass the use of aromatic dicarboxylic acids other than benzene -1,4-dicarboxylic acid and also to include certain aromatic polyetherester fibres.

polyethylene (fibre) (generic name)
A manufactured fibre composed of synthetic linear macromolecules of unsubstituted aliphatic saturated hydrocarbon. (See also Classification Table, p.401.)

polymer
A large molecule built up by the repetition of small chemical units.

polymer, addition
A polymer made by addition polymerization (see **polymerization, addition** under **polymerization**).

polymer, atactic
A linear polymer containing asymmetrically substituted carbon atoms in the repeating unit in the main chain, a planar projection of whose structure has the same substituents situated randomly to one side or the other of the main chain, e.g.,

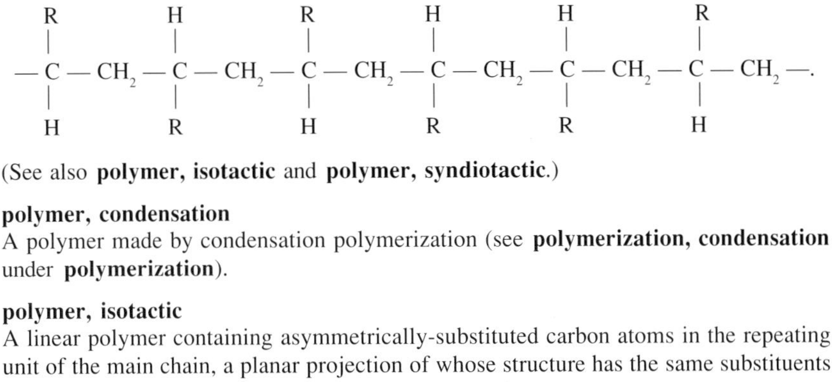

$$\begin{array}{ccccccccccccc} & R & & H & & R & & H & & H & & R & \\ & | & & | & & | & & | & & | & & | & \\ - & C & -CH_2- & C & -CH_2- & C & -CH_2- & C & -CH_2- & C & -CH_2- & C & -CH_2-. \\ & | & & | & & | & & | & & | & & | & \\ & H & & R & & H & & R & & R & & H & \end{array}$$

(See also **polymer, isotactic** and **polymer, syndiotactic**.)

polymer, condensation
A polymer made by condensation polymerization (see **polymerization, condensation** under **polymerization**).

polymer, isotactic
A linear polymer containing asymmetrically-substituted carbon atoms in the repeating unit of the main chain, a planar projection of whose structure has the same substituents situated uniformly on the same side of the main chain, e.g.,

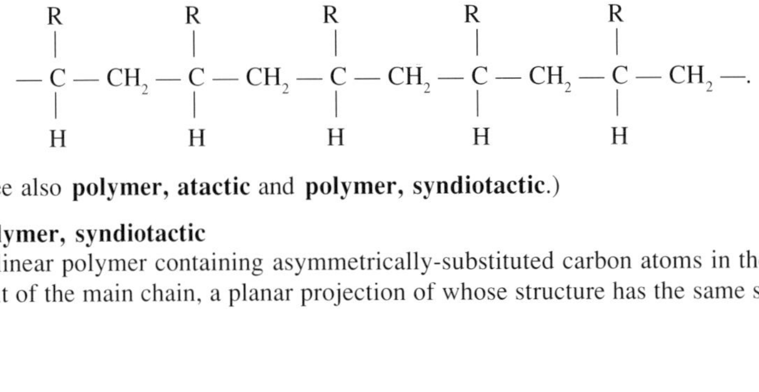

$$\begin{array}{ccccccccccc} & R & & R & & R & & R & & R & \\ & | & & | & & | & & | & & | & \\ - & C & -CH_2- & C & -CH_2- & C & -CH_2- & C & -CH_2- & C & -CH_2-. \\ & | & & | & & | & & | & & | & \\ & H & & H & & H & & H & & H & \end{array}$$

(See also **polymer, atactic** and **polymer, syndiotactic**.)

polymer, syndiotactic
A linear polymer containing asymmetrically-substituted carbon atoms in the repeating unit of the main chain, a planar projection of whose structure has the same substituents

polymer *(continued)*

situated alternately on either side of the main chain, e.g.,

$$
\begin{array}{ccccccccccccc}
 & H & & R & & H & & R & & H & & R & \\
 & | & & | & & | & & | & & | & & | & \\
— & C & —CH_2— & C & —CH_2— & C & —CH_2— & C & —CH_2— & C & —CH_2— & C & —CH_2—. \\
 & | & & | & & | & & | & & | & & | & \\
 & R & & H & & R & & H & & R & & H &
\end{array}
$$

(See also **polymer, atactic** and **polymer, isotactic**.)

polymer tape

A tape of synthetic polymer in unfibrillated form, that may be used as produced, or converted into a **fibrillated-film yarn**.

Note 1: Methods of production include the extrusion of flat narrow strips and the slitting of sheets or tubes of polymer. In either case, a hot-stretching process is usually included at some point to confer orientation and thus high longitudinal strength.

Note 2: The specific polymer used should be stated (e.g., polyolefin-tape yarns). It is customary in the industry to use, for example, 'polypropylene tapes' and 'polyethylene tapes' as complete terms.

polymerization

The process used to link small chemical molecules into a **polymer**.

polymerization, addition

The formation of a polymer by reaction of a compound or compounds without the formation of other reaction products of low molar mass, e.g.,

$$nCH_2 = CHCl \rightarrow \left[-CH_2 - CHCl-\right]_n.$$

polymerization, batch

A process for making polymer in batches.

polymerization, condensation

The formation of a polymer by reaction of one or more compounds with elimination of reaction products of low molar mass such as water, e.g.,

$$nH_2N(CH_2)_6NH_2 + nHOOC(CH_2)_4COOH \rightarrow$$

$$H\left[-HN(CH_2)_6NHCO(CH_2)_4CO-\right]_nOH + (2n-1)H_2O.$$

polymerization, continuous

A process for making polymer in which the reactants are fed continuously to, and the product is withdrawn continuously from, a vessel or series of vessels.

polynosic (fibre)

A type of regenerated cellulose fibre characterized by a high initial wet modulus of elasticity and a relatively low degree of swelling in sodium hydroxide solution.

Note: Such fibres fall within the ISO definition of **modal**. (See also Classification Table, p.401.)

polyolefin (fibre); olefin (fibre) (generic name U.S.A.)

A manufactured fibre in which the fibre-forming substance is any long-chain synthetic polymer composed of at least 85% by mass of ethene (ethylene), propene (propylene), or other olefin units. The term includes the ISO **generic names**: **polypropylene** and **polyethylene**. (See also

Classification Table, p.401.)
Note: The FTC generic name (U.S.A.) **olefin** excludes amorphous polyolefins listed under **rubber** 1.

polypropylene (fibre) (generic name)
A manufactured fibre composed of synthetic linear macromolecules having an aliphatic saturated hydrocarbon chain in which alternate carbon atoms carry a methyl group, generally in an isotactic disposition and without further substitution. (See also Classification Table, p.401.)

polystyrene (fibre)
A manufactured fibre made from a synthetic linear polymer of styrene. (See also Classification Table, p.401.)

polytetrafluoroethylene (fibre); PTFE (fibre)
A fibre made from a synthetic linear polymer in which the chief repeating unit is $—CF_2—CF_2—$. (See also **fluorofibre**.)

polyurea (fibre)
See **polycarbamide (fibre)**.

polyurethane (fibre)
A manufactured fibre composed of synthetic linear macromolecules having in the chain recurring aliphatic groups joined via urethane groups which together comprise at least 85% (by mass) of the chain.
Note: This name was formerly an ISO **generic name**. (See also Classification Table, p.401.)

poly(vinyl alcohol) (fibre); PVA
See **vinylal (fibre)**.

poly(vinyl chloride) (fibre); PVC
A type of **chlorofibre** in which the chlorine-containing repeating units are mainly:

$$
— CH_2 — \underset{\displaystyle Cl}{\underset{|}{CH}} —
$$

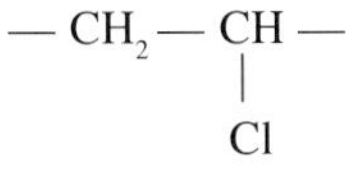

poly(vinylidene chloride) (fibre); PVDC
A type of **chlorofibre** in which the chlorine-containing repeating units are mainly:

$$
— CH_2 — \overset{\displaystyle Cl}{\overset{|}{\underset{\displaystyle Cl}{\underset{|}{C}}}} —
$$

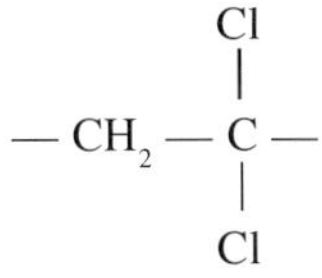

(See also **saran**.)

polyvoltine silkworm; multivoltine silkworm
A variety of silkworm producing several generations per year, more commonly found in the tropics.

pompadour
A silk term for small floral effects in a fabric.

pongee
Originally and traditionally a light-weight fabric hand-woven in China of wild silk in plain

pongee *(continued)*
weave. The term is now also applied to fabrics having a similar weight and appearance, power-woven, and made with yarns other than silk. If made from cotton, these fabrics are usually mercerized and schreinered.

ponte-Roma
See **punto di Roma**.

pony cloth
A term used for cut-pile fabric made in imitation of pony skin.

poplin
A plain-weave cotton-type fabric with weftway ribs and high warp sett, typically: 48x24; 16x18tex; K=19.2+10.2.

porcupine (lace machines)
A cylinder or shaft that draws the lace or net from the production zone. This is covered with either a special type of **card clothing** or brass sleeving having projections on the outer surface.

porcupine brass (lace machines)
Covering for the porcupine in the form of brass sleeving having projections on the outer surface.

positive drive (yarn winding)
A system in which a yarn package is driven mechanically at constant or controlled rotational velocity.
Note: The yarn speed varies with package diameter and is also dependent on angle of wind and cone taper angle if any.

positive feed (weft knitting)
The metered supply of a predetermined length of yarn to a given number of needles of a weft knitting machine. (See also **storage feed (weft knitting)**.)

positive let-off motion
See **let-off motion**.

positive shedding (weaving)
An operation in which the movement of the healds in both directions is under direct control (see **shedding**).

positive take-up motion
See **take-up motion**.

post bed
See under **bed (sewing machine)**.

post mercerization
See **mercerization**.

post-set yarn
See **stabilized yarn**, *Note 3*.

pot spinning
See **box spinning**.

potting
A finishing process applied mainly to woollen fabrics. The dyed fabric (which may have been **crabbed**) is batched on a roller and is then immersed in water. The temperature of the liquor and the duration of treatment depend on the effect desired. The fabric is cooled on the roller and rebatched end for end, and the process is repeated. The fabric is finally wound off the roller and dried.

poult
A plain-weave fabric woven from continuous-filament yarn with a rib in the weft direction. A construction satisfactory for a dress fabric was 80x22; 8.3x22.0tex; K=23.0+10.3.
Note: Poult belongs to a group of fabrics having ribs in the weft direction. Examples of this group arranged in ascending order of prominence of the rib are **taffeta**, **poult**, **faille**, and **grosgrain**. It was originally known as 'poult de soie'.

pouncing (hat manufacture)
Subjecting a felt **hood** or **body** to mechanical treatment with emery paper to produce a smooth finish.

powder bonding
A method of making **thermally-bonded nonwoven fabric** in which the fibre **web** or **batt** is bonded by activating heat-sensitive powder dispersed within it.

power loom
A loom which is driven by a source of power such as an electric motor.

power net, warp-knitted
An elasticated-net fabric produced from four half-set-threaded guide bars, the front two bars producing a net, the remaining two bars laying-in an elastomeric yarn. The lapping movements are as follows:

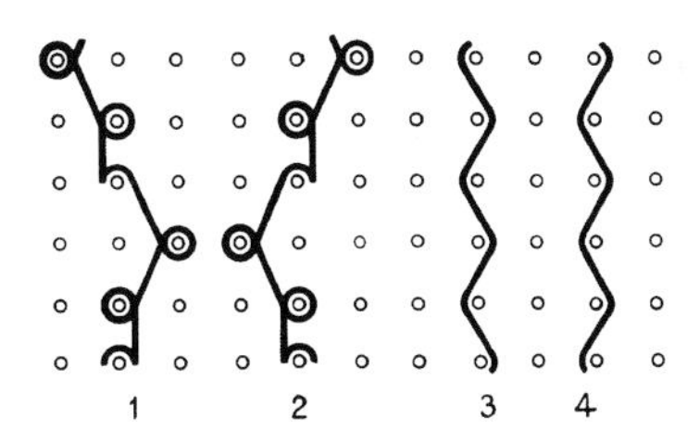

Power net structure

power stretch
See **stretch fabric**.

POY
See **partially oriented yarn**.

pre-condition (testing)
To dry a textile material to an approximately constant mass in an atmosphere with a relative humidity between 10% and 25% and a temperature not exceeding 50°C.

pre-sensitization
The treatment of a fabric with a reagent that will give stabilization of shape when the fabric, especially in garment form, is subsequently pressed.
Note: The term has been used, e.g., for (i) the application of a cross-linking agent and latent

pre-sensitization *(continued)*
catalyst to fabrics containing cellulosic fibres, so that the garments produced may be given durable shape by a heat treatment, and (ii) the application of a reducing agent, e.g., 2-hydroxyethylammonium hydrogen sulphite, monoethanolamine bisulphite (MEAS), to wool fabrics and particularly garments to accelerate setting.

pre-shrunk
A textile material that has been shrunk to predetermined dimensions in order to minimize shrinkage in use. Some fabrics are pre-shrunk by **compressive shrinkage**.

pre-spotting
Treatment of a soiled area before an article as a whole is dry cleaned or otherwise processed.

pre-treatment (testing)
An extraction procedure either using successively light petroleum, cold water and hot water, or a specified detergent solution, to remove **non-fibrous matter** before quantitative analysis of fibre mixtures.
Note: Some types of non-fibrous matter may be resistant to removal by this treatment.

preboarding
The operation of **boarding** carried out on garments or stockings, usually before they are scoured and dyed.
Note: Woven or warp-knitted articles are usually made from fabric that has been pre-set by either moist steam or dry heat according to established methods for stabilizing the fabric, and finished goods made from set fabric do not need to be preboarded. On the other hand, hosiery made from knitted fabric that has not been stabilized or fashioned depends to a great extent on the preboarding operation for conferring on it its final shape and ability to resist permanent creasing in such treatments as dyeing. Preboarding is confined mainly to fabrics or articles made from synthetic-polymer yarns (for example, nylon) and is done by submitting them to the action of steam under pressure or dry heat so that they shrink during the process into the desired shape. In addition to removing the stress introduced in the yarns during manufacture, the treatment stabilizes the dimensions of the articles so that they undergo no further change of shape during subsequent scouring and dyeing, provided that these processes are carried out under conditions less severe than those used during the preboarding. In general practice, stockings are drawn over stainless-steel formers and subjected to wet steam at 170 kPa for a predetermined time.

precision
The closeness of agreement between test results.

precision winding
A method of winding cones and cheeses in which the **wind** is constant and the **angle of wind** decreases as the diameter of the package increases. It can be used to build a very dense package.

precoating (carpet)
See under **substrate (carpet)**.

precrêping
1. The embossing of a fabric containing **crêpe yarns** with a design to influence the uniformity and fineness of the crêpe effect produced in subsequent treatment. This process is carried out by means of a precrêping calender.
2. See **crêpe embossing**.

president braid
A braid similar to **Russia braid** but with three cores, the centre core being larger than the two outer cores.

press felt (technical textiles)
A needled fabric incorporating an endless woven basecloth (commonly monofilament) which transports the paper sheet through the press section (where load is applied for mechanical dewatering) of a paper-making machine. Seamed press felts are suitable for some paper grades.

press finish
A finish given to fabrics or garments by **pressing**.

press ratio (alkali-cellulose)
The ratio of the mass of alkali-cellulose, after excess sodium hydroxide (caustic soda) solution has been pressed out, to the original mass of pulp.

press-off (knitting)
The accidental or intentional casting-off of the loops from the needles when the supply of yarn ceases while the knitting elements continue to perform their knitting cycle.

press-off narrowing (weft knitting)
See under **shaping (weft knitting)**.

presser (knitting)
A device used to close the needle beards on machines fitted with bearded needles.

presser foot (sewing machine)
An element which is used to constrain movement of the material during stitch formation by applying to it an adjustable downward pressure. It works in conjunction with the **throat plate** and **feed dog** to achieve controlled movement of the material and correct stitch formation. Different designs of presser foot are in use: the most common have a solid or hinged surface which controls the material.

presser-foot (weft knitting); stitch presser (weft knitting)
A device which presses downwards on to the stitches to prevent them rising with the needles. It thus enables a wider range of structures to be knitted than is normally possible with a conventional take-down arrangement. It consists of either a flexible wire of circular or angular cross-section, or a thin plate shaped to operate behind or between the needle bed(s), particularly in the case of vee-bed flat knitting machines.

pressing
The application of pressure, with or without steaming or heating, (i) to remove unintended creases and to impart a flat appearance to fabrics and garments, or (ii) to introduce desired creases in garments.

pressing (finishing)
A method of enhancing the appearance of fabrics by the application of pressure to increase their lustre and smoothness. (See also **calender**.)
Note: The equipment used varies with the type of fabric being finished, for example:
(i) Wool-type fabrics are pressed in (a) a hydraulic press in which layers of fabric formed by **cuttling** the material are interleaved with heated or cold press papers and the pile of fabric so formed is subjected to pressure, or (b) a rotary press in which fabric is passed continuously through the nip formed by a smooth rotating heated cylinder and a fixed curved metal plate, pressure between the two elements being applied hydraulically.
(ii) Garments and knitted goods are pressed by being placed on a perforated, fabric-covered steam bed and subjected to pressure by lowering a second, hinged steam-bed on to the material. Steam is blown through the material while under pressure and the fabric can be cooled by drawing air through it. This type of machine is often known as a Hoffman press and the process as Hoffman pressing.

pressing; press-finishing
See *Note* under **boarding**.

pressure boil
The **scouring** of cellulosic textiles with alkaline liquors in closed vessels under excess pressure, normally 140-210 kPa. (See also **kier boil**.)

pressure decatizing; pressure decating
A process for **setting** fabrics in an autoclave, which is a pressure vessel into which a trolley carrying a roll of fabric is placed. After the vessel is sealed, the air is evacuated and steam is introduced to give the appropriate setting temperature which is maintained for a designated time. To give wool fabrics a high degree of set, the fabric is wound on a roll between layers of cotton fabric and treated in the autoclave at up to 200 kPa (30 p.s.i.) gauge for 1-5 minutes.

pressure dyeing
The forced circulation of dye liquor through packages of fibre, yarn or fabric, without limitation of temperature. (See also **high-temperature dyeing**.)

pressure harness
A system of weaving damasks in which all the warp threads are drawn through a multi-eyed mail **jacquard harness** that controls them for pattern making only. Two to eight contiguous threads are drawn in each mail of the jacquard harness, and afterwards individually through separate deep-eyed mails in a set of heald shafts placed immediately in front of the jacquard harness that controls them for the ground or binding weave. A pressure harness combines the jacquard harness band heald shafts in a way which considerably increases the configuring capacity of the machine and saves cards.

pressure mark
An area of changed appearance of a fabric, often giving increased lustre, caused by irregularities of pressure during the finishing process. A **seam mark** is an example of a pressure mark.

prestress (technical textiles)
Tensile stress deliberately induced in a fabric structure to give it stiffness. It may be produced by jacking masts, pulling on cables or inflation.

primary backing (carpet)
See under **substrate (carpet)**.

primary cellulose acetate
See **cellulose acetate**.

primary strand
A **strand** which forms part of any of the ropes used as the **secondary strands** of a **cable laid rope**.

print bonding
A method of making **nonwoven fabrics** in which there is controlled application of adhesive to specific areas of the fibre **web** or **batt** by using printing techniques similar to those used for coloration. (See also **adhesive-bonded nonwoven fabric**.)

printed carpet
Carpet having coloured patterns applied by methods similar to those used for printing flat textiles, e.g., flat-bed screen printing, rotary screen printing, dye injection printing.

printed string
See **bolduc**.

printing
The production of a design or motif on a substrate by application of a colorant or other reagent, usually in a paste or ink, in a predetermined pattern.
Note: Since the 17th century the techniques employed have developed, from hand block printing to engraved roller printing and more recently to **screen printing**.

printing style
A concise, though not necessarily complete, indication of the method of production of a print in terms of the process or the class or classes of dye used (or both).

producer twist
The small amount of twist inserted during the production of multi-filament yarn by certain take-up systems such as pot, cap, or ring-and-traveller.

profile reed
A **reed** in which the front edges of the wires are shaped so as to form across the face of the reed:
(i) a channel in order to conserve the jet and guide the weft in a jet weaving machine;
(ii) superposed channels to locate the fells of the two ground fabrics in a double plush loom.

profile wire
See **wire (pile weaving)**.

profiled die (film-fibre technology)
An extrusion slot profiled in such a way as to produce film having longitudinal grooves that cause the film to separate almost completely into individual filaments when hot stretched.

projectile weaving machine
See **weaving machine**.

proof
1. Fully resistant to a specified agency, either by reason of physical structure or inherent chemical non-reactivity, or arising from a treatment designed to impart the desired characteristics.
Note 1: Proofing treatments are defined by specified limits ascertained by test, and the use of the term is related to the limiting conditions.
Note 2: 'Resistant', 'retardant' or 'repellent' are appropriate alternatives when the resistance is less than full.
2. Finishing process used in felt hat manufacture in which the **hoods** or **bodies** are treated with shellac or other suitable stiffening agents to improve shape retention of the hats produced.

proofed
Descriptive of material that has been treated to render it resistant to a specified agency.
Note: A designation of materials as 'proofed' should indicate that the material conforms to definite standards.

protective clothing
See **workwear**.

protein (fibre)
1. A natural fibre derived from animals (see **fibre, natural** under **fibre**).
2. A manufactured fibre obtained from natural protein substances by chemical regeneration (see **fibre, regenerated** under **fibre**).
Note: This name was formerly an ISO **generic name**. (See also Classification Table, p.401.)

prunelle twill
A 2x1 twill weave.

PTFE (fibre)
See **polytetrafluorethylene (fibre)**.

puckered selvedge
See **selvedge, slack**.

puckering
See **cockle (defect)**.

puffed sleeve
See under **sleeve (clothing)**.

pull; draw (sampling)
A sample of fibres abstracted manually from a bulk lot of raw material or sliver with a view to assessing the length and/or distribution of length of fibre within the sample.

pulled-in selvedge
See **selvedge, uneven**.

puller feed
See **feed mechanisms (sewing)**.

pulling (rag)
The operation of reducing rags and thread waste to a fibrous state.

pulling (wool)
The removal of wool from skins.
Note: Before removal, the fibres are loosened by treatment (see **skin wool**). The skins may be placed on a curved board, and, with ordinary skins, the wool is pushed or rubbed with the hands; with short-wool skins, a blunt knife, held with both hands, is used. When the puller is seated and pulls with his hands from the skin placed on his knees, the process is known as 'knee pulling'. The wool puller sorts the wool as he removes it from the skin.

pulling-back
See **unweaving**.

pulling-back place
An isolated narrow bar, running parallel with the picks of woven fabric, that starts abruptly and gradually shades away to normal fabric and is caused by pulling-back (see **unweaving**). The pick-spacing within this bar may be different from that of the normal fabric (see **pick bar** under **bar (woven fabric)**) or may be similar to it, but the effect will still be visible as a result of the greater degree of abrasion to which the warp has been subjected by being unwoven and woven again.

pulp (cotton)
Purified cotton **linters** usually in the form of standard sheets about 1mm thick.
Note: The preparation of the linters involves boiling under pressure with sodium hydroxide followed by hypochlorite bleaching, the severity and the duration of the boiling depending on the use to which the resultant material is to be put. The fibres are composed of glucose units to the exclusion of other sugars and only 1-2% of the cellulose is soluble in sodium hydroxide of 17.5% concentration at 20°C. Suitability for a specific purpose is determined by measurement

of the viscosity of a solution of the product under standard conditions, and different viscosity ranges are usually specified for material to be used for manufactured fibres, lacquers, etc., (see **fluidity** and **viscosity**). The material is also supplied in pressed bales.

pulp (wood)
Cellulose fibres isolated from wood by chemical treatments.
Note 1: The preparation of wood pulp involves the boiling of wood chips with alkaline liquors or solutions of acidic or neutral salts followed by bleaching with chlorine compounds, the object of these treatments being to remove more or less completely the hemicelluloses and lignin incrustants of the wood. The purified fibres are usually pressed into standard sheets about 1mm thick, and commercial material retains 4-12% of carbohydrates soluble in 17.5% soda at 20°C, the actual content depending on the severity of the purification treatments.
Note 2: Mechanical wood pulp is obtained by wet-grinding bark-free wood in stone or other mills. The material is used largely in mixtures with bleached pulp for newsprint and is quite different from wood pulp as defined above.

pulp, mechanical (wood)
See **pulp (wood)**.

pump capacity; pump delivery (manufactured fibre production)
The volume of liquid delivered by one revolution of a spinning pump (metering pump).

punching (wool industry)
A winding operation that prepares four-end balls of sliver for Noble combing.

punto di Roma; ponte-Roma
A non-jacquard **double jersey** fabric made on an interlock basis, using a selection of knitted loops and floats in the following sequence. (See also **double jersey, weft-knitted**.)
Note: The alternative 'ponte-Roma' is probably a corruption.

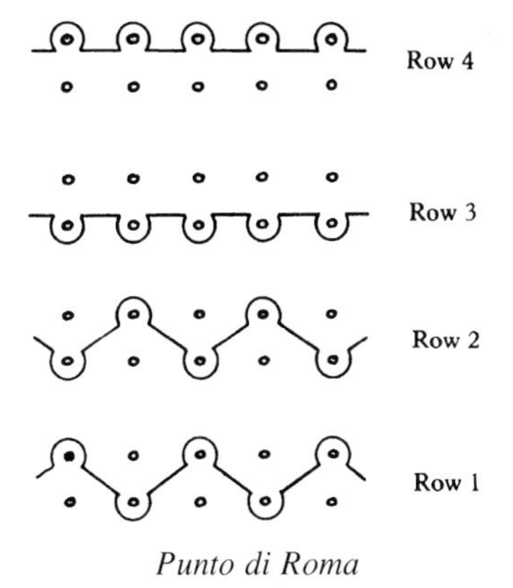

Punto di Roma

pure silk
See **silk, pure**.

purl fabric, weft-knitted
A fabric in which both back and face loops occur in some or all of the **wales**. The structure may be identified by the particular knitting sequence used, for example, (i) 1x1 purl fabric, in which a single course of back loops alternates with a single course of face loops. 2x2 purl and 3x3 are made in a corresponding way; and (ii) fancy purl, a general term used to describe patterned purl structures in which a design is formed from back and face loops; also known as 'links-links'.

purl gating; purl gaiting (weft knitting)
See **gating (knitting)**.

purl knitting machine (weft knitting)
A **knitting machine** having two aligned needle tricks and a set of double-headed needles which can knit in either of the two opposed needle beds under the control of sliders thus permitting the production of purl fabrics. (See also **flat knitting machine** and **double cylinder knitting machine**.)

purls (lace)
See **pearls**.

pussy willow
A plain-weave fabric characterized by fine horizontal lines and made from **net silk** yarns. The warp **sett** is much greater than the weft sett.

PVA (fibre)
See **poly(vinyl alcohol) (fibre)**.

PVC (fibre)
See **poly(vinyl chloride) (fibre)**.

PVDC (fibre)
See **poly(vinylidene chloride) (fibre)**.

quality
1. The totality of features and characteristics of a product or service that bear on its ability to satisfy stated or implied needs. (See also **colour quality**, **fibre quality index** and **lace quality**.)
2. A term, usually combined with a number or name, used to identify textile products.
3. A relative term used to indicate the perceived merits of similar products for the same end-use.

quality assurance
To carry out all those planned and systematic actions necessary to provide adequate confidence that a product or service will satisfy given requirements for quality.

quality control
The operational techniques and activities that are used to fulfil requirements for quality.

quantitative analysis
Method of determining the proportion of different substances in a sample, usually on a mass basis, e.g., the proportions of each fibre type in a blend; the proportion of impurity in raw fibres.

quarter
An abbreviation for a quarter of a yard (9 inches or 228.6 mm) used as a length unit in measuring the width of fabrics.

quartz (fibre)
A fibre of high-purity silicon dioxide glass, produced from mineral quartz or quartz sand. (See also **ceramic fibre** and Classification Table, p.401.)
Note: Although quartz is invariably crystalline, quartz fibres are not. For this reason, some authorities deprecate the use of this term, preferring 'fused silica fibres'.

queen's cord, warp-knitted
A two-bar construction made with full-set threading in both guide bars. The lapping movement

of the back guide bar involves underlapping three or four needle spaces, while the front guide bar chains continuously on the same needle.

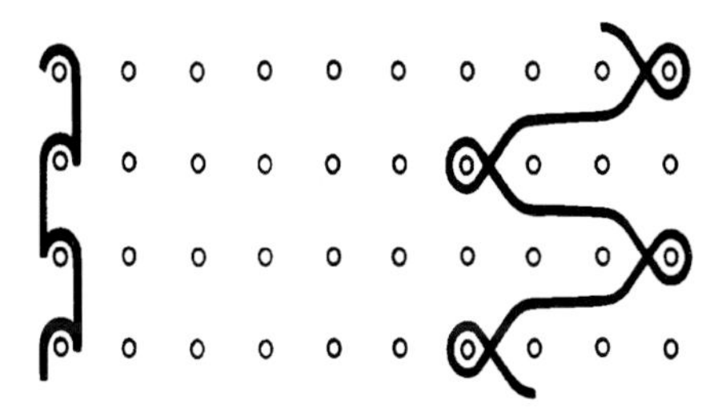

Queen's cord structure

quench
A cooling zone in which the temperature of melt-spun filaments is lowered very rapidly and/or at a controlled rate soon after extrusion. The two main types are generally referred to as water-quench and air-quench.

quetsch; quetch
The nip rollers of a padding machine (see **padding**) or size box.
Note: The term is also used to describe the whole machine, particularly in yarn sizing (see **sow box**).

quill (narrow fabrics)
1. A small **bobbin**, with or without flanges, on to which weft is wound. In use it revolves when mounted on the shuttle pin of a narrow fabric loom.
2. The weft package as described in 1 above.
3. See **pirn**.

quillings (lace)
A type of narrow lace of plain or spotted net.

rabbit fibre (hair)
Fibre from the rabbit (*Oryctolagus cuniculus*). (See also **angora fibre (hair)**.)

race board
That part of a **sley** in front of the **reed** and below the warp, over which the **shuttle** or **rapier** passes.

rack (lace machines)
An arbitrary number of **motions** of the machine, used as a basis for the calculation of machine speed, productivity, and **lace quality**.
Note: A Leavers rack is 1920 motions; furnishings rack, 720 full motions; bobbinet rack, 240 holes (12 or 20 motions per hole); warp rack, 480 motions.

rack (warp knitting)
A unit of 480 courses or complete knitting cycles.

rack (weft knitting)
See **shog (weft knitting)**.

rack and pinion motion (narrow fabric)
A positive system for controlling the movement of the **shuttles** of a narrow fabric weaving

rack and pinion motion (narrow fabric) *(continued)*
machine in which a pinion or cog meshes with teeth mounted on the shuttle(s) and is driven intermittently by a length of toothed rack.

racked stitch (weft knitting)
A rib structure in which a sideways-deflected loop lies across a loop formed in the same course on the opposite needle bed.

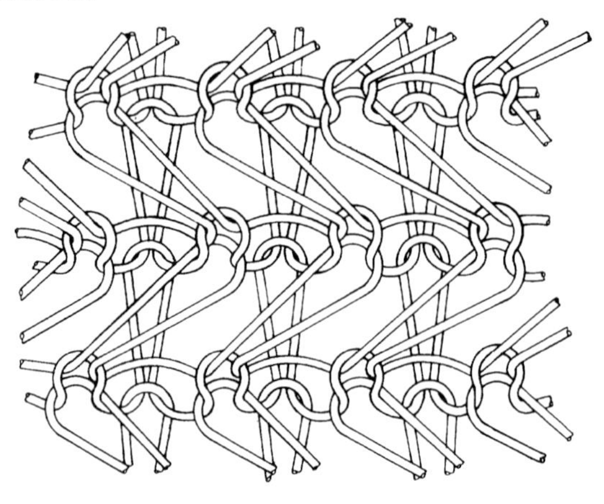

Racked stitch

racking (weft knitting)
The lateral movement of a needle bed or point bar across a predetermined distance on a flat knitting machine. (See also **shog (weft knitting)**.)

raffia
A fibre obtained from the leaves of the raffia palm *Raphia ruffia*.

raglan sleeve
See under **sleeve (clothing)**.

rags (new)
The waste fabric, whether woven or knitted, that is left after a garment has been cut out. The term also covers piece ends and discarded pattern bunches.

rags (old)
Fabrics from worn garments that have been discarded.

raised checks
A fabric figured with extra warp on a plain ground.

raising; napping
The production of a layer of protruding fibres on the surface of fabrics by brushing, teazling, or rubbing.
Note: The fabric, in open width, is passed over rotating rollers covered with teazles, fine wires, etc., whereby the surface fibres are pulled out or broken to give the required effect.

ramie
A bast fibre obtained from the stems of *Boehmeria nivea*, especially the variety *tenacissima*, belonging to the *Urticaceae* or nettle family. It usually reaches European markets in the form

of ribbons which are known as 'China Grass'.

random creeling
The exchange of individual supply packages during production whenever a supply package is empty or otherwise unsuitable. (See also **block creeling**.)

random dyeing
A form of **space dyeing**.
Note: It is so called because it can be used to produce random coloration in the final fabric.

random linking
linking in which, when stitching pieces of fabric together, no attempt is made to stitch through adjacent loops.

random range
See **pattern blanket**.

random winding
A method of winding cones and cheeses in which the **angle of wind** is constant and the **wind** decreases as the package diameter increases.

random-laid web
A term sometimes used to describe a **web** or **batt** produced by **air laying**.
Note: The orientation of the fibres is not usually random.

rapier; rapier loom
See **weaving machine**.

raschel lace
Lace fabric produced on a **raschel warp-knitting machine**.

raschel warp-knitting machine
A warp-knitting machine with one or two needle bars employing either latch or compound vertically mounted needles (see **needle (machine knitting)**). The fabric is supported on either one or two trickplates and is withdrawn from the knitting zone in a downward direction. (See also **knitting machine**.)

rat-tail cord; American cord
A tubular woven **cord** usually of satin construction.

ratch
1. See **reach** and **roller-setting**.
2. A slight additional drafting of the yarn which takes place towards the end of the **mule draw** (cotton system).

rate of dyeing
The rate at which a dye is absorbed by a substrate under specified conditions.
Note: It may be expressed quantitatively in several ways, such as the weight of dye absorbed in unit time, or the time taken for the substrate to absorb a given fraction of the amount of dye which it will absorb at equilibrium.

ratine
Originally a thick woollen fabric with a curled nap. This term or ratiné, the past participle of the French verb ratiner (meaning to cover with a curled nap), is also applied to a fabric, made

ratine *(continued)*
from any of a variety of fibres, with a rough surface produced either by using a fancy yarn in a fabric to which a special finishing technique may be applied or by using ordinary yarns in a fabric to which the special finish is applied.

ravel courses (knitting)
See **roving courses**.

ravensduck
A very heavy linen fabric in 2x1 twill weave, such as sailcloth.

raw silk
See **silk, raw**.

rayon (fibre)
1. (Generic name U.S.A.) A manufactured fibre composed of regenerated cellulose, as well as manufactured fibres composed of regenerated cellulose in which substituents have replaced not more than 15% of the hydrogens of the hydroxyl groups.
Note: The ISO **generic names** are **viscose**, **modal**, and **cupro**. (See also Classification Table, p.401.)
2. A term used in some manufacturing countries for any manufactured cellulose fibre, including, in some cases, fibres composed of cellulose acetate.

re-breaking
The shortening of fibres in a sliver or top by a process similar in principle to stretch breaking (see *Note* under **converting**). Re-breaking may be intended to shorten a limited number of overlength fibres and/or to reduce the average length.

reach; ratch; perry
The distance from the back heald to the back rest or back roller of a weaving machine. (See also **roller-setting**.)

reaching-in
The operation of selecting individual warp threads and presenting them for **drawing-in**. This may be done by hand and/or by machine.

reaction spinning (manufactured fibre production)
A process for fibre production in which polymerization is achieved during the extrusion of one set of reactants into another set of reactants.

reactive dye
A dye that, under suitable conditions, is capable of reacting chemically with a substrate to form a covalent dye-substrate linkage.

recommended allowance
The percentage that, in the calculation of commercial mass of textile material and of yarn linear density, is added to the oven-dry mass. The determination of this mass may or may not have been preceded by washing to remove natural or added oils and dressings. The recommended allowance is arbitrarily chosen according to commercial practice and includes the moisture regain. It may also include the normal finish that is added to impart satisfactory textile qualities to the material.

recovered wool
See **reused wool**.

redox potential
The potential developed when a bright platinum electrode is immersed in a solution containing an ionic species in two interconvertible oxidation (valency) states. The potential is dependent upon the ratio of the concentrations of the two oxidation states, e.g., iron II and iron III ions; 1,4-dihydroxybenzene (hydroquinone) and 1,4-benzoquinone (p-benzoquinone); vat dye and leuco vat dye.
Note: The standard redox potential, E, is that developed when the concentrations of the two oxidation states are equal to each other, and in vat dye systems indicates the difficulty with which the vat dye (the oxidized state) can be reduced.

reduction clearing
Removal of unfixed dyestuff (usually disperse) by an aqueous alkali/reducing system, usually sodium hydroxide/sodium dithionite.

reed; sley
1. A device, consisting of several wires closely set between two slats or **baulks**, that may serve any or all of the following purposes: separating the warp threads; determining the spacing of the warp threads; guiding the shuttle or rapier (if applicable); and **beating up** the weft.
2. To draw ends through a reed. (In the U.K., also known as: to sley, to bob the reed, or to enter the reed.)

reed, leasing
A reed constructed to permit warp ends passing through it to be separated into sheets suitable for **lease** formation. The usual construction consists of alternate open and blocked dents, but more complicated arrangements are sometimes used to aid segregation of particular ends, e.g., of one colour in a fancy warp.

reed, Scotch hook
A reed used in 'striking' a lease during the cotton-system sizing of manufactured continuous-filament warps. In simple form, each reed wire is provided with a small hook at the same side and at the same height in the reed. Warp ends passing through the dents may be engaged in the hooks or passed above them during formation of a **lease**. More complicated constructions make use of two hooks per wire and blocked dents to aid segregation of particular ends, e.g., of one colour in a fancy warp.

reed gratings
Transparent plates containing lines that are parallel and uniformly spaced in the cross-wise direction. By placing an appropriate grating on a reed it is possible to see if there is any irregularity in the spacing of the reed wires. Reed gratings can also be used for counting the number of reed wires per unit distance. (See also **parallel line gratings**.)

reed mark
A warpway crack in a woven fabric caused by a damaged or defective reed.

reed number
See **count of reed**.

reed-plate (tufting)
See under **tufting machine**

reediness
A noticeable grouping of warp threads due to the reed wires producing warpway **cracks**. It can be caused by the use of areed unsuited to the fabric construction employed.

reeding plan
See **denting plan**

reel
1. See **hank**.
2. The cylindrical former, usually flanged, suitable for use with domestic sewing machines, on to which **sewing thread** is wound.
3. The revolving drum, mill or swift of a **section warping** machine.
Note: One end of the reel is conical in order to accommodate the increasing diameter of each section as it traverses laterally. (See also **warping mill**.)

reeling machine; hanking machine
A revolving frame on to which a number of **hanks** or **skeins** or **reels** of yarn or roving are wound simultaneously. The frame normally comprises a number of parallel bars (some or all of which collapse to facilitate removal of the wound hank) mounted circumferentially on a cylindrical form. Traversing yarn guides provide a pattern which enables the hank to be **leased** so that it can be unwound without entanglement after storage or, e.g., wet processing.

refractive index
The ratio of the velocity of light in a vacuum to that in a given material. It may be used in fibre identification or as a measure of fibre orientation. (See also **birefringence**.)

regain
The ratio of the mass of moisture in a material to the **oven-dry mass**. The ratio is usually expressed as a percentage and is calculated as follows:

$$\frac{\text{Total moist mass} - \text{Oven-dry mass}}{\text{Oven-dry mass}} \times 100$$

regatta
A striped cotton-type fabric woven in 2/1 twill. The pattern consists of fast-dyed colour and white in warp stripes of equal width. The fabric has a white or undyed weft, typically: 30x25; 30x33tex; K=16.4+14.4.

regenerated cellulose fibre
See **rayon (fibre)**.

regenerated fibre
See **fibre, regenerated** under **fibre**.

regenerated protein fibre
See **protein (fibre)** and **fibre, regenerated** under **fibre**.

regina
A fine 2/1 twill of good quality, typically: 57x28; 7-10tex cotton type in warp and weft. Two-fold yarns can also be used and the fabric is around 100 g/m^2.

reindeer fibre (hair)
Fibre from the reindeer (Genus *Rangifer*).

relative humidity
The ratio of the actual pressure of the water vapour in the atmosphere to the saturation pressure of water vapour at the same temperature and same total pressure. The ratio is usually expressed as a percentage.

relaxation
The releasing of strains in textile materials.

relaxed elastic modulus
A term almost exclusively used to describe a property of highly sett nominally inextensible tapes and light conveyor belts. It is derived by determining the elastic modulus per unit width after cyclically straining the belt between known limits (usually 1% to 2%) for 500 cycles.

relief fabric, weft-knitted
A patterned rib-based fabric, the surface of which exhibits a characteristic relief or blister effect in which the number of loops in the relief portion is greater than in the surrounding area on the effect side and on the reverse side. The relief area may be of a different colour from the main ground and the ground may also be patterned. Two main types of structure are recognised: single relief or three-miss blister and double relief or five-miss blister. The latter has a greater preponderance of loops on the face of the fabric in the relief areas than the former. Also known as 'blister fabric' or 'cloqué fabric'.

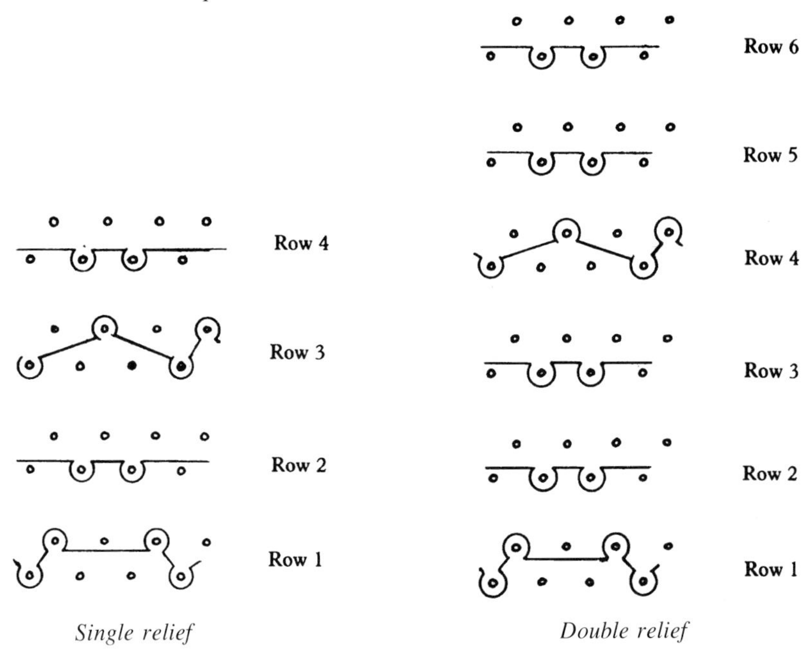

remnants
See **fents**.

repeatability (r)
The precision obtained under conditions where independent test results are obtained with the same method on identical test material in the same laboratory by the same operator using the same equipment within a short interval of time.

repeating tie
See **jacquard tie (weaving)** under **jacquard (mechanism)**.

repeating unit
A chemical group that recurs in the backbone of a polymer.

repp
A plain-weave fabric with a prominent weftway-rib effect, made from two warps and two wefts. Both the warp and the weft threads are arranged alternately coarse and fine. Coarse threads are raised above coarse picks and fine threads above fine picks, the rib effect being accentuated by different tensions in the warps.
Note: Less expensive fabrics are now often made with one warp and one weft but with the same general ribbed effect.

repping (defect)
The unintentional introduction into a woven fabric of a bar in which a prominently weftway-rib effect is evident.
Note: This fault is often associated with standing places (see **pick bar** under **bar (woven fabric)**) and is the result of differential relaxation of the upper and lower sheets of warp while the loom is standing.

reproducibility (R)
The precision obtained under conditions where independent test results are produced with the same method on identical test material in different laboratories with different operators using different equipment.

re-scutched tow
See **flax tow**.

residual shrinkage
The latent shrinkage of a fibre, filament, yarn, or fabric.

resist
1. A substance applied to a substrate to prevent the uptake or fixation of a dye in a subsequent operation.
Note: The substance can function by forming a mechanical barrier, by reacting chemically with the dye or substrate, or by altering conditions (e.g., pH value) locally so that development cannot occur. Imperfect preparation of the substrate may cause a resist as a fault.
2. In printing plate or roller making, a coating of, for example, light-hardened gelatine which protects from the action of the etching solution those areas of the plate or roller which are not required to be etched.

resist style
A method of printing in which undyed material is printed with resists (see **resist**) whereby, on subsequent dyeing or developing, a white or coloured pattern is obtained on a coloured ground.
Note: A coloured pattern (illuminated resist) is obtained by incorporating suitable dyes in the resist printing paste.

restraining agent
A substance that, added to a dyebath, decreases the equilibrium **exhaustion**.

resultant count
The actual **count** of a plied (folded) or cabled construction.
Note: This can be expressed in any count system.

retarding agent; retarder
A substance that, added to a dyebath, decreases the rate of dyeing, but does not substantially affect the equilibrium **exhaustion**.

retarding motion
See **cramming motion**.

retexturing
1. The passage of an already **textured yarn** through a further texturing process.
Note: The second texturing process may be the same type of process or different from the first texturing process.
2. A process for treating dry-cleaned garments to improve their handle.

retractive force (textured yarn)
The tension in a **textured yarn** due to the formation of crimp in the filaments under specified conditions of crimp development.

retting (flax)
The subjection of crop or deseeded straw to chemical or biological treatment to make fibre bundles more easily separable from the woody part of the stem. Flax is described as water-retted, dew-retted, or chemically-retted, etc., according to the process employed.

reused wool; recovered wool
Wool rags and manufactured waste, torn up and reprocessed into fibres again, and used for producing **shoddy** or **mungo** yarns.

reverse jacquard, weft-knitted
A rib-based fabric in which the design on the effect side is reversed on the other side by alternation of the two component threads between the two sides.

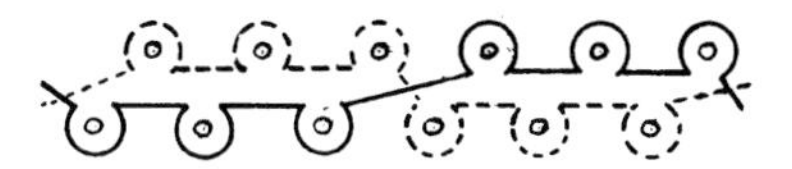

Reverse jacquard

reverse locknit, warp-knitted
A two-bar construction made with full-set threading. The front bar makes a 1x1 closed lap movement and the back bar a 2x1 closed lap movement in opposition.

reverse loop (weft knitting)
See **back loop** under **knitted loop (weft knitting)**.

reverse side (weft-knitted fabric)
See **plain fabric, weft-knitted**.

reverse toe (hosiery)
A form of toe in which the join between the toe and foot is on the underside of the foot.

reverse welt
See under **welt (knitting)**.

reverse-plated fabric, knitted
See **plated fabric, weft-knitted**.

reversible
Fabric that has pattern and appropriate finish on both sides, so that either side can be used as the **face** side.

rib fabric, weft-knitted
A fabric in which both back and face loops occur along the course, but in which all the loops contained within any single wale are of the same type, i.e., back or face loops.

broad rib fabric, weft-knitted
A rib fabric in which groups of three or more adjacent wales are of the same type, either face or back.

6x3 rib fabric, weft-knitted; Derby rib
A fabric in which all the loops of six adjacent wales are intermeshed in one direction and all the loops of the next three wales knitted at the same course are intermeshed in the opposite direction, and so on alternately.

1x1 rib fabric, weft-knitted; English rib
A rib fabric in which single wales of face loops alternate with single wales of back loops.

2x2 rib fabric, weft-knitted; Swiss rib
A rib fabric in which two adjacent wales of face loops alternate with two adjacent wales of back loops in a series.

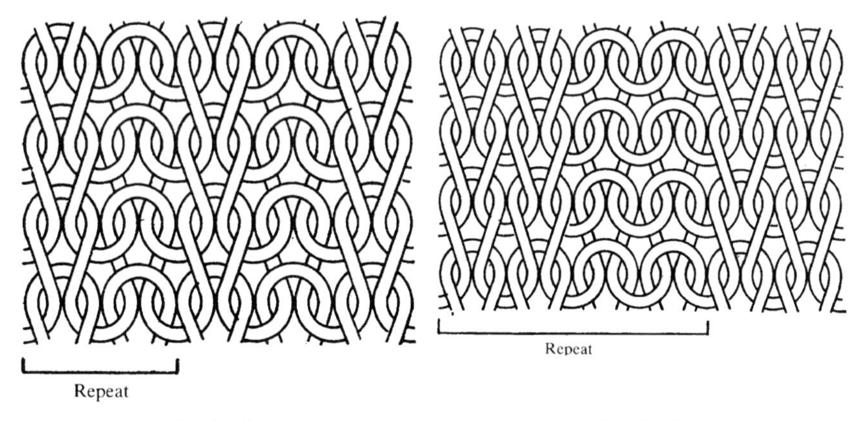

1 x 1 rib *2 x 2 rib*

rib fabric, woven
A fabric whose surface consists of warpway (weft rib) or weftway (warp rib) raised lines or ridges (see **warp rib** and **weft rib**).

rib gating; rib gaiting (knitting)
See **gating (knitting)**.

rib jacquard (weft knitting)
A patterned rib-based fabric the surface of which is essentially flat and exhibits a figure or design in differing colour or texture. The patterned surface is derived from the chosen arrangement of yarns, and of knitted and float loops. The back of the fabric may be either plain, **striped**, **birdseye** or **ladder backing**. (See also **double jersey, weft knitted**.)

rib transfer, straight bar (weft knitting)
The process of transferring a rib fabric by taking the loops of one course on to the needles of a plain machine in preparation for plain knitting.

rib-transfer stitch (weft knitting)
A stitch made by the transfer of the loop from a needle in one set to a needle in another set, the latter having a loop already on it. It may be used as an effect stitch.

ribbon
A **narrow fabric**, usually woven, and generally with a **continuous-filament warp** in a variety of styles for decorative and functional uses. Ribbon may also be made by slitting broad fabric. The product may have a fused edge. (See also **petersham ribbon (millinery), petersham ribbon (skirt), double satin ribbon, faille ribbon, galloon, lingerie ribbon, sarsnet ribbon, single satin ribbon, taffeta ribbon.**)
Note: The French term 'ruban' and the German 'Band' are incorrectly translated as 'ribbon' but are terms which embrace a wider range of woven narrow fabrics including **elastic**, **tape** and **webbing**.

ric-rac braid
A flat braid made by tensioning component threads differentially so that the fabric forms a zig-zag shape. (See also **van Dyke**.)

rice
See **swift**.

ridge cable (technical textiles)
Cable between two **fabric fields** which produces a kink in the surface of air-supported and tension-membrane structures.

rigging
The lengthways folding of fabric so that the folded material is half its original width.
Note: This term is used specifically in the wool textile industry.

ring doubling
See **ring twisting**.

ring spinning
A continuous system of spinning in which twist is inserted into a yarn by using a circulating **traveller**. The yarn is wound on to the package since the rotational speed of the package is greater than that of the traveller. (See also **spinning**.)

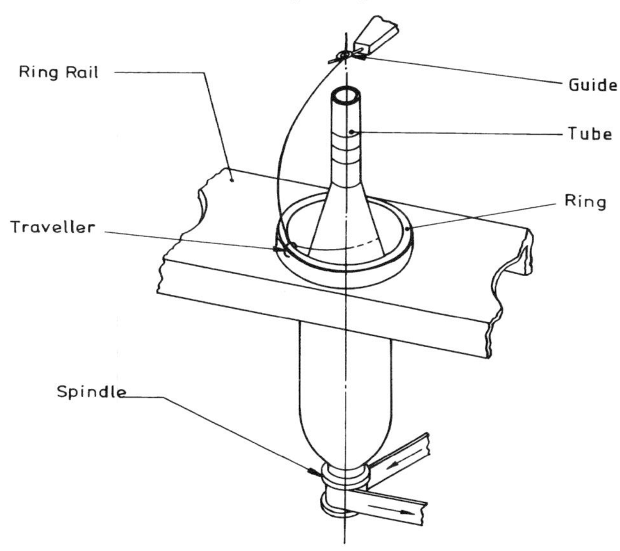

Ring spinning

ring twisting; ring doubling; downtwisting
A system of producing a **folded yarn** by twisting together two or more single yarns using ring-and-traveller as the twisting technique. The terms ring twisting and downtwisting are also used when the same technique is employed to increase or decrease the twist in a single yarn. (See also **uptwisting**.)

ripening
1. A process in the production of cellulose ethanoate (acetate) consisting in the splitting off of some of the ethanoic acid (acetic acid) and most of the combined catalyst present in the primary cellulose ethanoate (acetate).
2. A process in the manufacture of viscose in which, during storage prior to spinning, the xanthate slowly decomposes. The number of xanthate groups is reduced and some redistribution takes place. The process is time and temperature-dependent.

ripping
The operation of removing cotton or manufactured fibre linings from garments prior to the sorting and classification of rags.

ripple (weft knitting)

tuck ripple
A fabric, made on one set of needles, featuring raised effects that are developed by selective knitting and tucking, tuck loops generally being accumulated in alternate wales.

welt ripple
A held-loop fabric, made on two sets of needles by selective knitting and missing on one set and continuous knitting on the other and featuring roll or wave effects on the fabric side.

ripple shed
See **wave shed**.

ripples comb; ripple comb
A large comb used for removing the seed bolls or capsules from the flax crop by hand. (See also **roughing-out (flax)**.)

rocket package
A large version of a **supercop**, used as a means of supplying a coarse weft yarn to shuttleless weaving machines. It may be up to 800mm in length and 75mm in diameter.

rodier
See **double piqué, weft-knitted**.

roll welt
See under **welt (knitting)**.

roll-boiling
A comparatively short **potting** treatment at the boil.

rolled seam
See under **seam type**.

roller decortication
See **breaking (bast fibres)**.

roller locker machine
See **rolling locker machine** under **lace machines**.

roller-bed steamer
A **steamer**, in which fabric is carried in a relaxed state over a bed of individually driven rollers situated in the lower half of a steaming chamber.

roller-conveyor steamer
A **steamer**, in which fabric is supported and carried in a relaxed state on an endless conveyor made of stainless steel slats and driven by rollers at each end.

roller-embossed film (film-to-fibre technology)
A polymer film that has been indented to induce easy splitting during subsequent stretching.
Note 1: The indentations are produced by using a profiled embossing-roller and may be either longitudinal and followed by uniaxial stretching to produce individual filaments, or patterned and followed by biaxial stretching to produce a fine net.
Note 2: Polymer films may also be embossed for decorative purposes.

roller-setting; ratch; reach
The distance between the nips of the pairs of rollers in a roller-drafting or stretch-breaking system (see **converting**).
Note: Another common use of the term 'ratch' is to specify the distance between the front roller nip and the release point of the pins in a pin-drafting system.

rolling
The winding of finished fabric on tubes.

rolling (flax)
See **breaking (bast fibres)**.

rolling locker machine
See under **lace machines**.

rope
An article of **cordage**, more than approximately 4mm in diameter, obtained when: (i) three or more **strands** are laid (see **lay** 3) or plaited (see **braiding**) together; or (ii) a **core** is covered by a braided or plastic film sheath. Types of rope are:

> **braided rope; sennit rope; sinnet rope**
> A rope formed by **braiding** or plaiting the **strands** together.
>
> **cable laid rope**
> A rope formed by three or more ropes twisted to form a helix around the same central axis. The ropes that become the secondary strands are 'S' **lay** and the finished cable is 'Z' lay or *vice versa*.
>
> **combined rope**
> A rope in which the **strand** centres are made of steel and in which the outer portions of each strand are made from fibrous material.
>
> **double braided rope**
> A rope in which a number of **strands** are plaited to form a **core** and around which are plaited further strands to form a sheath. The core lies coaxially within the sheath.
>
> **8-strand plaited rope**
> A rope normally composed of 4 pairs of **strands** plaited in a double 4-strand round **sennit**.

rope *(continued)*

hard laid rope
A rope in which the length of lay of the **strands** and/or the rope is shorter than usual, resulting in a stiffer and less flexible rope.

hawser laid rope
A rope of three **strands** which are twisted to form helixes around the same central axis.

laid rope
A rope in which 3 or more **strands** are twisted to form helixes around the same central axis. (See also **ordinary lay**.)

shroud laid rope
A 4-strand rope with or without a **core** with the **strands** twisted to form a helix round the central axis.

soft laid rope
A rope in which the length of lay of the **strands** and/or the rope is longer than usual resulting in a more flexible rope which is easily deformed.

spring lay rope
A rope made with 6 **strands** over a main **core**, each strand of which has alternating wire and fibre components laid over a fibre core.

rope marks; running marks
Long **crease marks** in dyed or finished goods running approximately in the warp direction. They are caused during wet processing in the rope form and may be the result of (i) the formation of creases along which abrasion or felting may occur, or (ii) imperfect penetration or circulation of the processing liquors.

rope-form processing
The treatment of fabric that has been drawn into the form of a rope, often by passage through a ring (pot-eye) of appropriate diameter.

roping yarn
A yarn from which a **strand** is produced.

rotor spinning
A method of **open-end spinning** which uses a rotor (a high speed centrifuge) to collect and twist individual fibres into a yarn.
Note: The fibres on entering the rapidly rotating rotor are deposited around its circumference and temporarily held there by centrifugal force. The yarn is withdrawn from the rotor wall and, because of the rotation of the rotor, twist is inserted. (See also **spinning**.)

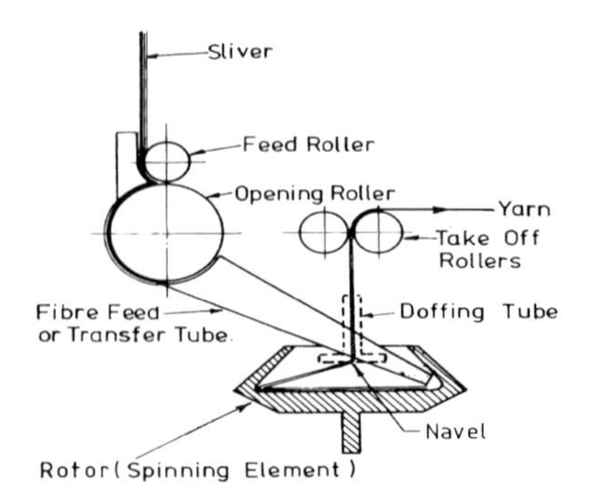

Rotor spinning

roughing (scutched flax)
A preliminary hand **hackling** operation involving the removal of tangled and short fibres by means of a roughing tool.

roughing-out (flax)
The rough separation of the seed from the chaff, short straw, weeds, and other extraneous material produced during deseeding. (See also **ripples comb**.)

round heel (knitting)
A fully fashioned hose heel made by continuous knitting across the whole width with widening or narrowing.

roving
A name given, individually or collectively, to the relatively fine fibrous strands used in the later or final processes of preparation for spinning.

roving courses; hand-hold; ravel courses; waste courses (weft knitting)
Additional **courses** used in the manufacture of knitted articles either as protective courses or to facilitate handling in subsequent operations. These courses are afterwards removed.

roving stop device
A roving clamp that automatically stops the flow of **roving** into the **drafting** zone of a spinning machine when the yarn breaks.

royal rib
1. A plain-weave fabric with a warp-way rib produced by **taped ends** and a high weft sett.
2. (Knitted) See **half cardigan rib**.

rubber (fibre) (generic name U.S.A.)
A manufactured fibre in which the fibre-forming substance consists of natural or synthetic rubber, including the following categories:
1. A fibre in which the fibre-forming substance is a hydrocarbon such as natural rubber, polyisoprene, polybutadiene, copolymers of dienes and hydrocarbons, or amorphous (noncrystalline) polyolefins.
2. A fibre in which the fibre-forming substance is a copolymer of acrylonitrile and a diene (such as butadiene) composed of not more than 50% but at least 10% by weight of acrylonitrile units

$$(-CH_2-\underset{\displaystyle CN}{\underset{|}{C}H}-).$$

The term **lastrile** may be used as a generic description for fibres falling within this category.
3. A fibre in which the fibre-forming substance is polychloroprene or a copolymer of chloroprene in which at least 35% by weight of the fibre-forming substance is composed of chloroprene units.

$$(-CH_2-\underset{\displaystyle Cl}{\underset{|}{C}}=CH-CH_2-).$$

(See also **elastodiene (fibre) (generic name)** and Classification Table, p.401.)

rubber-proofed sheeting
A woven sheeting fabric coated with a rubber compound and then vulcanized.

rubbing
1. A method of applying false twist to a **sliver**, **slubbing** or **roving** to effect the interlacing of

rubbing *(continued)*
fibres. The false twist is produced by the pressure applied to the fibrous materials as it passes between two surfaces of a machine, usually aprons, moving with a reciprocating sideways motion relative to the take-off direction of the material.
Note 1: The purpose of rubbing is to give the material sufficient strength to withstand the tensions imposed in the next process.
Note 2: As an alternative to inserting real twist, it enables faster operation of modern worsted drawing sets and removes the problems of **drafting** against twist.
2. See **crocking**.

ruche
A narrow woven or knitted heading (see **fringe** 2), usually having a very heavy multiple-thread weft passed through to form a skirt (see **loop ruche**), normally about 25mm wide over-all. It is used generally in lieu of piping round upholstery cushions. Other types consist of a web, the centre one-third of which consists of a pile weave or other raised effect, which is sewn around a central core. Ruches are usually flanged, the flanged portion being for insertion into the seam of the article to which it is to be sewn.

Cauliflower ruche

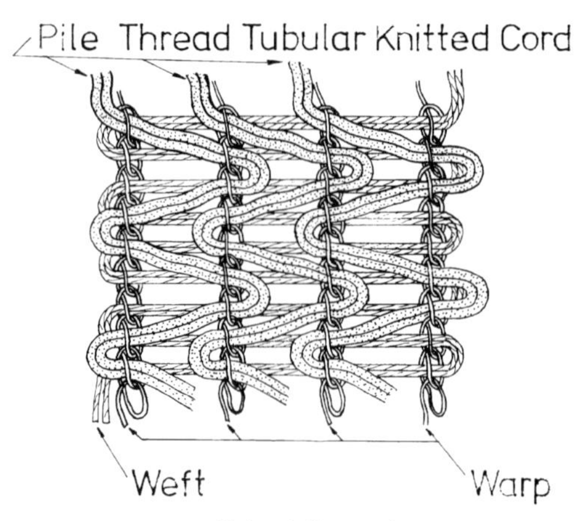

Knitted flat ruche

Knitted flat ruche: end section

Loop ruche

cauliflower ruche
A **woven flat ruche**, the weft of which forms a heavy uncut-pile effect on one side. The cross-section of a cauliflower ruche is almost semi-circular. It is sometimes called 'half-round ruche'.

crimped loop ruche
A **loop ruche**, the weft for which has been passed through a crimping machine.

cut crimped ruche
A **cut ruche**, the weft for which has been passed through a crimping machine.

cut ruche
A **ruche** woven or knitted in double width and cut down the middle.

knitted flat ruche
A **ruche** consisting of a loosely constructed warp-knitted web with an uncut-pile effect on one surface.

loop ruche; tape ruche
A **ruche**, the skirt of which is in the form of an uncut pile

pile ruche
A **ruche** made as a **woven flat ruche**, but with pile-forming elements.

woven flat ruche
A **ruche** made from a narrow woven tape with warp patterning, but without pile forming elements.

ruched fabric, warp-knitted
A three or four-bar fabric in which the front one, two, or three guide bars are part-set threaded and mis-lap while their warps are stopped so producing a discontinuous pleat in the form of small raised areas, the size, shape, and arrangement of which depend on the threading and lapping movement of the guide bars.

ruffle; beam ruffle
The cylindrical surface integral with or attached to a **beam** that serves as a brake drum in the systems of control of **warp** tension.

ruffling
See under **gathering**.

rug; mat
A textile floorcovering of predetermined shape and limited dimensions.

rug, braided
A textile floorcovering of braided cords sewn together.

rug (flax)
1. The partly scutched short straws that fall below the two compartments of a **scutching** machine after the **shives** have been shaken out of it.
2. The waste made during the production of scutched flax on a wheel.

rug tow
See **flax tow**.

rug wool
A wool yarn, generally woollen-spun, twisted six-fold, the single thread before plying being not finer than 350 tex.

run-in (warp knitting)
The length of yarn used in warp knitting 480 courses, generally measured on one end of warp thread of each warp sheet. On lace designs it is generally the length of yarn per pattern repeat. (See also **rack (warp knitting).**)

runnage (cordage)
specific length especially of cords and ropes.

runner
A long narrow length of textile floorcovering finished at both ends.

running loom efficiency
See **loom efficiency (running)**.

running marks
See **rope marks**.

running-on (weft knitting)
The operation of placing a series of loops on to **points** or needles on a loop-to-point or loop-to-needle basis, preparatory to further knitting, or to joining fabrics together by **linking**.

running-on course
See **slack course (weft knitting)**.

Russia braid; soutache braid
1. A narrow braid with two cores side by side, covered with fine yarns, which go backwards and forwards over one core and under the other as a continuous figure 8.
2. Two or more adjacent braids, interlaced to form stripes.

Russian cord
A colour-striped shirt or dress fabric in which cord stripes are produced by **leno weaving**. One end is made to cross a number of ends in an alternate crossed and open shed sequence.

sable fibre (hair)
Fibre from the sable (*Mustela zibellina*).

sacking
A general name applied to coarse fabrics used chiefly for the making of bags or sacks. They are often made of jute, hemp, flax or polyolefin, and the number of threads per centimetre may vary from 2 to over 12.

sailcloth
Originally a tightly woven cotton or linen canvas used in the manufacture of ship and yacht sails. It is now more common for these fabrics to be manufactured from nylon for spinnakers, and polyester or aramid for foresails and mainsails. Newer developments include laminated constructions which give greater dimensional stability.

salt figure; salt index; salt number
The concentration of an aqueous sodium chloride solution in g/100 ml required to produce incipient coagulation of viscose under standard conditions. (See also **Hottenroth number**.)

salt sensitivity
1. The extent to which the dyeing properties of a dye are affected by the addition of a neutral electrolyte to the dyebath.
Note: This term is usually only applied in the dyeing of cellulosic fibres.
2. The susceptibility of coloured material to change in colour when spotted with aqueous solutions of neutral electrolytes.

sample
A separate unit or part representative of the lot, consignment or design.

sample swatch
See **swatch**.

sand crêpe
A fabric with an irregular surface texture made from silk or manufactured fibres; it is heavier and has a rougher, harsher handle than **crêpe de chine**. A typical plain-weave construction: 40x18; 15.6tex pigmented acetate warp x 22.2tex viscose rayon crêpe weft inserted 2S:2Z; woven 1.1m wide for 0.91m finish; 136 g/m^2. A crêpe weave may be used in which case the crêpe weft would not necessarily be picked 2S:2Z, and the warp need not be pigmented, but the fabric would be finished to give the appearance of sand.

sandfly net, warp-knitted
A fine **net** produced with two half-sett threaded guide bars each making four course atlas-type lapping movements in opposition to each other.

sandwash finish
A characteristic soft handle conferred on silk materials which have been subjected to the gentle abrasion of a sand and water mixture.

sandwich-plated fabric, knitted
See **plated fabric, weft-knitted**.

Sanforizing
A controlled compressive shrinkage process.
Note: The trade mark Sanforized® is owned by Cluett-Peabody Inc. and can be applied to fabrics which meet defined and approved standards of washing shrinkage.

sansevieria
A fibre obtained from the leaves of various species of plants of the Genus *Sansevieria*.

saponification
Hydrolysis of ester groups by an alkali. (See also **deweighting** and **S-finish**.)

saran (fibre) (generic name U.S.A.)
A manufactured fibre in which the fibre-forming substance is any long-chain synthetic polymer composed of at least 80% by weight of 1,1-dichloroethene (vinylidene chloride) units. (See also Classification Table, p.401.)
Note: saran is a **chlorofibre**.

sarsnet ribbon
A ribbon constructed entirely in plain weave of very fine warp and weft and with high density, the weft density being higher than that of the warp.
Note: A true sarsnet is made wholly of silk.

sateen
1. (Weave) A weft-faced weave in which the binding places are arranged with a view to producing a smooth fabric surface, free from **twill**.
Note: Since there is confusion in the use of this term, it is safer to qualify it by 'weft'.
2. (Fabric) A fabric made in sateen weave.
Note: In North America, this is a strong, lustrous, cotton fabric generally made with a five-harness satin weave in either warp or filling-face effect.

satin
1. (Weave) A warp-faced weave in which the binding places are arranged with a view to producing a smooth fabric surface, free from twill.
Note: Since there is confusion in the use of this term it is safer to qualify it by 'warp'.
2. (Fabric) A fabric made in satin weave.
Note: In North America, this is a smooth, generally lustrous fabric with a thick close texture made in silk, manufactured or other fibres in a satin weave for warp-face or sateen weave or filling face effect.

satin (warp knitting)
A two-bar warp-knitted fabric in which the front-bar underlaps are arranged with a view to producing a smooth surface. Typical front-bar laps are given by the notation:

/1-0/3-4/ or /1-0/4-5/.

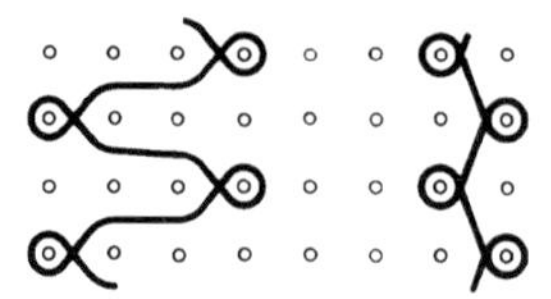

Front bar *Back bar*
Warp-knitted satin structure

satin drill
See **drill**.

satinet weave; satinette
A four-end irregular **satin** or **sateen** weave.

saturation (of a colour)
The nearness of a colour in purity to the associated spectral colour.

saturation bonding
A method of making **nonwoven fabrics** in which a fibre **web** or **batt** is treated by immersion in an adhesive in liquid form. (See also **adhesive-bonded nonwoven fabric**.)

saturation value
The maximum quantity of a dye which can be absorbed by a substrate under defined conditions.

saturator
A machine for thoroughly wetting fabrics with an aqueous solution, or allowing interchange of liquor in wet-on-wet processing.

saxony
A high-quality fabric, made of wool of 60s quality or finer, spun on the woollen system.

scaffolding yarn
That component of a plied yarn that is used to support a weaker component through further processing until it is satisfactorily introduced into a fabric.
Note: The scaffolding yarn may then be removed by solvent or other chemical action or, alternatively, be retained in the fabric to make it more durable.

scale harness
See **split harness**.

scallop
Curves or indentations along the edge of a fabric.

schappe silk
Originally, yarn spun from fibre degummed by **schapping**, but nowadays the term is increasingly used as an alternative to **spun silk**.
Note: The change in meaning reflects the greatly decreased use of fermentation processes for degumming.

schappe-spun
Originally used to describe a silk yarn from fibre degummed by the **schapping** process, but now used in Europe as a term synonymous with **silk-spun**.

schapping
A method of degumming applied to silk waste, which removes part of the gum by a fermentation process. Up to 10% of gum may remain on the fibre.

Schiffli embroidery machine
An embroidery machine consisting of a multiplicity of lockstitch sewing elements working on a basic net or fabric that is attached to a frame movable vertically and horizontally according to the requirements of the pattern.

schreiner
1. Descriptive of a calender with two or three **bowls** in which one (the middle one in a three-bowl calender) is of highly polished steel engraved with very fine parallel lines (grooves) running at an angle of approximately 20° to either the bowl axis or cross direction.
2. (Finish) Descriptive of a finish obtained by passing a fabric through a schreiner **calender**. The object of the process is to enhance the **lustre** of the fabric.
3. (Bowl) The engraved bowl of a schreiner calender.
Note 1: The number of lines on the bowl may vary from 5-24 per millimetre, but is usually in the range 9-14.

schreiner *(continued)*
Note 2: The angle of inclination of the lines is chosen to ensure good cover of the fabric and minimum fibre damage, e.g., a low inclination of 15-20° is recommended for weft **sateen** fabrics. Optimum effects are obtained when the lines slightly cross the direction of the surface yarn twist.
Note 3: In use, the engraved bowl is heated, usually to 60-120°C for finishing cotton fabrics.

Scotch beaming; Scotch dressing
See under **dressing (warp preparation)**.

Scotch carpet
See **ingrain carpet**.

scouring (washing)
The treatment of textile materials in aqueous or other solutions in order to remove natural fats, waxes, proteins and other constituents, as well as dirt, oil and other impurities. (See also **solvent scouring**.)
Note: The treatment varies with the type of fibre.

scray
A curved trough in which fabric accumulates, either during a dwell period in a process, or whilst awaiting treatment.

screen printing
A design reproduction process, developed from stencilling, in which print paste is forced through the unblocked areas of a mesh, in contact with the substrate. The mesh may be a woven fabric or a metal screen, flat or cylindrical (rotary screen). Pressure is applied to the paste by a squeegee (blade or roller), which is moved when the screen is stationary or stationary when the rotary screen is rotating.

scrim
A general term, irrespective of structure, for a lightweight **basecloth**.

scrimp roller; scroll roller
Rollers or bars (rails) characterized by grooves or projections inclined at equal and opposite angles to the centre line on each half and used for removing folds or creases during finishing operations.

scrimp(s)
A printing defect in which a lengthwise portion of the fabric is not printed because it is creased as it passes through the printing zone.

scroll (tufting)
See under **tufting machine**.

scroll gimp
A woven figured narrow fabric having two series of wefts and a warp. Each series consists of

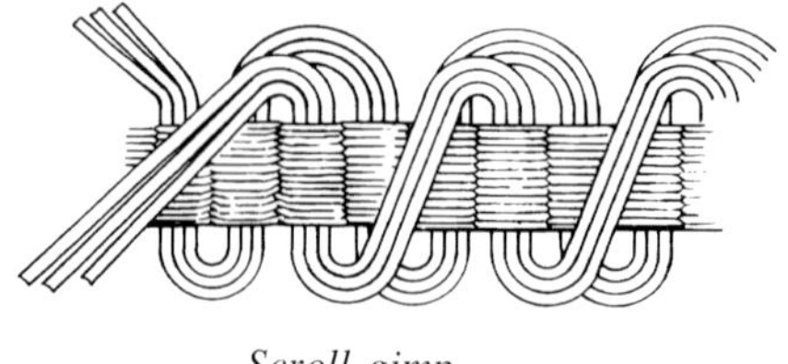

Scroll gimp

three **gimp** cords laid flat. The ground series projects at one edge to form a triple loop; the figure series passes through the warp and returns over the warp alternately to form a loose scroll on the surface. The overall width of the fabric, which is in plain weave, is about 16mm.

scroll roller
See **scrimp roller**.

scroop
A rustling sound produced when a material, which usually has a characteristic 'dry' handle, is compressed by hand. Scroop is usually associated with silk but is also produced in certain cellulosic fibres, yarns, or fabrics by suitable finishing treatments.

sculptured pile
See **pile (carpet)**.

scutcher (finishing)
A machine for continuously opening fabric which has previously been in rope form.

scutching (flax)
The operation of separating the woody part of deseeded or retted flax straw (see **retting (flax)**) from the fibre.

scutching; picking (cotton)
The final operation in the cotton system preparation line, in which the cotton **flocks** are opened mechanically, cleaned, and formed into a **lap** of specified mass per unit area, for feeding to a **carding** machine.
Note: Scutching (picking) machines are being extensively replaced by the combination of opening and cleaning machine units followed by **chute feeding**.

scye (clothing)
The armhole.
Note: Possibly a corruption of 'arm's eye'.

Sea Island cotton
Gossypium barbadense, an exceptionally fine, long-staple type of cotton grown formerly in Sea Island, off the coast of Georgia, U.S.A.

seal fibre (hair)
Fibre from the seal (Family *Pinnipedia*).

sealed edge
See under **selvedge, woven**.

sealed sample
A **sample** confirmed by specified persons and durably designated as the standard for a product or one or more of its features, for example, colour, style, make, dimensions, fabric, or trimmings.

seam
The join between two or more plies or pieces of material. Seams are usually formed by sewing but welding, adhesives or other joining means are sometimes used.

seam allowance; turnings
A pre-determined amount of material between the edge of component parts of the garment and the **seam line**. (See also **bight**.)

seam binding
A woven **narrow fabric**, usually of cotton or manufactured fibres, made in plain weave, usually with contrasting selvedges, and intended for use in covering or strengthening seams or edges in clothing.

seam damage
Damage to a seam leading to a reduction in seam performance or appearance. It is invariably associated with a change in the physical condition of one or more of the components of the seam. (See also **stitching damage**.)

seam failure
The effective breakdown of a seam due to rupture of the sewing thread, rupture of yarns in the fabric, excessive **seam slippage**, or any combination of these. (See also **seam strength**.)

seam fraying
Descriptive of fabric yarns running parallel to the seam in the turning (see **seam allowance**) becoming loose and falling out from a raw edge.

seam grin; grinning; seam gape; seam gap
The unwanted opening of a **seam** whilst under transverse stress, which reveals the sewing thread spanning the gap. (See also **grin**.)
Note 1: Typical causes of seam grin are low sewing tensions and excessively extensible threads.
Note 2: If a seam gap develops due to the displacement of the warp or weft threads parallel to the seam, then this may be additionally described as **seam slippage**.

seam line
The line along which pieces of material are joined.

seam mark
A particular form of **pressure mark** in a fabric produced by the thickness of a seam during scouring, dyeing, or finishing operations.

seam pucker
Unwanted waviness of the material along the **seam line**. This may be apparent immediately after sewing or it may develop later in use.

seam slippage
The unwanted displacement of fabric yarns parallel to a **seam line** arising from transverse stress, resulting in a partial or complete loss of seam integrity.

seam strength
The load required to cause **seam failure**, either in a sample removed from a made-up product or in a sample made to simulate a seam in a product, normally measured at right angles to the **seam line**.

seam type
A formal description of the configuration of a seam.
Note 1: Four classes of seam are defined in ISO 4916:1991 *Textiles - Seam types: classification and terminology*. This standard designates seam types using an alpha-numeric coding system, with the following general categories:

Class SS Superimposed seams
Class LS Lapped seams
Class BS Bound seams
Class FS Flat seams

Other seam configurations which are not load-bearing are classed as:

Class EF Edge finishing stitchings

Class OS Ornamental stitchings

Note 2: For formal definitions and illustrations, reference should be made to the ISO documents. The seam types below are a selection of constructions with common-usage definitions. Many of these seam types require specialised sewing machines but reference to this machinery is not essential to the definition. However, some terms in common use fail to distinguish between stitch, seam and type of sewing machine and may be ambiguous, such as lockstitch.

bound seam
A seam having its material edges covered with a strip of additional material.
Note: May be used for neatening, protecting and as a design feature.

butt seam
A seam with the two material edges abutting each other.

cup seam
See **cup seaming**.

flatlock seam
A **butt seam** formed using a flatlock stitch. The edges of the material are trimmed during stitching.

French seam
A type of flat folded seam with only one line of stitching visible. Two plies of material are superimposed back to back and stitched together producing a seam with a narrow **bight**. The plies are then opened out, folded face to face, and stitched again at a distance from the fold sufficient to enclose the edges of the material.
Note: Formation of this seam is often accompanied by trimming, to maintain the narrow bight and to remove frayed threads.

overlock seam
Two or more superimposed plies of material, aligned along their edges, are joined together, edge-trimmed and oversewn in one operation, with overedge stitches having two or more threads.

plain seam; flat seam
Usually a lockstitch seam, formed by a line or lines of stitches joining two pieces of material together face to face.

rolled seam
A seam in which the two edges of the material to be joined are rolled over together and secured by a single line of stitching.
Note: This construction is also used for edge finishing single plies of fabric.

taped seam
A seam stabilised and reinforced by a tape.

seat angle
A variable angle to the vertical of the centre back seam used when drafting bifurcated garments.
Note: It is a means of achieving ease of movement in the finished garment.

seat belt webbing
A woven webbing, usually of polyester, with special physical characteristics used for body restraining purposes.

secant modulus
The ratio of change in stress to change in strain between two points on a stress-strain curve,

secant modulus *(continued)*
particularly the points of zero stress and breaking stress.

secondary backing (carpet)
See under **substrate (carpet)**.

secondary cellulose acetate
See **cellulose acetate**.

secondary strand
A rope forming one of the **strands** of a **cable laid rope**.

seconds
Textile products which, owing to some fault or imperfection, do not reach an agreed standard of quality.

section marks
Individual warp stripes in woven or warp-knitted fabrics that occur at regular intervals across part or all of a fabric width, the distribution coinciding with the width of the warping section, and the stripes being the result of tension or individual package differences within the section during warping.

section warping
1. Yorkshire and Scotch warping and silk-system warping; horizontal-mill warping. A two-stage-machine method of preparing a warp on beam, consisting in:
(i) winding a warp in sections on a reel (drum, mill, swift);
(ii) beaming-off the complete warp from the reel on to a warp beam. The procedure is as follows:

(a) Ends in closely spaced sheet form (approximately loom-warp sett), withdrawn from a warping creel, are wound on the machine reel to loom warp length.

(b) Each such sheet of ends is called a 'section'.

(c) Stability of the yarn-build of the first section on the reel is obtained by moving the section sheet laterally at a regular rate as winding proceeds, so that its outer edge is supported by a fixed or adjustable incline fitted at one end of the reel.

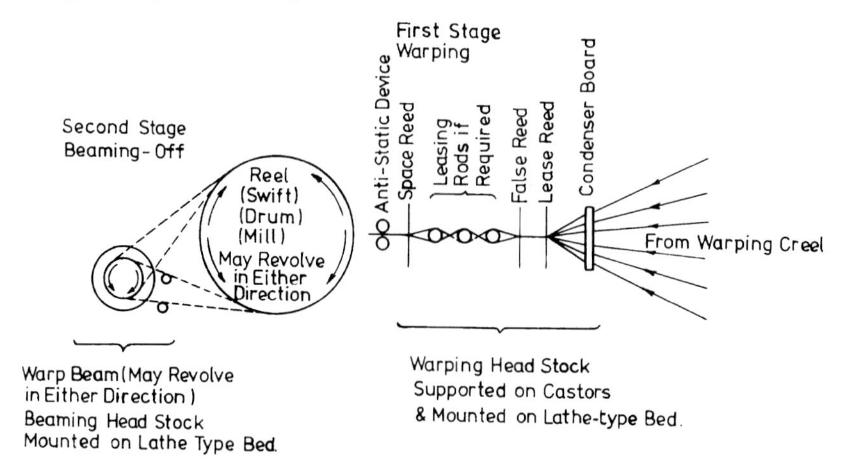

End view of horizontal-mill section-warping 1 machine

(d) When the first section is complete, the other edge of its build on the reel is a replica of the original incline. Sections to the number required for the complete warp are wound similarly.

(e) Fancy warps are prepared by 'dressing' yarn packages in the creel for each section in conformity with the warp pattern plan.
(f) Section sheets are attached to a beam and withdrawn simultaneously from the reel by rotation of the beam. This unrolling is controlled to provide suitable warp tension for winding the warp on to the beam. As beaming-off proceeds, the warp beam is moved laterally at the same rate but in the direction opposite to that of the section during warping. This ensures that the complete warp sheet runs from the reel to between the beam flanges without the need for lateral deviation.

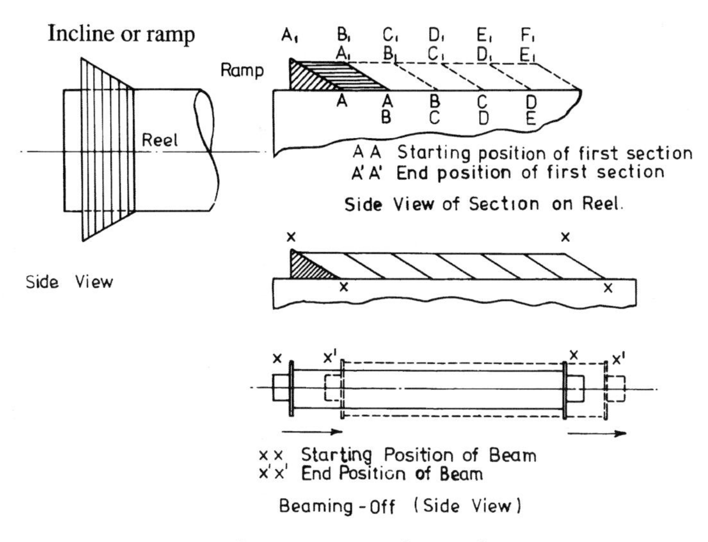

Section-warping 1 procedure

Note: The efficiency of a section-warping machine is normally of the order of 25-30% because warping ceases during the beaming stage. Single-stage section-warping machines (without a beaming mechanism) are designed with interchangeable reels to improve warp output. The reels may be either of normal capacity (one warp of loom-beam length) or of large capacity (several warps of loom-beam length). Warp from either type may be transferred to warp beams in a beaming machine or on to loom beams in a warp-sizing machine. In each method, a reel is traversed at the same rate as during warping but in the opposite direction.
2. A two-stage-machine method of preparing a warp on beam, consisting in:
(i) winding 'section' beams;
(ii) assembling section beams in warp-beam form. The procedure is as follows:
(a) Ends in closely spaced sheet form (approximately loom-warp sett), withdrawn from a warping creel, are wound on to a beam to loom-warp length. The width of the beam between flanges is equal to the loom warp-sett width of the section sheet. The number of equal-length section beams wound is determined by the respective number of ends in the section and the complete warp.
(b) Section beams (with or without flanges) are assembled side by side on the 'shaft' or barrel of a suitable warp beam.
(c) Section sheets are then fed simultaneously to a sizing machine (or dressing frame) and taken-up on a loom beam (beam-to-beam-sizing). This method of warping is used for making fancy warps in which the number of ends in a pattern repeat is greater than the number of ends in the warping creel. All sections of the same end-pattern in a warp may be made in succession and afterwards assembled in correct order. Planned asymmetrical stripes may be obtained by reversing the winding-off direction of alternate section beams of one pattern. By such means, the amount of creel dressing is reduced in comparison with that required for section warping 1.

seed cotton
Cotton which has been harvested but not ginned, so that the fibre is still attached to the seed.

seed hair
Fibres growing from the surface of seeds or from the inner surfaces of fruit cases or pods. Such fibres (seed hairs) are formed by the marked elongation of epidermal cells. The most important example of seed hair is cotton.

seersucker
A fabric characterized by the presence of puckered and relatively flat sections, particularly in stripes, but also in checks.
Note: The effect may be produced in a variety of ways, e.g.,
(i) by weaving from two beams, with the ground ends at a higher tension and the ends for the puckered stripes woven at lower tension;
(ii) by treatment of cellulosic fabrics, particularly linen and cotton, with sodium hydroxide (caustic soda) solution, which causes the treated parts to contract;
(iii) by using yarns having different shrinkage properties; these yarns maybe combined in the warp and/or the weft.

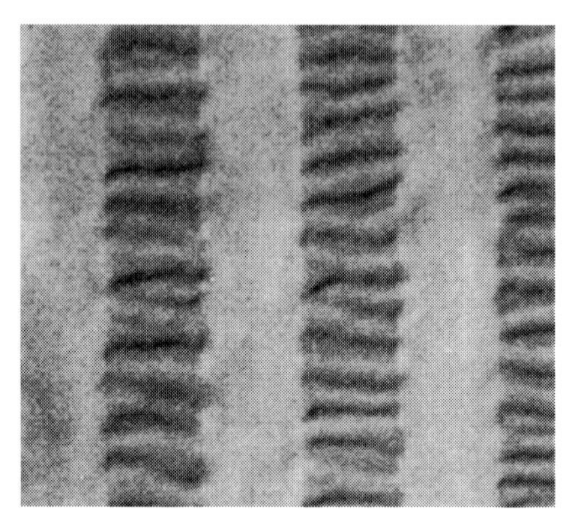

Seersucker fabric

self-stitching
See **double cloth**.

self-twist spinning
A method of spinning a plied yarn by **false-twisting** two drafted fibrous strands in order to insert alternating S and Z twist along their lengths. The strands are then brought into contact along their lengths so that their untwisting torques coincide and cause the strands to twist around each other producing a plied yarn termed a **self-twist yarn** or ST yarn.
Note: The alternating-twisted strands are termed AT-strands.

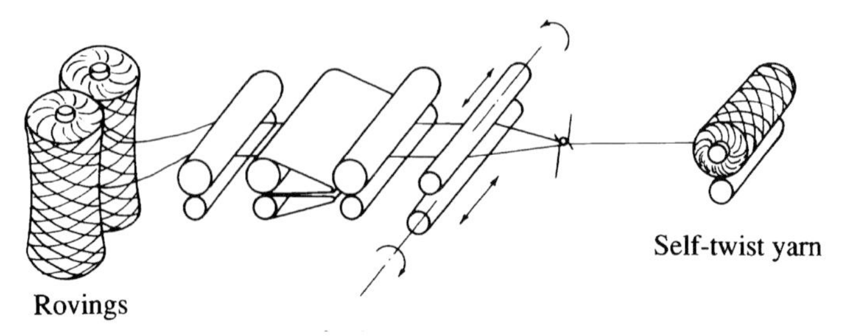

Self-twist spinning

self-twist yarn; ST yarn
An inherently twist stable, 2-ply yarn structure having alternating S and Z ply twist strands (see **twist direction**). The two strands are themselves twisted with alternating S and Z twist prior to being plied.
Note 1: The alternating twist present in the strands and in the resulting yarn takes the following change of order: Z and S in the strands give S and Z respectively in the self-twist yarn.
Note 2: The repeated pattern in the self-twist yarn is S twist-no twist-Z twist. The total length of this pattern is called the cycle length. The no-twist length is shorter than the twist lengths and is called the twist-change-over zone.
Note 3: The lengths of S and Z ply twist are essentially equal.

selvage
Synonym of **selvedge**.

selvedge (knitting)
When used without qualification, the longitudinal edge of a fabric or a garment panel produced during knitting. The term can also be applied to fabric in which the yarn is cut rather than turned at the end of a course of loops.

selvedge, cockled
See **selvedge, slack**.

selvedge, cracked
A tight selvedge in which the warp threads have been broken during processing.

selvedge, distorted
A selvedge that does not itself vary in width, but which is not straight as a result of variations in the fabric width.

selvedge, dog-legged
See **selvedge, uneven**.

selvedge, dressing (lace)
A band of fabric on the longitudinal edges of a piece providing a means of holding the lace during the process of dressing. It is removed after dressing. In plain net, a cord edge is provided.

selvedge, puckered
See **selvedge, slack**.

selvedge, pulled-in
See **selvedge, uneven**.

selvedge, slack
A selvedge that is slacker than the adjacent fabric owing to incorrect balance of fabric structure between the ground and the selvedge or owing to insufficient tension when selvedge ends are being woven.

selvedge, tape
A selvedge produced from a **selvedge warp** with ends weaving in pairs.

selvedge, tight
A selvedge that is tighter than the adjacent fabric owing to incorrect balance of fabric structure between the ground and the selvedge or owing to excessive tension when selvedge ends are being woven.

selvedge, uneven
A selvedge that varies in width. This should not be confused with distorted selvedge (see **selvedge, distorted**).
Note: Variations in weft tension or lack of control of the warp ends within the selvedge may result in such unevenness. Pulled-in selvedges are caused by pulling-in of the edges by isolated tight picks; dog-legged selvedges are the result of the characteristic gradual change in weft tension that occurs as some types of weft pirn are unwound (see **cop-end effect**), regular changes in selvedge width being present at each pirn change.

selvedge, woven; list; listing
When used without qualification, the term refers to the longitudinal edges of a fabric that are formed during weaving, with the weft not only turning at the edges but also passing continuously across the width of the fabric from edge to edge.
Note: Selvedges are often up to 20mm wide and may differ from the body of the fabric in construction or weave or both, or they may be of exactly the same construction as the body of the fabric and be separated from it by yarns of different colour. Selvedges may contain fancy effects or may have brand names or fabric descriptions woven into or printed on them but their main purpose is to prevent fraying of the outside ends from the body of the fabric and to give strength to the edges of the fabric so that it will behave satisfactorily in weaving and subsequent processes. Examples of different types of selvedges are:

> **leno edge**
> A set of threads interlacing with a gauze weave either at the edge or in the body of a fabric. In the latter case, it prevents ravelling when the fabric is severed in the direction of the warp. When in the body of the fabric, a leno edge is often referred to as a 'centre selvedge'. (See also **splits**.)
>
> **sealed edge**
> The cut edge of a fabric produced from yarns having thermoplastic properties that has been treated by heat or chemical means to prevent ravelling of the edge.
>
> **shuttleless weaving machine edge**
> 1. In some types of shuttleless weaving, either one, or in some cases, both edges differ from the normal woven selvedge in that the weft is held in position at the turn by threads other than the warp threads, e.g., by the use of an independent thread to lock the weft in position at the edge, or by the interlocking of the weft threads.
> 2. In some shuttleless weaving machines, an edge formed by severing the weft near to the edge of the fabric, and tucking the cut end into the shed formed on the next pick.
>
> **twist selvedge; helical selvedge**
> A set of threads making a half or complete revolution around one another between picks at the edge of a fabric.
>
> **narrow fabric selvedge (needleloom)**
> A double pick is inserted into each successive shed from one side of fabric. The selvedge at the side of the loom from which the double pick is inserted is secured by the structure of the fabric (the outside end must be made to change sheds) and at the opposite selvedge successive loops of weft are locked by one of the following methods: (i) knitting successive loops together; (ii) knitting a catchthread through the loops; (iii) knitting a catchthread through the loops together with a locking thread; (iv) passing a catch thread through each loop by means of a small shuttle.

selvedge guider
An electronic or mechanical device for presenting fabric to stenter pins or clips.

selvedge uncurler
A device for straightening selvedges and fabric edges which have rolled during processing.

selvedge warp
A number of threads intended to form a selvedge, wound on to either a beam or a bobbin, and delivered in the form of a sheet.

selvedge widening (weft knitting)
See under **shaping (weft knitting)**.

semi-collapsed balloon spinning; semi-suppressed balloon spinning
A system of **ring spinning** in which the rotating yarn balloon is greatly reduced at the start of an empty bobbin and is allowed to expand later when the bobbin is becoming filled with yarn. The small-balloon condition is achieved by allowing contact between the rotating balloon and the spindle top. As the bobbin fills, contact is lost and the yarn **balloon** is allowed to expand.
Note 1: Special spindle top extensions are used to obtain the required effect.

semi-milled finish
A finish on lightly milled fabrics containing wool.

semi-open shedding
A type of shedding in which threads that are to remain in the top shed line for the next pick are lowered a short distance and are then raised again. The other threads are raised and lowered as in **open shedding**. (See also **shedding**.)

semi-permanent set
See **setting**.

semi-suppressed balloon spinning
See **semi-collapsed balloon spinning**.

semi-worsted spun
A term applied to yarn spun from sliver produced by carding and gilling in which the fibres are substantially parallel, the carded sliver not having been condensed or combed. Alternatively, the yarn may be produced from a roving.
Note: The above definition is descriptive of processing technique and not of fibre content.

sennit rope
See **braided rope** under **rope**.

separating course (weft knitting)
A course of knitted loops separating one garment or garment part from another, that, on removal, permits the separation of articles that are knitted as a succession of interconnected units. Separating courses may be cut, unroved, or dissolved, and several such courses may be made consecutively.

sequential knitting (weft knitting)
The knitting of shaped garment pieces in a fixed sequence, for example, a front, a back and two sleeves.

sequestering agent
A chemical capable of reacting with metallic ions so that they become part of a complex anion. The principle is used to extract calcium and magnesium ions from hard water, iron (II) and copper ions from peroxide bleach liquors and various metallic ions from dyebaths, by forming a water-soluble complex in which the metal is held in a non-ionizable form.

serge
A piece-dyed fabric of simple twill weave (usually 2/2) of a square or nearly square construction

serge *(continued)*
and with a clear finish. It was originally made of wool, but now is sometimes made of other fibres, or of blends of wool with other fibres.

sericin
See **gum**.

set (testing)
Strain remaining after complete release of the load producing the deformation.

set twist
See **twist set**.

set yarn
See **stabilized yarn**.

set-in sleeve
See under **sleeve (clothing)**.

set-out (lace machines)
The selected width of repeat on a Leavers lace machine, determined by the threading of the warp and beam yarns in the steel bars. It is defined in terms of the number of carriages in the width.
Note: Narrow laces may be made on sub-divisions of a set-out, but the bars will not normally be rethreaded to correspond with these sub-divisions.

sett; set; pitch
1. A term used to indicate the density of ends or picks or both in a woven fabric, usually expressed as the number of threads per centimetre. The state of the fabric at the time should be described, e.g., loomstate or finished.
2. See **count of reed**

sett, square
A fabric in which the number of ends per centimetre and the number of picks per centimetre are approximately equal. For practical reasons, the linear densities of warp and weft would normally be approximately the same in such a fabric.

sett, unbalanced
A fabric in which there is an appreciable difference between the numbers of ends and picks per centimetre.

setting
The process of conferring stability of form upon fibres, yarns, fabrics, or garments, usually by means of successive heating and cooling in dry or moist conditions.
Note 1: The different levels of setting may be described as:
(i) temporary set: set which is destroyed by relatively minor treatments such as light mechanical stress or immersing the set material in water or by ordinary usage. 'Cohesive set' is a term used in the wool industry for set that is removed easily, especially by water immersion.
(ii) semi-permanent set: set which is resistant to ordinary effects experienced in use but which can be destroyed by more severe treatment.
(iii) permanent set: set in which a structural change takes place that cannot be reversed. More severe setting treatments may cause further irreversible changes of structure.
Note 2: The term is sometimes used in conjunction with a description of the particular characteristic to be stabilized (e.g., crêpe setting, crimp setting, flat setting, twist setting) or of the setting medium (e.g., heat setting, steam setting, thermal setting).

settle (hat manufacture)
Mechanical treatment of hardened **forms** in a dilute solution of inorganic acid to effect further felting and **hardening**.

sew-knit fabric
See **stitch-bonded fabric**.

sewing damage
See **stitching damage**.

sewing thread
A thread intended for stitching material either by hand or by machine.
Note 1: A variety of thread structures and linear densities are used for sewing. Structures include cabled, braided, core-spun, textured and mono-filament yarns; the very finest sewing threads are often monofilaments; the coarsest include thick cabled threads used in sewing heavy canvasses for tarpaulins, etc., and for saddlery; in knitted garments, especially in linked seams, the knitting yarn may also be used as a sewing thread.
Note 2: To facilitate sewing, a lubricant, wax or other finish is often applied to the thread and some thread types may be polished.
Note 3: Many sewing thread manufacturers classify threads by means of ticket numbers which relate indirectly to ranges of nominal thread linear density (decitex) values. The sewing thread ticket number or 'thread size' determines the choice of sewing needle size.
Note 4: Although sewing threads are usually employed for the seaming or edge-binding of fabrics or other sheet materials or for attaching other components (e.g. buttons), they may also be used for decoration only. However, threads or yarns used in embroidery are usually called embroidery threads, yarns or silks. (See also **shirring thread**.)

S-finish
A finish produced on cellulose ester textiles by alkaline hydrolysis causing surface **saponification**.

shade
1. A particular depth or intensity of **hue**.
2. To bring about relatively small modifications in the colour of a substrate by adding a further small amount of colorant with the object of matching a given pattern more accurately.

shade bar
See under **bar (woven fabric)**.

shade cloth
An open fabric used in agriculture to control shade and temperature, reduce evaporation, and reduce damage to plants and crops from wind, rain, hail and frost. Shade cloth creates a local microclimate and may also keep out birds and insects.

shaded; shading
General terms referring to variations in the shade of dyed textile materials, excluding **ending**. (See also **listing** 1.)

shading (carpet)
A change in the appearance of a textile floorcovering owing to a difference in light reflection because of localized alterations in the orientation of the fibres, tufts, or loops. Shading can occur as follows:
(i) temporary shading: a reversible localized change in orientation described as a normal characteristic of certain cut-pile floorcoverings.

shading (carpet) *(continued)*
(ii) permanent shading: an irreversible localized change in orientation of the pile of textile floorcoverings (sometimes known as 'water-marking', 'pooling', or 'pile reversal').
(iii) tracking: a gradual change in appearance of a textile floorcovering from edge to middle of a narrow band caused by repeated walking over the same area. It may result in a localized change in pile orientation and may be irreversible.
Note: The boundary delineating the shaded area often, if not always, assumes a rough, matted texture where the pile is trying to lie in the two different directions, i.e., normal and reversed. This effect is part of the shading phenomenon.

shading (warp knitting)
Transverse defects caused by structural distortion in warp-knitted fabric.

shadow check
See **shadow stripe, woven**.

shadow stripe fabric
See **atlas fabric, single bar**.

shadow stripe, woven
An effect, caused by different reflections of light, produced in woven fabrics by employing yarns of different characteristics, usually of S and Z-twist, in warp or weft (or in both, when it becomes a 'shadow check').

shadow welt
See **after-welt**.

shafty wool
Strong, dense and well-grown wool with good length and spinning characteristics.

shag pile
See **pile (carpet)**.

shaker bar (tufting)
See under **tufting machine**.

shaker motion
See **leno-weaving**.

shalloon
A 2/2 twill-weave fabric made from crossbred worsted yarns, used as a lining for coats, liveries, etc.

shantung
A plain-weave silk dress fabric exhibiting random yarn irregularities resulting from the use of yarn spun from tussah silk.

shantung-type yarn
An irregular yarn made from fibres other than natural silk to imitate the yarn used for making shantung.

shaping (weft knitting)
Descriptive of the process used to shape a knitted product during knitting by changing the number of stitches per **course**, **wale**, or unit area in the fabric. The various methods which may be used separately or in combination may be identified as follows:

full-fashioned; fully fashioned (weft knitting)
Terms applied to knitted fabrics and garments that are shaped wholly or in part by widening and/or narrowing by loop transference to increase or decrease the number of wales. This increase or decrease is brought about by increasing or decreasing the number of needles operative at either or both selvedges without alteration to the character of the stitch whilst maintaining a secure selvedge.

press-off narrowing (weft knitting)
A method of shaping a garment panel by pressing-off loops at the selvedge of the fabric and thereby reducing the total number of needles knitting.

selvedge widening (weft knitting)
A method of shaping a garment panel by introducing additional needles at one or both **selvedges** in a particular sequence designed to increase the width of the fabric over a given number of courses whilst maintaining a secure selvedge and without changing the structure.

stitch shaped
A garment shaped wholly or partially by change of stitch length, or structure, or both, e.g., from 1x1 rib to half cardigan rib.

sharkskin fabric
A generic name used to describe a woven or warp-knitted fabric, the characteristic of which is a firm construction and a rather stiff handle. One type of woven sharkskin fabric is made from continuous-filament yarn of high linear density in plain weave. Others may have several ends and/or picks woven as one to give a softer handle. Spun yarn may be used. The usual warp-knitted sharkskin fabric is a two-bar construction made with full-set threading in both guide bars, the lapping movement of the back guide bar involving the underlapping of three or four needle spaces, and that of the front guide bar of one needle space in the opposite direction.

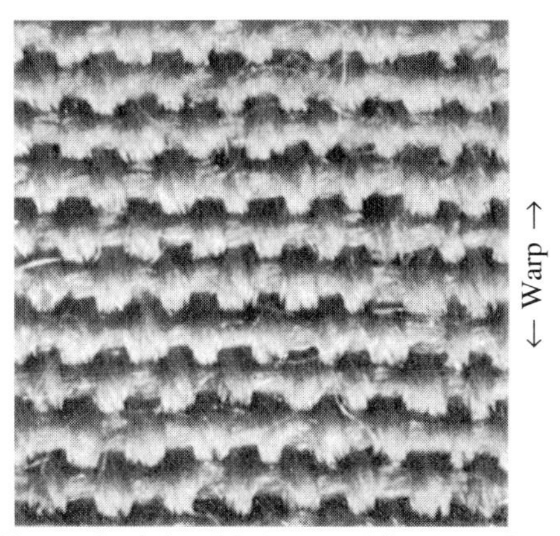

Woven sharkskin fabric (magnification x 5)

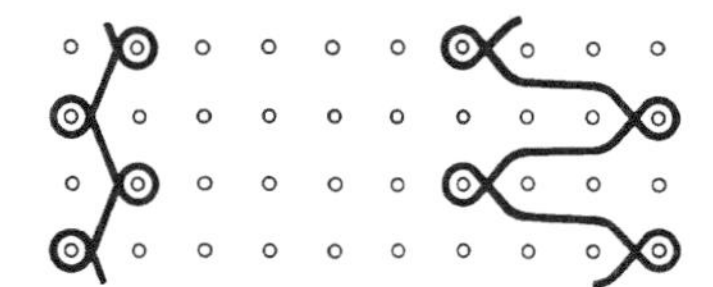

Warp-knitted sharkskin structure

shear
1. To cut the fleece from a sheep.

shear *(continued)*
2. To cut a nap or pile to uniform length or height (also called **crop**).
3. To cut loose fibres or yarn from the surface of a fabric after weaving (also called **crop**).

sheared terry pile
See **velour, woven** 3.

shed; warp shed
The opening formed when warp threads are separated in the operation of weaving.

shed efficiency
See **loom efficiency (overall)**.

shedding
A primary motion in weaving: the separation of warp threads, according to pattern, to allow for weft insertion or **picking** prior to **beating up** (see **open shedding**, **semi-open shedding** and **closed shedding**).

sheeting
A medium-weight, closely woven, plain-weave or 2/2 twill-weave fabric, made from yarns of medium linear density and used primarily for bed coverings. Condenser-spun weft may be used. Typical construction:
(i) plain weave: 25x23; 35x33tex; K=14.8+13.2;
(ii) 2/2 twill weave: 24x28; 21x37tex; K=11.0+17.0;
(iii) plain weave - raised: 28x18; 30x9tex; K=15.3+5.4.

shell-stitch fabric
A warp-knitted fabric having a raised shell-like surface produced by holding loops on certain needles while knitting takes place on others at each course (see **held loop**). The arrangement of the held loops in pattern formation gives the shell-like effect.

shepherd's-check effect
A small check effect developed in black and white, or in contrasting colours, generally by groups of four, six, or eight threads of the two colours and in twill weaves, commonly the 2/2 twill. The description applied to the effect probably originates from the traditional ¼ inch wide check pattern developed in black and white and in 2/2 twill weave as featured on the plaids worn at one time by shepherds in the hills of the Scottish Borders. Dogstooth or **houndstooth check** is a particular form of shepherd's-check effect.

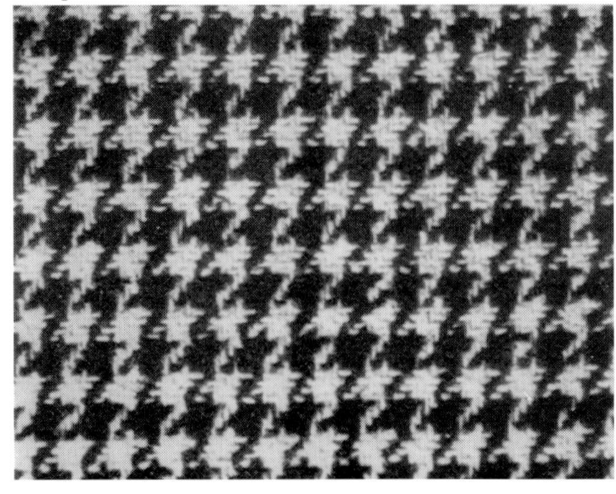

A shepherd's-check effect

Shetland
1. (Original usage) Descriptive of a yarn spun by hand in the Shetland Islands from the wool of sheep bred and reared in those islands.
2. (Common usage) Descriptive of a yarn, spun on the woollen system from 100% new wool, of a quality capable of imparting to a fabric the handle attributed to the products formerly made exclusively from wool from the Shetland breed of sheep.
Note: In current trade usage, the term Shetland is qualified by the adjective 'genuine', 'pure', 'real', or any similar description. This implies that the wool actually originated in the Shetland Islands. This usage is recognised by the International Wool Textile Organisation.

shiner
A warp or weft thread, usually of a continuous-filament yarn, that is more lustrous (and generally tighter) than its neighbours (see **tight end (defect)** and **tight pick (defect)**).

shingosen
Advanced synthetic fibres and fabrics, usually the product of a combination of progressive technologies, with some speciality in **handle** and high aesthetic appeal.
Note 1: This Japanese expression (literally, 'new synthetic fibre') is used in English as a plural noun or as an adjective qualifying a singular or plural noun.
Note 2: Shingosen fabrics are almost exclusively of polyester though small amounts of other fibres, e.g., rayon may occasionally be included.

shires
See **weft crackiness**.

shirring
See **gathering**.

shirring thread
A covered elastomeric yarn used in sewing to produce ruched and pleated effects and for elastication.

shives (flax)
Short pieces of woody waste beaten from straw during **scutching**.

shivey wool
Wool that contains small particles of vegetable matter other than burrs.

shoddy
Fibrous material made in the woollen trade by **pulling** new or old knitted or loosely woven fabric in rag form. (See also **mungo**.)

shoddy shaker; Issit's shaker
A machine used for shaking **shoddy** reclaimed from under carding engines to remove dirt.

shog (lace machines)
A lateral movement, usually of a specific number of gaits (see **gait (lace machines)**) imparted to certain of the bars of a machine, e.g., the guide bar, point bar (bobbinet and furnishing), and comb bar (bobbinet).

shog (warp knitting)
See **underlap**.

shog; rack (weft knitting)
The lateral movement of the needle-bed on a flat knitting machine or the angular displacement

shog; rack (weft knitting) *(continued)*
of the dial relative to the cylinder on a circular machine.

shogging (warp knitting)
The lateral movement of a guide bar relative to the needle bar over a predetermined number of needle spaces. (See also **overlap (warp knitting)** and **underlap (warp knitting)**.)

shoot; shute
See **weft**.

short-staple carding
See under **carding**.

short-staple spinning
The use of cotton spinning machinery to produce staple yarns from cotton or any other type of fibre possessing similar length and fineness characteristics (see **cotton-spun**).

shot
See **pick**.

shot effect
A colour effect seen in a fabric woven with a warp of one colour and a weft of a contrasting colour by using dyed yarns or by cross-dyeing. The effect is usually associated with fabrics of plain or 2/2 twill weave.

shottage (carpet weaving)
The number of pick insertions per row of pile woven. For example, three weft units inserted per pile row denotes a 'three-shot' construction.

shrink-resistant; shrink-resist
Descriptive of textile materials that exhibit dimensional stability conforming to specified standards based on tests designed to simulate normal conditions of usage.
Note: This property may be an inherent property of the textile material or may be conferred by physical or chemical processes or both.

shrink-resistant finish
A treatment applied to a textile material to make it **shrink-resistant**.

shrinkage
Reduction in length (or width) of a fibre, yarn, fabric or other textile, induced by conditioning, wetting, steaming, chemical treatment, wet processing as in laundering, dry heat or mechanical action.
Note: In commercial practice and in the literature the following specific terms have been used to describe the shrinkage which occurs in testing procedures: **residual shrinkage**, relaxation shrinkage, felting shrinkage, and consolidation shrinkage. (See also **compressive shrinkage**.)

shroud laid rope
See under **rope**.

shuttle (lace machines)
1. (Schiffli embroidery machine) A boat-shaped yarn-package holder travelling in a slide in such a manner that it passes through the loop formed in the needle thread thus forming the back thread of the lock stitch. The shuttle-yarn package is a coreless **cop** and tension is applied by means of a spring over the thread hole.
2. (Lace furnishing machine) A term used in Scotland for the **carriage**.

shuttle (weaving)
A yarn-package carrier that is passed through the shed to insert weft during weaving. It carries sufficient weft for several picks.

shuttle box
A compartment at each end of the loom sley for retaining the shuttle in the required position before and after picking.

shuttle checking
The action of arresting the shuttle in the **shuttle box** after picking.

shuttle loom
A term loosely used to describe a **hand loom**, **treadle loom**, or **weaving machine** that uses a **shuttle** to insert the weft.

shuttle marking; shuttle tapping
Warp bruising usually caused by abrasion of the lower shed between the under-side of the shuttle and the **race board**. This is often the result of faulty shuttle flight.
Note: Excessive frictional heat produced by the shuttle in its flight can, as it abrades a warp of yarns made of synthetic fibre, cause filaments to melt. This is often associated with faulty shuttle flight and/or faulty shedding.

shuttle trap
See **trap**.

shuttleless weaving machine
See **weaving machine**.

shuttleless weaving machine edge
See under **selvedge, woven**.

siddo rags
See **syddo rags**.

side weft fork
See **weft fork**.

side-by-side bicomponent fibre
See **conjugate fibre** under **bicomponent fibre**.

sides (weft knitting)
See under **knitted loop (weft knitting)**.

sighting; tinting
Temporary coloration of textile materials either (i) for visual identification of a particular fibre quality or type of yarn, or (ii) to enable a printer to see the pattern when applying colourless substances.

sighting colour
See **fugitive tint**.

silesia
A lining fabric with a smooth face. Originally it was a plain-weave fabric but now 2/1 or 2/2 twill weaves are chiefly used. The fabric may be piece-dyed, colour woven in stripes, or printed.

silicone finish
A polymeric **finish** containing combined silicon applied to the surface of textile materials to confer water repellency and/or softness.

silk
1. The protein fibre forming the cocoons produced by silkworms.
Note: The natural fibre is covered by sericin (silk gum), which is usually removed in processing.
2. Descriptive of yarns, fabrics, or garments produced from silk. The gum may or may not be present.

silk, pure
Silk in which there is no metallic or other weighting of any kind, except that which is an essential part of dyeing.

silk, raw
Continuous filaments or strands containing no twist, drawn off or reeled from silk cocoons.
Note: The term raw silk is often incorrectly used to describe a silk-noil woven fabric.

silk, wild
Fibres extruded by insect larvae other than *Bombyx mori*.
Note: The most important of these larvae is tussah (tussore) (see **tussah silk**). Other types include: eri (*Phylosamia ricini*), muga (munga) (*Antheraea assamensis*), anaphe (*Anaphe*), kuriwata, and *Gonometa postica*.

silk lap (warp knitting)
A lapping movement produced on a Milanese machine in which the yarn traverses two wales per course. (See also **milanese fabric, warp-knitted**.)

silk noils
Fibres extracted during silk dressing or combing that are too short for producing **spun silk**. These fibres are usually spun on the condenser system to produce what are known as silk-noil yarns.

silk waste
The fibres remaining after drawing off, reeling, or throwing net silk, and fibres obtained from damaged or unreelable cocoons.

silk-spun
A term applied to staple yarn produced by dressing or combing and spinning on machinery originally designed for processing waste silk into yarn (see **spun silk**). (See also **schappe-spun**.)
Note: Whenever the term silk-spun is used, it is qualified by the name of the fibre and fibres from which the material is made.

simili binding
See under **binding**.

simili mercerizing
A calendering process for increasing lustre. The effect is similar to that obtained by mercerizing but is not permanent.

simple rib, weft-knitted
A fabric in which loops of single wales intermeshed in one direction are separated by loops of one or more wales intermeshed in the other direction.
Note: A common example is 3 and 1 rib.

simplex fabric
A double-faced fabric usually made on two needle bars of a bearded-needle warp-knitting machine: the two sets of warp threads are meshed together successively on each needle bar to produce a fabric that normally has the same appearance on both sides.

simplex warp-knitting machine
A double needle bar machine using bearded needles mounted vertically or nearly so, in which the fabric is supported and controlled by sinkers. The fabric is removed from the knitting zone in a downward direction. (See also **knitting machine**.)

singe
To remove, in a flame, or by infra red radiation, or by burning against a hot plate, unwanted surface fibres. The operation is usually performed as a preliminary to bleaching and finishing.

singed yarn
See **gassed yarn**.

single bar vandyke fabric
See **atlas fabric, single bar**.

single jersey, weft-knitted
A generic name applied to knitted fabrics made on one set of needles. The characteristics may be varied to achieve the desired end-use or patterned effects. (See also **single-jersey jacquard, weft-knitted; single-jersey tuck jacquard, weft-knitted;** and **plain fabric, weft-knitted**.)

single lift (weaving)
A term applied to lever dobbies (see **dobby**) and jacquard mechanisms in which a single knife or **griffe** is used to effect the **lift**. A **closed shed** is produced.

single marl yarn
See **worsted yarns, colour terms**.

single mottle yarn
See **worsted yarns, colour terms**.

single phase weaving machine
See **weaving machine**.

single piqué, weft-knitted
A non-jacquard **double jersey** fabric made on an interlock basis using a selection of knitted and tuck loops in the following sequence. (See also **double jersey, weft-knitted**.)
Note: The fabric is sometimes referred to as **cross tuck**.

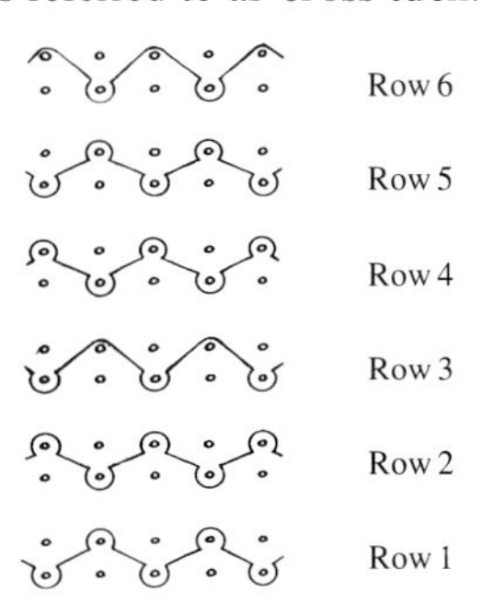

Single piqué

single satin ribbon
A ribbon woven from continuous-filament yarns, formerly also in silk, having a satin face on one side only and a contrasting edge.

single tie
See **jacquard tie (weaving)** under **jacquard (mechanism)**.

single yarn; singles
A thread produced by one unit of a **spinning machine** or of a silk **reel**.

single-head loom
A narrow-fabric loom that weaves one **piece** only.

single-jersey jacquard, weft-knitted
A patterned **single-jersey** weft-knitted fabric, usually made from two or more yarns of differing colour or texture to give a construction that consists essentially of knitted and float loops, but may incorporate tuck loops. The surface pattern is derived from the chosen arrangement of the yarns and of the knitted and float loops. The inclusion of tuck loops into the construction eliminates long lengths of floating threads from the back of the fabric. (See also **accordion fabric** and **single-jersey tuck jacquard, weft-knitted**.)

single-jersey tuck jacquard, weft-knitted
A patterned **single-jersey** weft-knitted fabric usually made from two or more yarns differing in colour or texture in construction that consists of knitted and tuck loops. The surface pattern is derived from a chosen arrangement of the yarn and of the knitted and tuck loops.

singles
See **single yarn**.

sinkage
1. Loss of mass in wool cleansing, usually expressed as a percentage.
2. Unaccounted or 'invisible' loss of mass in processing, usually expressed as a percentage.

sinker (weft knitting)
A blade that works in conjunction with knitting needles and assists with (i) loop formation, and (ii) fabric holding-down.

sinker loop (weft knitting)
See under **knitted loop (weft knitting)**.

sinker timing (weft knitting)
See **timing (weft knitting)**.

sinker top machine (knitting)
A **circular knitting machine** in which a set of needles, mounted vertically in a needle cylinder, co-operates with a corresponding set of horizontal **sinkers** mounted radially in a sinker ring. (See also **knitting machine**.)

sinker wheel machine
A bearded-needle circular weft **knitting machine** in which knitting takes place on a set of horizontal radially mounted needles, the yarn being manipulated with the aid of **sinkers** mounted in a wheel.

sinnet rope
See **braided rope** under **rope**.

sisal
A pale cream fibre obtained from the leaf of the sisal plant (*Agave sisalana*). The fibre from other *Agave* plants, and particularly from henequen (*Agave fourcroydes*) resembles sisal very closely and indeed is sometimes also termed sisal. (See also **hemp, true**.)

sival machine
See under **lace machines**.

six-by-three rib fabric, weft-knitted
See under **rib fabric, weft-knitted**.

size
1. A gelatinous film-forming substance, in solution or dispersion, applied normally to warps but sometimes to wefts, generally before weaving, to protect the yarns from abrasion in the healds and reed and against each other; to strengthen them; and, by the addition of oils and fats, to lubricate them.
Note: The main types are carbohydrates and their derivatives, gelatin, and animal glues. Other substances such as linseed oil, polyacrylic acid, and polyvinyl alcohol are also used.
2. See *Note* under **count of yarn**.

size concentration
The number of kilograms of oven-dry size solids in 100 kilograms of size liquor.

size percentage
The number of kilograms of oven-dry size solids on 100 kilograms of oven-dry yarn.

size pick-up
The number of kilograms of size liquor on 100 kilograms of oven-dry yarn as it leaves the nip of the squeeze rollers.

sizing, beam-to-beam
The method of machine sizing in which a warp is transferred from a warp beam to a loom beam. The procedure is as follows:
(i) Warp in sheet form, withdrawn from a warp beam (e.g., beamed-off a **section-warping** machine reel), is passed through a **sow box** and the **quetsch** squeezing-rollers of a sizing machine. Application of size solution by immersion or by contact with a partially immersed roller, penetration of the yarn by size solution, and removal of the surplus size solution occur at this stage.
(ii) The sized warp is dried by hot air or by contact with steam-heated cylinders (cans) *en route* to the loom beam.

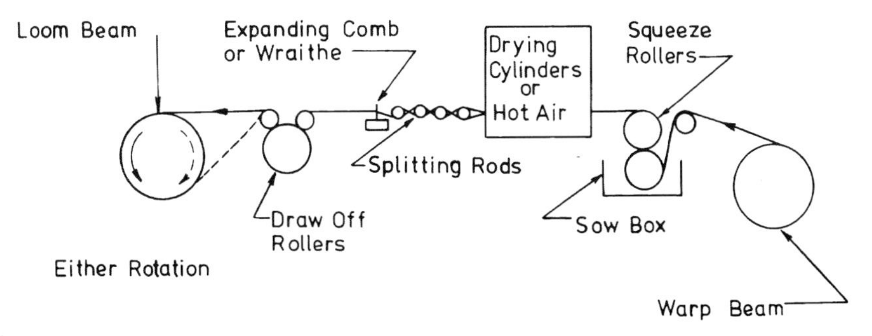

Beam-to-beam sizing (silk system)

sizing, cake
See **cake sizing**.

sizing, crêpe
The application of size to continuous-filament yarns intended for twisting with crêpe twists.

sizing, dry
The application to textile yarns of sizing materials compounded with liquids volatile at temperatures below 100°C (212°F).

sizing, single end
The application of size to: (i) a thread supplied from a single package: the thread receives a unidirectional side-traverse as it is wound on to a large drum on which it forms an endless sheet of yarn. This sheet is then cut and wound on to a beam under tension to produce a sample (or short length) warp; (ii) a low density sheet of yarn in which the adjacent ends do not touch one another: the yarn is usually supplied either from a cone creel or a single warper's beam, and the resulting beam is usually combined with other similar beams in a **dry taping** process, to produce a weaver's beam.

skein
See **hank**.

skein break factor
See **Lea count-strength product**.

skein sizing
See **hank sizing**.

skew
A fabric condition in which the warp and weft yarns, although straight, are not at right angles to each other. The effect is due to the fabric's structure and is not a distortion imposed during processes subsequent to weaving. (See also **drawn piece** and **off-grain**.)

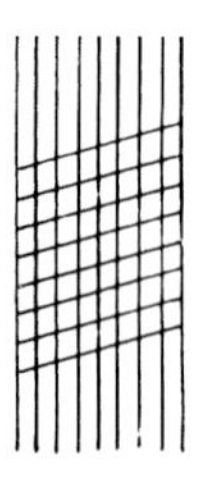

Skew

skin wool
Wool removed from the skins of slaughtered sheep (see **slipe**).
Note: There are three methods of removal: (i) lime-steeping; (ii) sweating (by bacterial action); (iii) painting the flesh side of the skin with, for example, sodium sulphide (sometimes combined with (i)).

skirt (narrow fabric)
See **fringe** 2.

skirting (wool)
1. The removal of wool, different from the main bulk, from the edges of a fleece. (See also **wool**

classing.)
2. A wool sorting term for stained parts of the fleece, such as the legs and the whole edge of the fleece.

skitteriness
An undesired speckled effect arising from differences in colour between adjacent fibres or portions of the same fibre.

slack course (weft knitting)
A **course** of knitting made with loops longer than normal for a special purpose, e.g., linking, running-on, etc.

slack end
A warp thread or part of a warp thread that has been woven into fabric at a lower tension than the adjacent ends which are tensioned normally.

slack mercerization
See **mercerization**.

slack pick
A weft thread or part of a weft thread that has been woven into fabric at a lower tension than the adjacent picks which are tensioned normally.
Note: Such a pick may include a **weft loop (defect)**.

slack selvedge
See **selvedge, slack**.

slasher sizing (cotton system)
A process in which warp yarns are sized during transfer from warper's beams to loom beams. Two or more size boxes may be used in parallel and/or in tandem if the warp sheet is too dense for effective sizing in one box, or if it contains yarns with different fugitive tints. Slasher sizing is also known as 'slashing'.
Note: The term appears to originate in the action of cutting through the combined warp sheet to separate loom-beam warps, as each beam is completed.

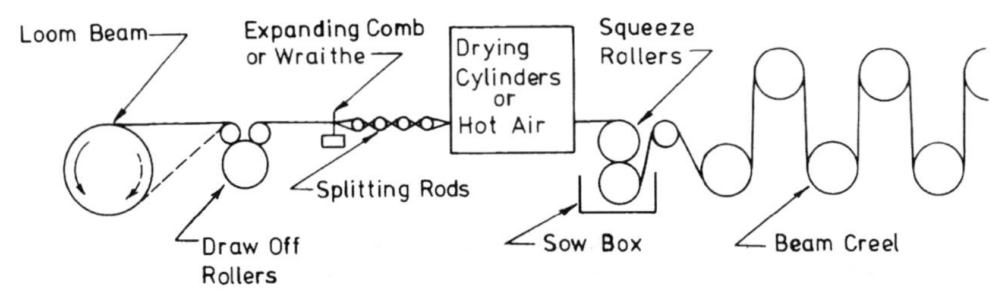

Slasher sizing

slashing
1. A synonym for **slasher sizing**.
2. A term that has been adopted to indicate the process used to reduce the extensibility of rayon yarns, particularly yarns used in the manufacture of tyres. The process consists in stretching yarn in the wet state and then drying it while maintaining the stretched length.

slat (tufting)
See under **tufting machine**.

sleeve (clothing)
That part of a garment that covers all or part of the arm.

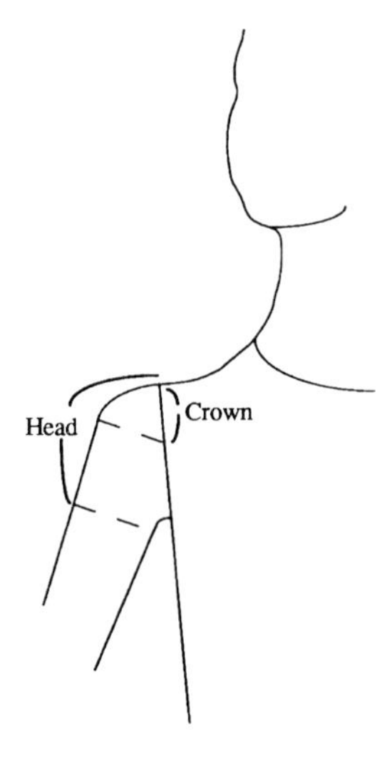

Sleeve crown and sleeve head

kimono sleeve
Generally used to describe any loose fitting, wide sleeve cut in one with the body of the garment. The term derives from the traditional garment of Japan.

magyar sleeve
A sleeve which is integral with the body of the garment.
Note: For close-fitting garments made from a woven fabric, a **gusset** may be included.

puffed sleeve
A short sleeve which finishes above the elbow and has **fullness** gathered at either the **sleeve crown** or the sleeve hem, and frequently at both.

raglan sleeve
A sleeve with the armhole line extending from the front and back of the **scye** to the neckpoint so that the shoulder section is joined to the **sleeve crown**, eliminating the conventional shoulder and **sleeve head** seams.

set-in sleeve
A sleeve which is sewn into the armhole.

sleeve crown
The top section of the **sleeve head** (approximately the upper half) (see diagram).

sleeve head
The upper part of the sleeve, that which is inserted into the armhole (see diagram).

sley; slay
1. That oscillating part of a weaving machine, positioned between the **healds** and the **fell** of the cloth, which carries the **reed**.
Note: Also known in the U.K. as lay, lathe, batten, going-part, fly-beam.
2. See **reed**.

sley (lace machines)
A device consisting of several wires, or a woven wire mesh, or plates drilled with holes at regular intervals, used for separating and determining the spacing of threads of jacquard strings on lace machines and ancillary winding equipment.
Note: In Scotland, the jacquard-string sley is known as a 'holey board'.

sleying plan
See **denting plan**.

slide fastener tape
See **zip fastener tape**.

sliding needle bar (tufting)
See under **tufting machine**.

slip
See **leno weaving**.

slip drafting
1. A technique commonly employed in yarn production in which **drafting** takes place between two pairs of rollers and within the drafting zone some system of controlling fibre movement is utilized. The most common method is to utilize double **apron** control with the top apron purposely slightly separated from the bottom apron so that long fibres which are gripped by the front rollers can slip through whilst shorter fibres are held back. This is usually achieved by having a recessed top apron roller.
2. The withdrawal of fibres during drafting from a pair of rollers of which the top one is self-weighted or is purposely slightly separated from the bottom roller.

slipe
Lime-steeped wools (see **skin wool**).

slipper satin
A heavy, smooth, high-quality satin made from continuous-filament silk or manufactured fibre yarns, suitable for wedding gowns, evening dresses, evening shoes, and jockeys blouses, traditionally with 120 ends/cm and enough picks to ensure a well-constructed fabric. A typical construction using acetate yarns would be: 130x38; 67x44dtex; K=33.6+7.6; 136 g/m^2; 8-end satin weave.

slit-film yarn
See **tape yarn**.

sliver
An assemblage of fibres in continuous form without twist.

sliver high-pile, weft-knitted
A single-jersey fabric in which untwisted staple fibres in **sliver** form are normally knitted at each loop to form a pile surface on the technical back of the jersey structure.

slop padding
See **padding** 1.

slotted wire
See **wire (pile weaving)**.

slub
An abnormally thick place in a yarn.

slub repp
A **repp** fabric in which the coarse weft is a slub yarn (see **slub yarn** under **fancy yarn**).

slub yarn
See under **fancy yarn**.

slubbing
The name given, individually or collectively, to the fibrous strands produced during the stages of preparation for spinning, and also to strips of web from a **condenser card** that have been consolidated into a circular cross-section by rubbing (see **roving**).

slubby yarn
A yarn that contains unintentional **slubs**.

slurgalling (weft knitting)
A fabric fault caused by variation in loop length between successive courses on a straight-bar knitting machine.

slurry steeping
A process in the manufacture of viscose rayon in which a pulp is dispersed in a solution of sodium hydroxide in the preparation of alkali-cellulose. (See also **steeping**.)

smash
The breakage of a large number of warp ends, usually as the result of a shuttle trap (see **trap** 1). Other weaving machine parts, e.g., the **shuttle**, the **reed** etc., may also be damaged.

smooth-drying
See **easy-care**.

snag (knitting)
A pulled thread course-wise in weft knitting or wale-wise in warp knitting. In weft knitting a small snag is known as a 'fisheye'.

snagging test
A test by which woven or knitted fabrics are characterised for their tendency to form undesirable surface loops when caught by sharp points.

snarl
See **warp snarl**.

snarl yarn
See under **fancy yarn**.

snarly yarn; lively yarn
Yarn that shows an excessive tendency to twist round itself if held with insufficient tension. (See also **twist liveliness**.)

snow
An accumulation of yarn debris generated by abrasion, especially during false-twist texturing by friction-twisting of synthetic-fibre yarns.

snubber pin
A stationary pin or guide that induces a localized change in yarn tension. A **draw pin** is a type of snubber pin.

soft laid rope
See under **rope**.

soft-flow jet dyeing machine
A **jet-dyeing machine** in which the fabric is transported from the dyeing chamber over a driven reel (which provides the motive force) and is then transported back to the dyeing chamber by the gentle action of the jet.

softening
The application of a chemical agent and/or a mechanical process, e.g., calendering, to give fabrics a soft handle and frequently a smooth appearance. A number of chemical softening agents also confer a fullness of handle.

soil release finish
A textile finish which, when applied to a textile, makes it easier to remove soiling or stains by ordinary domestic washing.

solid colour yarn
See **worsted yarns, colour terms**.

solid woven belting
See under **belting**.

solvent bonding
A method of making **nonwoven fabrics** in which a solvent is used to soften the fibre surfaces in a **web** or **batt** and hence cause bonding between fibres.

solvent dyeing
Dyeing carried out from a continuous non-aqueous phase.
Note: Water may be added to assist the dyeing process.

solvent finishing
The treatment of textile materials with reagents, other than dyes, dissolved in organic solvents.

solvent scouring
The treatment of fabrics in organic solvent media to remove impurities such as lubricating oils and spin finishes.

solvent-assisted dyeing
Dyeing carried out from an aqueous dyebath containing a small proportion of an organic solvent in solution, which normally acts to accelerate dyeing.

souple
Partially degummed silk.

sour
To treat textile materials in a bath of dilute acid.

soutache braid
See **Russia braid**.

sow box; quetsch
Primarily the container (trough, pan) of the size solution of a warp-sizing machine, often steam jacketed and/or provided with open or closed steam piping for heating the size solution.
Note 1: The term is also used loosely to indicate the assembly of trough, immersion, and sizing rollers of a slasher-sizing machine.
Note 2: The tendency is to restrict the use of sow box to the above primary meaning, and the term quetsch to indicate the complete assembly. The use of the terms quetsch-box and quetsch-trough should be noted. (See also **quetsch**).

space dyeing
The production of multicolour yarns by application of various colorants at intervals along a yarn by printing or other methods.
Note: Alternative processes include 'knit-deknit' where a knitted tube of fabric is printed, fixed and deknitted; the injection of dye-liquor into the inner layers of wound packages (Astro-dyed method); and blanking off portions of wound packages before treatment with dye-liquor (Frauchiger method).

span length
See under **fibre length**.

spandex (fibre) (generic name U.S.A.)
A manufactured fibre in which the fibre-forming substance is a long-chain synthetic polymer comprised of at least 85% of a segmented polyurethane (see **elastane**). (See also Classification Table, p.401.)

SPC
See **statistical process control**.

specific stress
The ratio of force to the linear density expressed in mN/tex or N/tex.

specification
A precise statement of a set of requirements to be satisfied by a material, product, system or service that indicates the procedures for determining whether each of the requirements is satisfied.

spectral colour
The colour produced by light of a single wavelength.

spectral tristimulus values
See **colour measurement**.

spin winder; spin assembly winder
A spinning machine in which the outputs from two adjacent spinning units are assembly wound on to one package. These are usually subjected to a subsequent twisting operation.
Note: An example of this technique is **fasciated yarns** which are combined before being twisted together to produce a two-fold yarn.

spin-draw-texturing
A process for making **textured yarns** in which extrusion, drawing and texturing stages are integrated on a single machine.
Note: At the present time, spin-draw-texturing is restricted to the hot-fluid jet method of textured yarn production (see **textured yarn**, *Note 1* (iii)).

spin-drawing
A process for spinning partially or highly oriented filaments in which most of the orientation is introduced between the first forwarding device and the take-up, i.e., spinning and drawing are integrated sequential stages. (See also **draw-spinning**.)

spin-stretch ratio; draw-down; extrusion ratio
In manufactured filament extrusion, the ratio of take-up or haul-off speed to the average speed of the spinning fluid at the exit from the spinneret hole.

spindle tape
A woven narrow fabric of width normally not greater than 50mm , usually of high warp density and designed for the transmission of power to spindle **wharves** of small diameter, e.g., on textile spinning and twisting machinery.

spinneret; spinnerette
1. (Entomology) The small orifices on filament-spinning insects, in particular, on the lower lip of the silkworm and at the rear of the abdomen in the spider, through which thread-forming material is extruded in the formation of a cocoon, web, or other filamentous structure.
2. (Manufactured fibres) A nozzle or plate provided with fine holes or slits through which a fibre-forming solution or melt is extruded in fibre manufacture.

spinners' double; double end; married yarn (defect)
Two ends inadvertently running on to one spindle during spinning operations. This is usually caused by the breakage of one end followed by its subsequent entanglement with an adjacent end after which the two continue to run in married form.

spinning
The present participle of the verb 'to spin' used verbally, adjectivally, or as a noun, meaning the process, or the processes, used in the production of yarns or filaments.
Note 1: The term may apply to:
(i) the drafting and, where appropriate, the insertion of twist in natural or staple manufactured fibres to form a yarn (see **cap spinning, continuous yarn felting, double roving spinning, flyer spinning, friction spinning, hollow-spindle spinning, jet spinning, mule spinning, open-end spinning, ring spinning, rotor spinning, self-twist spinning, twistless spinning**, and **wrap spinning**);
(ii) the extrusion of filaments by spiders or silkworms; or,
(iii) the production of filaments from glass, metals, fibre-forming polymers or ceramics.
Note 2: In the spinning of manufactured filaments, fibre-forming substances in the plastic or molten state, or in solution, are forced through the holes of a spinneret or die at a controlled rate (extrusion). There are five general methods of spinning manufactured filaments, but combinations of these methods may be used (see **dispersion spinning**, **dry spinning**, **melt spinning**, **reaction spinning**, and **wet spinning**).
Note 3: In the **bast** and leaf-fibre industries, the terms 'wet spinning' and 'dry spinning' refer to the spinning of fibres in the wet state and in the dry state respectively.

spinning bath
1. A coagulating bath into which a solution or dispersion of a fibre-forming polymer is extruded during the processes of **wet spinning** or **dispersion spinning** respectively.
2. A bath containing one set of reactive components into which a solution containing another set of reactive components is extruded during the process of **reaction spinning**.

spinning frame
A machine consisting of a number of spinning positions for converting **slivers**, **slubbings**, or **rovings** into **yarn**. (See also **spinning**.)

spinning pack
See **pack (manufactured fibre extrusion)**.

spinning pump
A small pump, usually of the gear-wheel type, used to provide a uniform flow of a spinning solution or molten polymer to a spinning jet.

spinning solution
A solution of fibre-forming polymer as prepared for extrusion through a **spinneret**.
Note: A spinning solution is often referred to as 'dope', a term historically associated with the use of cellulose ethanoate (cellulose acetate) solutions as varnishes.

spiral yarn
See under **fancy yarn**.

spirality
See **course spirality (weft knitting)** and **wale spirality (weft knitting)**.

splicing (knitting)
The reinforcement of areas of knitted goods by the knitting of another yarn, known as splicing yarn, along with the main thread.

splicing (yarn joining)
The joining of two yarns by fibre entanglement, ideally to produce a tailless joint of short length and of similar diameter to the yarns, and therefore smaller than the knot which is the usual alternative.

split (defect)
See **crack**.

split cam (weft knitting)
A cam consisting of two or more parts disposed at different distances from the needle bed or cylinder, at least one of these parts being movable. The parts are used in conjunction with knitting elements having butts of different lengths to move some or all of them into a different path.

split end
A continuous-filament warp thread that has lost some of its filaments, usually as a result of abrasion during weaving, and has woven as a thin end.

split film; split tape
See **fibrillated yarn**.

split harness; scale harness; bannister harness
1. A jacquard harness that has a knot in each double harness cord below the comber board and above the mail so as to form a loop long enough to allow a proper depth of shed. A rod is passed through the loops of each long row of harness cords so that each is capable of lifting all the ends in this row independently of the figuring cords. The jacquard lifts the required ends to form the figure and the rods lift to form the ground weave. The harness is used in weaving jacquards with a finely sett warp and more open weft.
2. See **divided harness**.

split reed (narrow fabrics)
A special reed (used in some high-speed single-head shuttle looms) consisting of upper and

lower parts, which separate to allow the weft to move from behind the reed to the fell prior to **beating-up**.

splits
Fabrics woven two, three, or more in the width and later separated, one or two empty dents usually being left to indicate the cutting line. Fraying of the edges may be prevented by the use of a leno edge or by other suitable means (see **leno edge** under **selvedge, woven**).

splitting
1. The arrangement, prior to dressing, of differently coloured warp threads in the order required in a fabric.
2. The separation of the several warp sheets in the **slasher-sizing** machine.

sponge weave
Any one of a variety of weave arrangements that groups ends and picks together in order to form a cellular structure and to create a soft spongy effect in the fabric. Examples include spot weaves, diamond effects, honeycombs and sateen-based structures with lifts added. (See diagram for example.) Uses include shawls, counterpanes, drapes, bathing wraps and dress fabrics.

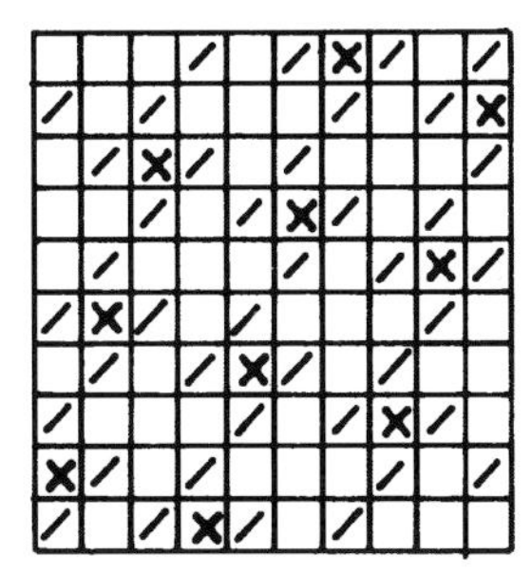

X indicates sateen base / indicates added lifts

Sponge weave

sponging
The sequence of steaming, shrinking, pressing and reconditioning of finished, or partly finished wool piece goods, before use in garment manufacture (see **London shrinking**).

spool (Axminster and gripper-spool)
A double-flange bobbin in which a number of threads of pile yarn are wound in a predetermined order for use in spool-Axminster and gripper-spool looms. The yarns from a spool form all or part of a row in the carpet design.

spool (lace machines)
A cylindrical barrel with flanges at each end designed to fit on the wires of the spool board of a lace machine. The spools carry the yarns that form the main patterning threads on the lace furnishing and string warp lace machines.

spool (machine sewing)
See **bobbin (machine sewing)** 1.

spool Axminster
See **Axminster carpet**.

spool-gripper Axminster
See **Axminster carpet**.

sports surface, artificial
Any manufactured surface, including fabrics, on which a sporting activity takes place.
Note: Sports surfaces fall into three classes: those used by a player(s) only; those used with a ball only; and those used by a player(s) and ball. Fabrics are normally canvas-like, felt-like, or pile fabrics. Pile fabric, sometimes referred to as 'artificial turf', may be used with the pile almost filled with sand.

spray bonding
A method of making **nonwoven fabrics** in which droplets of adhesive are sprayed on to the fibre **web** or **batt**. (See also **adhesive-bonded nonwoven fabric**.)

spray dyeing
Application of colorant to a substrate using a spray gun with the object of producing **ombré** effects.

spray printing
A form of **stencil printing**.

Spray rating test
See **water repellency**.

spread loop (weft knitting)
A needle loop expanded over two or more wales.
Note: Applied to stockings, the term spread loop refers to expansion over two wales and the stockings are described as 'mesh' or (technically) 'half-point transfer'. The stitch has ladder-resistant properties.

spreader (machine sewing)
A hooked stitch-forming element used to expand the upper/needle thread loop or to position a loop of thread to facilitate the formation of the desired stitch. (See also **looper**.)

spring frame (lace machines)
A heavy-duty grid frame held on rods that project from the end of a Leavers lace machine at the level of the well. It serves as an anchor for the springs which tension the steel bars (see **guide bars**, *Note 3*) against the pull of the jacquard.

spring lay rope
See under **rope**.

spring needle (machine knitting)
See under **needle (machine knitting)**.

sprit (flax)
Small pieces of woody epidermal tissue adhering firmly to flax fibre strands.

spun glass calender
A machine similar to a **Schreiner calender** except that the steel bowl is engraved with 'vertical' (circumferential) lines and acts as a friction bowl at the same time as it embosses the design on the fabric.

spun silk
1. Yarn produced by dressing or combing processes from silk waste that has been 'boiled off' to remove most of the gum.
2. Descriptive of fabrics produced from spun silk.

spun yarn
See **spinning**, but commonly used to describe (i) a yarn that consists of staple fibres held together usually by twist, and (ii) a manufactured continuous filament yarn that has not yet undergone a drawing process.

spun-dyed; spun-coloured
Descriptive of yarns produced with **mass coloration**.

spunbonded fabric
See **spunlaid fabric**.

spunlaced fabric
See **hydroentangled fabric**.

spunlaid fabric; spunbonded fabric
A **nonwoven fabric** made by the extrusion of filaments which are then laid down in the form of a **web** and subsequently bonded.

spyndle
1. A length of 14,400yds of dry-spun flax or jute yarn consisting of 48 leas or cuts of 300yds each.
Note: The count is the weight in pounds of one spyndle.
2. In the Alloa system of woollen yarn numbering, a length of 11,520yds of yarn.
Note: The Alloa count is the number of spyndles in 24 pounds of yarn. (See also Table, p.401.)

square
See **sett, square**.

square heel (knitting)
A fully fashioned hose heel composed of two square tabs with their extremities linked together to form a pocket. Where the heel tabs meet the sole of the stocking, the wales in the foot are at right angles to those in the heel.

stabilized finish
A treatment applied to a textile material in order to increase its resistance to dimensional changes in laundering and use.

stabilized yarn; set yarn
Yarn which has been subjected to a heating and cooling or other **setting** treatment, usually in order to reduce its tendency to shrink, contract, twist, snarl, or stretch.
Note 1: Stabilization or setting may be carried out either as a batch process, e.g., by steaming packages of yarn in an autoclave, or continuously, by heating and cooling a running yarn.
Note 2: In the case of staple yarns and untextured continuous- filament yarns, stabilization or setting is usually carried out to reduce shrinkage in subsequent processing or use, and/or to reduce **twist liveliness**.
Note 3: In the case of certain types of **textured yarn** a second setting or stabilization (post-setting) is employed to reduce the stretch or retractive properties of the yarn, while retaining most of its bulkiness (see **textured yarn**, *Note 2*). In the case of air-textured yarn (see **textured yarn**, *Note 1* (viii)), stabilization may be used to modify and set the loop structure.

stabilizer (hydrogen peroxide bleach baths)
A chemical compound which, when added to an alkaline peroxide bleaching liquor, controls the rate and nature of decomposition of the peroxide, thus providing a controlled process of bleaching with a minimum of tendering of the cellulose substrate.

stage twisting
See **two-stage twisting**.

stainblocker
A chemical substance which, where applied on or in a textile substrate, imparts partial or total resistance to staining, especially that caused by colorants in foodstuffs.

staining
1. An undesirable local discoloration.
2. In fastness testing of coloured textiles, the transfer of colorant from the test material to the adjacent materials (see *Standard Methods of Test for the Colour Fastness of Textiles and Leather*, Bradford: The Society of Dyers and Colourists, 5th Edition, 1990).
3. In textile printing, the soiling of whites in the washing process.

standard atmosphere for pre-conditioning (textiles)
An atmosphere with a relative humidity of between 10% and 25% and a temperature not exceeding 50°C.

standard atmosphere for testing (textiles)
1. Standard temperate atmosphere: an atmosphere at the prevailing barometric pressure with a relative humidity of 65% and a temperature of 20°C (68°F).
2. Standard tropical atmosphere: an atmosphere at the prevailing barometric pressure with a relative humidity of 65% and a temperature of 27°C (81°F).

standard condition for physical testing (textiles)
A textile material is in standard condition (or is 'conditioned') for physical testing when, after having been dried to approximately constant weight in an atmosphere with a relative humidity not higher than 10%, it has been kept in the **standard atmosphere for testing** until it has reached equilibrium.
Note: In cases where the textile material is not likely to lose volatile matter other than water, or to change dimensions, the preliminary drying may be carried out in an oven at 50-60°C situated in the standard atmosphere for testing, which is a convenient way of achieving a relative humidity of about 10%. When the oven is located in a standard tropical atmosphere, an oven temperature of 60-70°C is required. Equilibrium with the standard atmosphere for testing may be assumed when successive weighings, at intervals of not less than two hours, show no progressive change greater than 0.25% in the weight of the textile material.

standard hairweight
The hairweight of a cotton fibre divided by its maturity ratio. It is usually denoted by H_S. (See also **hairweight**.)

standard heald
See **leno weaving**.

standing bath
A liquor in which batches of material are processed in sequence.
Note: In order to obtain reproducible results after each batch the bath must be restored to its original state.

standing place
See **pick bar** under **bar (woven fabric)**.

staple
1. A lock or tuft of fibres of uniform properties. Hence a lock or tuft prepared to demonstrate

fibre length.
2. A mass of fibres having a certain homogeneity of properties, usually length.

staple diagram
See **fibre diagram** under **fibre length**.

staple fibre
A fibre of limited and relatively short length.
Note: Natural staple fibres range in length from a few millimetres (e.g., cotton linters), up to about a metre (e.g., some bast fibres). Manufactured staple fibres are produced over a similar range of lengths. They are normally prepared from extruded filaments by cutting or breaking into lengths suitable for the subsequent processing system or end-use; they may also be crimped.

staple fibre waste
See **waste**.

staple length
See under **fibre length**.

star tape
A cotton **tape** typically R42tex x 17tex: 22.5 ends x 11 double picks per cm.

starch
A carbohydrate component extracted from certain plants and used in sizing (see **size**) and **finishing**. Its use in these operations depends on its adhesive or film-forming properties.

starting place
See **pick bar** under **bar (woven fabric)**.

static loading test
A test intended to give information on the response of a floorcovering to loads applied for a period of time, a standard load being applied to a specimen of carpet for a standard time. It is intended to simulate the effect when, e.g., furniture rests on a carpet.

statistical process control; SPC
The comparison of measured variables within a process against pre-selected statistical limits, as a means of assessing whether the process variables are inside acceptable confidence limits and the taking, if necessary, of appropriate steps to bring the process within the limits of the acceptable confidence interval.
Note: Departures from normality are normally associated with either engineering or control malfunctions or the input of raw materials which are out of specification.

stay binding
See under **binding**.

stay tape
A woven **narrow fabric** in plain weave, usually with linen warp and cotton weft, generally used for stabilising garments.

steady dial linking machine (knitting)
A **linking machine** that has a point ring continuously driven, usually by worm and worm wheel.

steam setting
See **setting**.

steamer
A chamber for treating textiles with steam. Depending on the moisture content of the steamer atmosphere, steaming is known as wet steaming or dry steaming.

steel bars (lace machines)
See **guide bars**, *Note 3*.

steeping
1. (General) The treatment of textile material in a bath of liquid, usually, though not necessarily, without continuous or intermittent agitation. The term is also applied to processes whereby the materials are impregnated with a liquor, highly squeezed, and then allowed to lie.
2. In viscose manufacture, the process of immersing the dissolving pulp in a solution of caustic soda of mercerizing strength (17-20%). The purpose of this treatment is twofold: (i) to produce alkali-cellulose; (ii) to remove soluble impurities from the pulp. The operation is controlled by time and temperature.
3. The process of **retting** flax straw by immersion in an aqueous liquor.

stencil printing
The application of colorant to a substrate by brushing-on or spraying through a stencil usually cut in thin sheet metal or waterproofed paper.

stenter; tenter; frame
1. An open-width fabric-finishing machine in which the selvedges of a textile fabric are held by a pair of endless travelling chains maintaining tension.
Note 1: Attachment may be by pins (pin stenter) or clips (clip stenter).
Note 2: Such machines are used for: (i) drying; (ii) heat-setting of thermoplastic material; (iii) fixing of dyes and chemical finishes; (iv) chain mercerizing; (v) controlling fabric width.
2. To pass fabrics through a **stenter**. (See also **jigging stenter**.)

step number
See **move number**.

stepped cam (weft knitting)
A cam (usually of a fixed type), used in circular or flat knitting machines, for moving knitting elements that have butts of different lengths into different paths.

sticky cotton
Cotton which sticks to roller surfaces, especially at a card or drawframe, causing difficulties in processing or even making the material impossible to process without special precautions. Causes of cotton fibre stickiness range from contamination with cotton seed oil or the pesticides and defoliants used during cotton growing, to the presence of bacteria and fungi or of different types of sugars. The best known cause of sticky cotton is **honeydew**.

stiffness
The resistance offered by a material to a force tending to bend it. (See also **bending length** and **flexural rigidity**.)

still point (warp knitting)
See **point (milanese machines)** under **point (knitting)**.

stitch; binding point (weaving)
A special form of thread interlacing used, for example, to join the layers of compound cloths (see **double cloth, woven**), or to consolidate single structures (see **hopsack**).

stitch; float (defect)
Local incorrect interlacing of warp and weft threads in a woven fabric. This is often due to some interference with the opening of the shed, one or more ends being prevented from following the movement of the healds carrying them.
Note: Such interference may be caused by a broken end, a large knot or waste in the shed. Stitching may also arise near the selvedges if a clean, open shed is not formed before the shuttle enters, owing, for instance, to lack of sufficient warp tension. Additionally the defect may be caused by a **wrong lift**.

stitch (knitting)
An intermeshed loop.

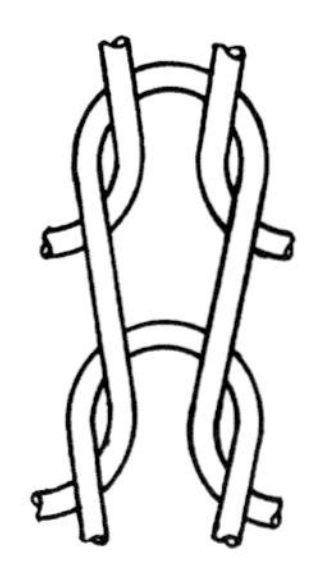

Stitch (knitting) (plain)

stitch (Schiffli embroidery)
1. An interlacing of the needle thread with the cop thread.
2. The distance between two adjacent interlacings.

stitch cam (weft knitting)
That part of a knitting machine that is used for actuating needles or **sinkers** to draw new loops.

stitch density (knitted fabric)
The product of **wale density** and **course density.**

stitch finish
A **finish** applied to yarns or fabrics, or both, to facilitate the movement of sewing thread and/or the penetration of a needle through the material.
Note 1: The object is to reduce damage to fabrics and sewing threads or to avoid overheating of the sewing needle.
Note 2: Stitch finishes involve the application of lubricants. They are frequently applied to closely woven fabrics such as collar cloth and shoe canvases, to fabrics containing filling material of an abrasive nature, such as metallic oxides; and to fabrics which may be embrittled by chemical or other finishing treatments.

stitch holding (weft knitting)
The retention of a stitch by a needle for the purposes of patterning or **shaping**.

stitch length (knitting)
The length of yarn in a knitted **loop**.

stitch shaped
See under **shaping (weft knitting)**.

stitch transfer (weft knitting)
The transfer of a stitch from the needle on which it was formed on to another needle for the purpose of patterning or **shaping**.

stitch type (sewing)
A formal description of the configuration of recurring stitches used in sewing.
Note: Eight classes of machining stitch type are defined in ISO 4915:1991 *Textiles - Stitch types: classification and terminology*. These are summarised as follows:

Class 100 Chainstitch
Class 200 Handstitch (simulated)
Class 300 Lockstitch
Class 400 Multi-thread chainstitch
Class 500 Overedge chainstitch
Class 600 Flat seam stitch
Class 700 Single thread lockstitch
Class 800 Combination stitch

stitch-bonded fabric; sew-knit fabric
A multi-component fabric, one component of which is a series of interlooped stitches running along the fabric length. The other components may be fibre **web** or **batt**, yarns, or pre-formed fabric. Examples of stitch-bonded fabrics are:
(i) a fibre web or batt bonded by stitching yarns (Arachne, Maliwatt);
(ii) cross-laid yarns with or without machine-direction yarns bonded by stitching yarns (Malimo);
(iii) a structure in which either the sewing yarns or other threads lying in the machine direction are taken over sinkers to form loop pile (Araloop, Malipol); and
(iv) a structure in which the stitching loops are formed from the fibres of the web or batt (Arabeva, Malivlies).

stitched hopsack
See **hopsack**.

stitching
See **cobwebbing**.

stitching damage; stitch damage; needle damage; sewing damage
Damage occurring to the material during the stitching/seaming process. There are many causes, attributable to the sewing conditions, the characteristics of the sewing needle, and the material being sewn. (See also **seam damage.**)

stitching thread
See **binding thread.**

stockinette
A plain knitted fabric.
Note: The term is no longer in common use.

Stoddard solvent
A petroleum solvent adopted in 1928 in the U.S.A., as the standard dry-cleaning solvent of the white spirit type. It has a relatively high flash point (38°C).

stone-washed finish
A finish obtained by vigorous tumbling of wet garments, usually made from denim, with pebbles.

stop line; stopping line (warp knitting)
A defect produced in warp-knitted fabrics whenever the knitting action of the machine is stopped. It appears as a horizontal line and consists of a number of courses that are different in

stitch length from the main part of the fabric. It is attributed largely to the changes in warp tension that take place during the deceleration and acceleration of the machine.

storage feed (weft knitting)
A yarn-furnishing device that is designed to maintain a uniformly wound store of yarn between supply package and needles, and from which yarn is withdrawn as required by the needles during knitting.
Note: The object is to improve the regularity of the fabric, especially by eliminating the effects of tension variations arising from package unwinding.

store curtain (lace)
A patterned lace curtain made as a panel of definite size, traditionally 36-72 inches (90-180cm) wide, 2yds (180cm) or more long. It is usually decorated by a single large design. Provision may be made for the insertion of a curtain rod or wire.

stoving
bleaching of wool, silk, hair, or other proteinaceous materials in a moist condition with sulphur dioxide in an enclosed chamber. (Wet stoving is the treatment of a material with a solution of a sulphite or bisulphite.)

straight of grain
See **grain line.**

straight-bar machine (weft knitting)
A **knitting machine** having bearded needles fixed in a movable straight bar or bars, used to produce fashioned or fully fashioned goods.

strand
1. A single, two or multi-fold yarn used as a component of a folded or cabled yarn or of a rope.
2. Linear textile material generally.

straw tow
Flax straw in tossed and broken condition, resulting from threshing a flax crop too poor for normal processing.

straw yarn; flat yarn
Extruded monofilament yarns that have the cross-section and appearance of natural straw.

stress
The intensity at a point in a body of the forces or components of forces that act on a given plane through the point. It is expressed in units of force/unit cross-sectional area, i.e., N/mm^2.

stress relaxation
The time-dependent decrease in stress in a solid under given constraint conditions.
Note: Stress-relaxation tests are usually made under given constant total strain or deformation conditions.

stretch breaking
See **converting.**

stretch fabric
A fabric characterized by greater than normal capacity for stretch and recovery from stretch.
Note: The term is used for materials with greater extension and recovery properties than traditional woven or knitted structures from conventional yarns and implies the use of **stretch**

stretch fabric *(continued)*
yarns, elastomeric threads, or finishing treatments. Such fabrics may have different degrees of extensibility and recovery specified for particular uses. Stretch fabrics, so defined, are used particularly for body conforming garments (comfort stretch) and for figure controlling purposes (power stretch). Where the prime requirement is power of recovery, the term **elastic fabric** is preferred.

stretch ratio
See **draw ratio.**

stretch spinning (manufactured fibre production)
A process of spinning whereby the extruded filaments are substantially stretched at some stage between extrusion and collection.
Note: The term is normally applied specifically to the application of substantial stretch in the production of **cupro (fibre)** yarns.

stretch yarn
Yarn capable of a pronounced degree of stretch and recovery from stretch.
Note: Stretch yarn may consist of: (i) conventional yarn treated by certain texturing processes (see **textured yarn**); (ii) **elastomer** in continuous-filament form.

strick
A small bunch of scutched flax (see **scutching (flax)**) of a size that can be held in the hand.
Note: In the jute industry, the corresponding term is **strike** which refers to a bunch of jute similar to a **head** but smaller, usually 1 to 2 kg.

strike
1. The uptake of dye by a substrate from a dyebath in the early stages of dyeing.
2. The result of the first period of dyeing, especially in wool dyeing.
3. See **strick.**

strike-back
The flow of the adhesive on a fusible **interlining** from the coated to the uncoated surface during the fusing process.

strike-through
1. (Hygiene products) The rate at which a liquid passes through the **coverstock** of an absorbent pad.
2. (Interlinings) The flow of the adhesive on a fusible **interlining** on to the outer surface of the top fabric during the fusing process.

striking jack (weft knitting)
An intermediate lever on a straight-bar machine, acted on by a slurcock to transmit motion to **sinkers.**

string warp machine
See under **lace machines.**

string yarn (hosiery)
Coarse mercerized cotton yarn used for the manufacture of gloves.

stripe yarn
See under **fancy yarn.**

striped backing (weft knitting)
The reverse side of a **rib jacquard** fabric characterized by successive courses of horizontal stripes of each of the yarns used to form the pattern. The effect is obtained by knitting on all the needles in the set opposite to that used to form the pattern.

Colour 'B'

Colour 'A'

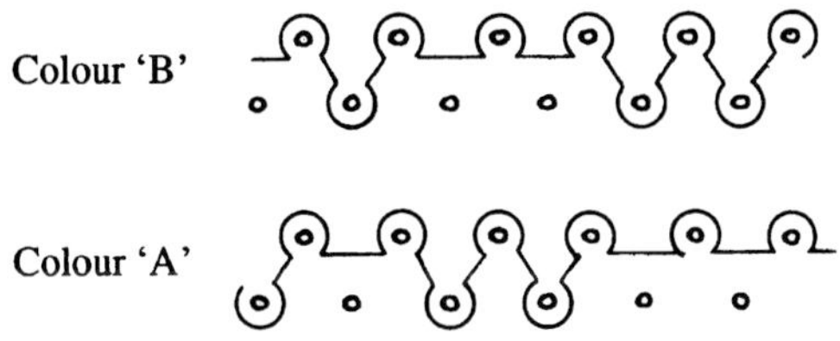

Example of a striped backing for a two colour rib jacquard fabric

stripiness (warp knitting)
Longitudinal defects caused by yarn variation or structural distortion in warp-knitted fabric.

stripiness (weft knitting)
See **barré (fault)** 2.

striping finger (weft knitting)
One of a number of yarn feeder guides mounted in a block which can be individually selected to guide a particular yarn into the needle hooks.

striping unit (weft knitting)
A mechanism comprising a number of **striping fingers** that allows two or more weft yarns to be used selectively at a **feeder**.
Note: Means for cutting and trapping threads may also be included.

striping yarn; striping end; striping pick
Any yarn, plain or fancy, from any textile raw material and used singly or as a small number of threads to produce stripe effects in a fabric.

strippers
Wire-covered rollers that form part of a roller and clearer card and are used to perform the action of transferring fibres from the **workers** to the swift.

stripping
Destroying or removing dye or finish from fibres or fabrics.

stroll test
A test to assess the electrostatic propensity of a carpet when it is subjected to walking. Using standard shoes a person walks on the carpet under test, and the body voltage generated is measured.

structured needlefelt
A pile fabric formed by subjecting a previously needled **web** or **batt** to a further punching operation with forked, single barb, or side-hook needles. Rib, velour and pattern structures may be produced.

strusa
See **frisons.**

stubble
broken filaments, whose length corresponds to the warp floats in rayon satin. They are

stubble *(continued)*
generally caused by filament breakage at the beat-up.

stud breaker
See **breaking machine.**

stuffer; stuffer yarn
A warp yarn used to give stability to the foundation of a carpet and to support the pile. The stuffer lies between, and separates, the upper and lower shots of weft.

stuffer box
A crimping device consisting of a confined space into which a tow, a converted tow, a sliver, a yarn or a similar assembly of filaments or fibres is injected by feed rollers or other means such as a fluid jet and in which the fibre assembly is packed and compressed so that the individual filaments or fibres buckle and fold.
Note 1: Multifilament tows normally pass through a stuffer box after extrusion and drawing in order to improve the coherence of the tow. Tows which have been converted are similarly crimped both for the same reason and also to improve the subsequent processibility of the fibre.
Note 2: A stuffer box is used in the production of some types of textured yarn (see **textured yarn**, *Note 1* (ii)).

stuffer thread
See **wadding thread.**

stuffer-box textured yarn
See **textured yarn**, *Note 1* (ii).

stumping
Final mechanical treatment of **hoods** or **bodies** to obtain the desired size.

S-twist
See **twist direction.**

ST yarn
See **self-twist yarn.**

style, printing
See **printing style.**

sublimation printing
A form of **transfer printing** employing dyes that sublime readily and have **substantivity** for the substrate to which they are applied.

substantivity
The attraction between a substrate and a dye or other substance under the precise conditions of test whereby the latter is selectively extracted from the application medium by the substrate.

substrate
A material to which dyes and chemicals may be applied.

substrate (carpet); carpet backing
A construction, integral with the use-surface and composed of one or more layers, which serves as a support for the use-surface and possibly stabilizes the dimensions and/or acts as a cushioning layer.

foam backing
A mechanically or chemically foamed polymer, applied to the back of a textile floorcovering to act as an integral underlay.

impervious backing
A **secondary backing** consisting of a polymeric material, e.g., rubber, PVC, etc.

precoating
The operation of depositing a bonding agent on the back of a textile floorcovering to ensure its penetration into the substrate and pile root without reaching the use-surface.

primary backing
Material which acts as a carrier for the use-surface and which is often additionally used as an anchor for other parts of the substrate.

secondary backing
A fabric which forms an additional stabilizing layer in the substrate of a textile floorcovering and which often forms the final coating or layer.

unitary backing
A heavy back-coating of a polymer material, e.g., latex, to impart dimensional stability to a carpet. Unitary backed carpets are normally installed using adhesive-based systems.

sueded cloth
A fabric finished in such a way as to imitate suede leather.

sueding
See **emerizing.**

suint
Excretion from sweat glands of sheep, which is deposited on wool fibres.

sulfar (fibre) (generic name U.S.A.)
A manufactured fibre in which the fibre-forming substance is a long-chain synthetic polysulphide in which at least 85% of the sulphide (-S-) linkages are attached directly to two aromatic rings. (See also Classification Table, p.401.)

sulphur dye
A dye, containing sulphur both as an integral part of the **chromophore** and in attached polysulphide chains, normally applied in the alkaline soluble reduced (leuco) form from a sodium sulphide solution and subsequently oxidized to the insoluble form in the fibre.

sunn
A bast fibre obtained from the plant *Crotolaria juncea.* (See also **hemp, true.**)

sunray pleats
See under **pleats.**

supercop
1. A wooden or plastic conical base on which weft is wound for use in a shuttle loom.
2. The weft package produced by winding yarn on to the base defined above.

suppressed balloon spinning
See **collapsed balloon spinning.**

suppression
The creation of shape by the elimination of surplus fabric from the surface area of a garment

suppression *(continued)*
using darts, tucks, seams or gathers.

surface flash
Rapid spread of flame over the surface of a material without ignition of its basic structure.
Note 1: However, if the latter occurs simultaneously or sequentially with surface flash, it is not considered as a part of surface flash.
Note 2: Surface flash is of particular relevance to pile fabrics.

surface pile density
The ratio of mass to volume of the pile of a carpet above the backing.

surface pile mass; surface pile weight
See **pile mass.**

surfactant
An agent, soluble or dispersible in a liquid, which decreases the surface tension of the liquid.
Note: A contraction of 'surface-active agent'.

surgical dressing
See **wound dressing material**.

suspender web
See **brace web.**

suture (surgical)
Thread used to close a wound by sewing. Monofilament and multifilament sutures are used, with multifilament sutures being twisted or braided in construction. Sutures may be supplied with curved needles swaged on to the ends of a cut length. A variety of natural and synthetic materials are used to impart particular absorption rates in the body or to render the suture nonabsorbable. (See also **ligature.**)

suture line (weft knitting)
A line within a knitted fabric at which the wales are caused to change direction collectively. It is formed either by knitting more loops in certain wales than in others and/or during the knitting of pouches.

swab (surgical)
A pad consisting of several layers of absorbent fabric, used to prepare the site of an operation or a wound, to absorb and remove excess blood and body fluids from an incision, and for packing internal body cavities during an operation.
Note: Swabs were traditionally made from woven cotton, but other structures and cellulosic fibres are finding application as national regulations permit. All swabs are constructed in such a way as to eliminate fraying or linting. Swabs may incorporate X-ray opaque yarns or tags to enable them to be located by X-ray examination should they be left in the body.

swansdown
A general term applied to various soft, raised fabrics. A typical weave for a cotton swansdown fabric is the five-end **sateen** with an extra riser added. Typical construction: 25x50; 30x25tex; K=13.7+25.0. (See also **fustian.**)

Swansdown weave

swatch; sample swatch
Fabric for display, test, or record purposes, in the form of a single sample or an assembly of small samples, the latter being sometimes called a 'bunch'.

swealing
1. Migration of dye into the angles of folds and creases during fabric drying.
2. Partial transfer of colour, dirt or grease into the surrounding fabric, caused by unsatisfactory removal of stains by hand from a fabric when using an aqueous or solvent treatment.
Note: The resulting mark is frequently referred to as a 'sweal-mark', 'halo', or 'ring'.

swell ratio; bulge ratio
In fibre manufacture, the ratio of the maximum diameter of the extrudate as the solution or melt emerges from the spinneret to the orifice diameter. (See also **die swell.**)

swelling agent
A substance that causes the total liquid take-up of a fibre to increase.
Note: A swelling agent may be used in a dyebath or a printing paste to promote coloration by accelerating the diffusion of dyes into a fibre.

swift; hank swift; rice
The revolving frame on to which a **hank** is mounted when it is unwound.

swift (carding)
See **cylinder**.

swift (warping mill)
A large diameter reel which may be horizontal or vertical, on to which warp yarns are wound during the process of **section warping.**

swing needle machine; zig-zag machine.
A sewing machine with a needle bar which swings laterally during sewing to produce a zig-zag formation of stitches.

Swiss bar (lace machines)
The middle guide bar on a **lace furnishing machine** equipped with three guide bars. It is usually clothed with threads from the bottom spool board.

Swiss double piqué
See **double piqué, weft-knitted.**

Swiss lace
A furnishing lace obtained by contrasting two densities of clothing. The lighter densities consist of V-ties made from the **Swiss bar** between two or more **pillars**. The heavy density is made from the back bar in a complementary class of work.
Note 1: In single-tie Swiss work, the Swiss bar throws on every full motion, but jacquard control determines whether each thread ties or pillars.
Note 2: In two-gait Swiss work, each V-tie consists of one spool thread throwing to the adjacent pillar and back again.
Note 3: In three-gait, single-tie Swiss work, each V-tie crosses one pillar and ties to the next and back again, thus forming overlapping double ties.
Note 4: In three-gait, double-tie Swiss work, the Swiss bar throws to the left on the back motion and to the right on the front motion. Jacquard control determines whether each thread ties to the left or to the right or pillars.

Swiss rib
See under **rib fabric, weft-knitted.**

swivel weaving
A fabric in which figure is achieved by the introduction of additional weft threads into a base fabric to produce spot effects. The figuring yarn is fed from a series of small shuttles mounted over the top of the weaving surface.

syddo rags; siddo rags
Rags consisting of interlinings from garments. The best types are produced from fabrics made from yarns of hair, or blends of hair with wool, made on the worsted system.

synchronised needle timing (weft knitting)
See **timing (weft knitting)**.

syndet
A detergent that is not a soap: a contraction of 'synthetic detergent'.

syndiotactic polymer
See **polymer, syndiotactic** under **polymer.**

syntan
A name for synthetic tanning agents. (See also **backtanning.**)

synthetic fibre
See **fibre, synthetic** under **fibre.**

tab; footing
The starting point of a weave. This is seen in diagrams A and B which are said to be plain weaves 'on opposite tab'. Tab is probably derived from 'tabby', especially when used in relation to plain weave.
Note: When two or more weaves are combined to form a stripe, check, or figured design, a better fit of the weaves and a neater edge to the figure is obtained by ensuring that the weaves used are 'on the tab' or on the correct footing relative to each other. (See also **mock.**)

tabaret; tabourette
A finely woven, yarn-dyed furnishing fabric that has alternate warp stripes of satin and plain weave.

tabby
See **weave, plain.**

tablet weaving
A method of producing woven plain or patterned narrow fabrics. The warp shed is controlled by tablets made of thin, stiff material, e.g., cardboard, plastic, bone etc. Tablets are usually about 5cm to 10cm square, although other shapes, e.g., triangles, hexagons, etc., are also employed. Each tablet has a hole at each corner through which the warp yarns are threaded. Rotating the tablets controls the rise and fall of the warp yarns. If the tablets are continually turned in one

direction, the yarns threaded through the same tablet twist around each other giving a warp twisted structure, but by frequently reversing the turning direction, the twist can be obviated and normal warp weft interlacements are produced. This, combined with the turning of tablets individually, not altogether as a pack, gives a large range of structure and design possibilities. In all these cases, the result is a warp-face fabric, the weft normally being visible only at the selvedge. Rotation of the tablets, weft insertions and beat-up are normally hand operations.

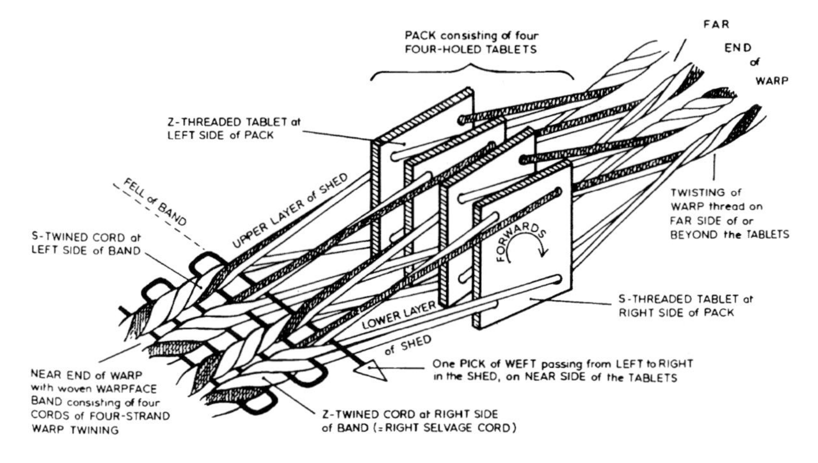

Tablet weaving

tabourette
See **tabaret.**

tacking
See **bagging.**

tackspun fabric
A material made from a polymer film with a backing substrate. The film is melted by a roller to which it adheres. As the film and roller separate a fibrous pile is formed.

taffeta
A plain-weave, closely woven, smooth and crisp fabric with a faint weft-way rib, produced from filament yarns. The rib effect is produced by making the warp end density greater than the pick density. The warp and weft yarns are of similar linear density.
Note 1: Taffeta belongs to a group of fabrics that have ribs in the weft direction. Examples of this group, arranged in ascending order of prominence of the rib are: taffeta, poult, faille, and grosgrain.
Note 2: The term 'wool taffeta' is often applied to a plain-weave, light-weight fabric produced from worsted yarns.

taffeta ribbon
A **ribbon** of continuous-filament yarn in plain weave, with a relatively high warp density and a very fine, almost imperceptible rib, generally with a selvedge of contrasting weave.

tail
A length of yarn wound on to a cheese or cone at the commencement of winding that protrudes from the main build-up of yarn and can later be attached to the free end of yarn of a second package during magazine creeling.

tailing
A dyeing fault consisting of a gradual change in colour along a length of material to which colorant has been applied by padding or other continuous techniques.

take-up lever
A lever on a sewing machine, located at the front of the horizontal **arm**, which takes up excess needle thread after stitch formation and draws more thread through the tension discs at the same time as completing the interlacing of the **stitch**.

take-up motion
A mechanism for controlling the winding-forward of fabric during weaving. There are two main types:
(i) Positive take-up motions in which the **take-up roller** is gear driven, a change wheel or variable-throw pawl and rachet being provided to allow the required rate to be obtained, so determining the pick spacing.
(ii) Negative take-up motions in which the take-up roller is rotated by means of a weight or spring, this roller only rotating when the force applied by the weight or spring is greater than the warp-way tension in the fabric. The take-up rate is controlled by the size of the force applied by the weight or spring and/or the warp tension.

take-up roller
A roller whose speed of rotation relative to the picking rate determines the pick spacing during the weaving (see **take-up motion**). Its surface is covered with one of a variety of materials designed to grip the fabric firmly so that it can be drawn forward at the required rate.

taker-in; licker-in
A saw-tooth wire covered **cylinder** of a card used for the **opening** of the fibrous lap or batt feed.

tape (textile)
1. A single ply **narrow fabric**, usually of plain-weave, sometimes knitted, used in non-loadbearing applications or the reinforcing of fabrics to resist wear and deformation. (See also **webbing.**)
2. A long narrow flat structure with textile-like properties made from thermoplastic polymer, paper, or other appropriate material.

tape, oriented
A tape made by extruding a thermoplastic polymer, usually a polyefin, in the form of a sheet or film, slitting the sheet and hot-stretching to induce molecular orientation and hence high longitudinal strength. In some cases the slitting may follow the hot-stretching process.

tape ruche
See **loop ruche** under **ruche.**

tape yarn
A yarn which comprises a tape with a large width-to-thickness ratio, and which has an apparent width not exceeding an agreed limit (e.g., 5mm or 8mm).
Note: Such yarns are usually of paper or are formed by slitting a wide film of (usually) polyethylene or polypropylene polymer into individual tapes, with hot-stretching either before or after slitting to induce high longitudinal strength. The draw ratio in hot-stretching is kept low enough to avoid excessive longitudinal **fibrillation.** The tape yarn so produced is suitable for weaving.

taped ends (weaving fault)
Two or more warp yarns drawn through the healds and reed as one as a result of being stuck together after the sizing process.

taped seam
See under **seam type.**

taper line gratings
Transparent plates containing lines more widely spaced at one end than the other. By selecting the appropriate taper line grating and placing it parallel to a set of threads in a woven fabric, it is possible to ascertain the number of threads per unit length (cm or inch) as a result of a star form created. These gratings can also be used to determine the number of courses per unit length in weft-knitted fabrics, or the number of dents per unit length in a reed.

tapered cone
See **biconical package.**

tapestry
A closely woven figured fabric of compound structure in which a pattern is developed by the use of coloured yarns in the warp or in the weft or both. A fine binder warp and weft may be incorporated. It is normally used for upholstery.
Note: Originally the term was applied to furnishing fabrics in which the design was produced by means of coloured threads inserted by hand as required. Modern tapestry fabrics are woven on jacquard looms, coloured threads being used to produce the required pattern. There are various fabric structures in which two or more warps and wefts of different materials may be used. The face of the fabric is usually of uniform texture, the design being developed in various colours, but in some tapestry fabrics figures of the brocade type formed by floating some of the threads are also to be found.

tapestry carpet
A patterned carpet woven by the single-pile Wilton process, in which a warp, printed before weaving, is used to produce the design. When the pile is cut, the carpet is known as **tapestry velvet.**

tapestry velvet carpet
A cut-pile carpet woven from a printed-pile warp or single frame of yarn. It was traditionally woven on a tapestry-carpet loom with bladed wires.

taping
1. A term for **slasher sizing**.
2. Groups of ends stuck together by size to give a tape-like appearance.

tappet fabric
Fabrics of a simple weave structure that may be woven on a cam or tappet loom.

tappet shedding; cam shedding
The control of the movement of heald shafts in weaving simple constructions by means of cams or tappets. Positive tappets raise and lower the heald shaft. Negative tappets move the heald shafts in one direction only and require another mechanism, usually springs, to return them.

tartan
Originally a woollen fabric of 2/2 twill woven in checks of various colours and worn chiefly by the Scottish Highlanders, each clan having its distinct pattern. Other materials and weaves are now used.

tear
The ratio of **top** to **noil** produced in combing.

tear drop (defect)
See **cannage.**

teariness (defect)
See **cannage.**

tearing force; tearing strength
The average force required to continue a tear previously started in a fabric.

teazer
See **willey.**

teazle; teasel; teazel
The dried seed-head of the plant *Dipsacus fullonum* (Fullers' thistle) used to raise a pile or nap on certain fabrics. The machine used for this purpose is known as a 'teazle gig'.

teazle gig
See **teazle.**

technical back (weft-knitted fabric)
See **plain fabric, weft-knitted.**

technical face (weft-knitted fabric)
See **plain fabric, weft-knitted.**

technical textiles
Textile materials and products manufactured primarily for their technical performance and functional properties rather than their aesthetic or decorative characteristics.
Note: A non-exhaustive list of end-uses is: aerospace, industrial, marine, medical, military, safety and transport textiles, and geotextiles. (See also **industrial textiles.**)

teg wool
See **hog wool.**

temperature-range properties (of a dye)
The extent to which the depth produced by a dye under specified application conditions is affected by a change of temperature.

temple
A device used in weaving to hold the fabric at the fell as near as possible to the width of the warp in the reed.

temple cutting
Fracture of the warp or weft yarn, or both, by temple pins during weaving.

temple marking
Disturbance of the fabric surface as it passes through the temple during weaving.

temporary set
See **setting.**

temporary shading (carpet)
See **shading (carpet)**.

tenacity
The tensile force per unit linear density corresponding with the maximum force on a force/ extension curve.
Note 1: In testing textile fibres and yarns, tensile force is normally measured in Newtons (or multiples or submultiples thereof) and linear density in tex (or decitex). Thus:

$$\text{tenacity} = \frac{\text{maximum tensile force (N)}}{\text{linear density (tex)}}$$

Note 2: For the purposes of this definition, the linear density is that measured before a test piece is extended, not that applying at the point of maximum force. Because of the relatively high extensibility of most textile materials, the extension at maximum force should always be stated.
Note 3: The expression 'ultimate tensile stress' is sometimes used in textiles as a synonym for tenacity but, in some other disciplines, 'ultimate tensile stress' and tenacity relate to the maximum tensile force per unit cross-sectional area of a test piece.

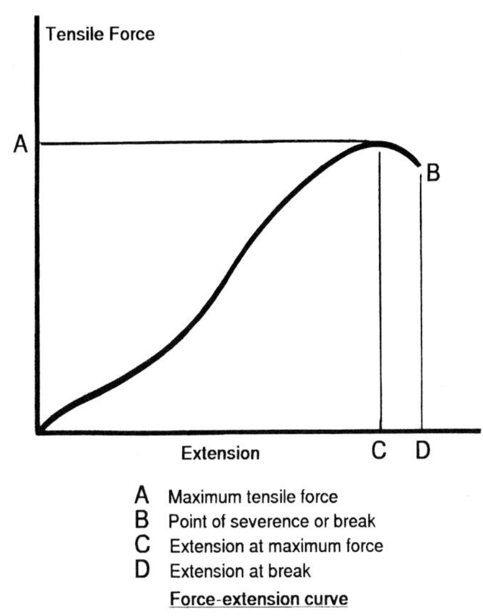

Force-extension curve

tennis ball cloth
Traditionally a heavy weight (600-760 g/m^2) **melton** type fabric in a sateen weave, produced with a cotton warp and wefts of various fibre types, e.g., wool, wool nylon, viscose nylon or nylon. It can also be produced as a **needlefelt.**

tensile strength
See **breaking strength.**

tensile strength at break; breaking force
The tensile force recorded at the moment of rupture.
Note: The tensile strength and the tensile strength at break may be different if, after yield, the elongation continues and is accompanied by a drop in force resulting in tensile strength at break being lower than tensile strength.

tensile test
A test in which the resistance of a material to stretching in one direction is measured.
Note: The tearing test is not regarded as a tensile test.

tension
A uniaxial force tending to cause the extension of a body or the balancing force within that body resisting the extension.

tension bar
See under **bar (woven fabric)**.

tenter
See **stenter.**

terry
Uncut loops in fabric. The term has become widely synonymous with woven Turkish towelling.

terry fabric, warp-knitted
A fabric generally produced with a continuous-filament yarn for a ground construction and cotton or similar yarn for a pile. The terry loops are generally made by forming loops on alternate needles and then pressing these loops off. Other less common methods of forming terry loops include over-feeding of the pile yarn, pressing and mis-pressing.

terry fabric, weft-knitted
A term sometimes used as a synonym for weft-knitted plush (see **plush, weft-knitted**).

terry fabric, woven
A warp-pile fabric in which loops are created, without positive assistance, by varying the relative positions of the fell and the reed. A high tension is applied to the ground warp and a very low tension to the pile warp.
Note: The loose picks (L) are not beaten up to the conventional cloth fell, but when a fast pick (F) is beaten up all three picks are forced to the furthest forward position to form the pile loops on the face and back.

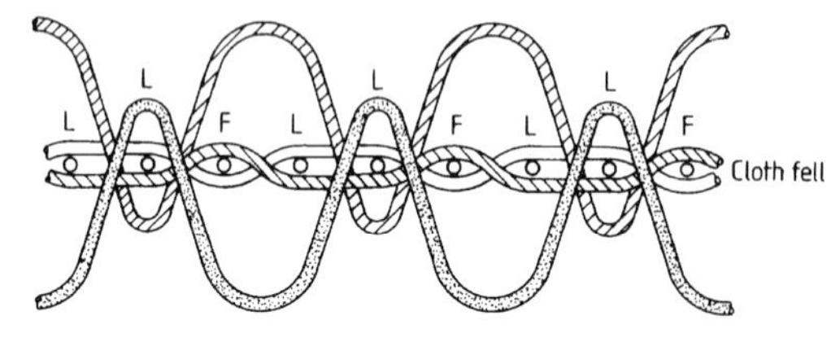

Three-pick terry, pile on both sides

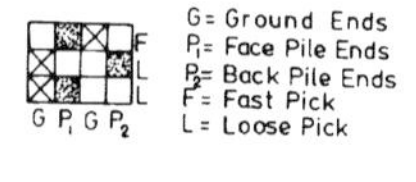

Three-pick terry weave, pile on both sides

terry velvet
See under **velvet, woven.**

tex
The basic unit of the **tex system.**

tex system
A system of expressing linear density (mass per unit length) of fibres, filaments, slivers, and yarns, or other linear textile material. The basic unit is tex, which is the mass in grams of one kilometre of the product. Multiples and sub-multiples recommended for use in preference to other possible combinations are: kilogram per kilometre, designated kilotex (ktex); decigram

per kilometre, designated decitex (dtex); and milligram per kilometre, designated millitex (mtex). Tex is a recognised SI unit (see Tables, p.396-397).

texipiqué (weft knitting)
A non-jacquard **double-jersey** fabric made on an interlock basis and consisting of a selection of knitted and tuck loops in the following sequence. (See also **double jersey, weft-knitted.**)

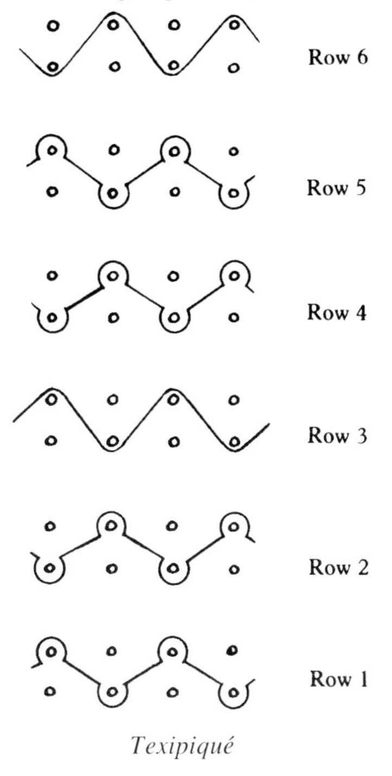

Texipiqué

textile
1. A **textile** was originally a woven fabric, but the terms **textile** and the plural **textiles** are now also applied to fibres, filaments and yarns, natural and manufactured, and most products for which these are a principal raw material.
Note: This definition embraces, for example, fibre-based products in the following categories: threads, cords, ropes and braids; woven, knitted and nonwoven fabrics, lace, nets, and embroidery; hosiery, knitwear and made-up apparel; household textiles, soft furnishings and upholstery; carpets and other floorcoverings; technical, industrial and engineering textiles, including geotextiles and medical textiles.
2. Descriptive of textiles as defined in 1 above and of the raw materials, processes, machinery, buildings, craft, technology, personnel used in, and the organisations and activities connected with, their manufacture.

textile agent
An individual or organisation acting on behalf of a principal who establishes contact with potential customers for the purpose of making sales of textile goods supplied by the principal. The agent does not take title to the goods. (See also **textile merchant** and **merchant converter.**)

textile film
A manufactured textile material in film form within which molecular orientation is predominantly in the longitudinal direction.

textile film *(continued)*
Note: Polymer films for non-textile use are commonly unoriented or bi-axially oriented, but uni-axial orientation is present in some cases.

textile floorcovering
A product having a use-surface composed of textile material and generally used for covering floors.

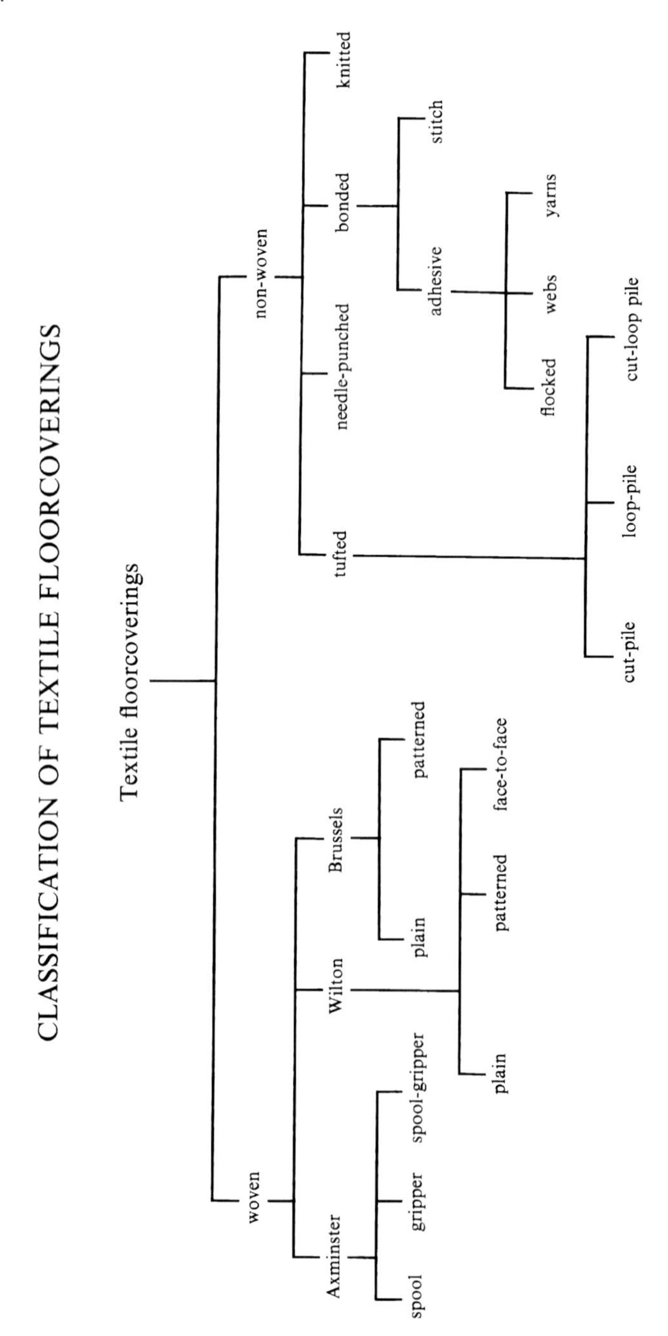

textile merchant
An individual or organisation that locates a supplier and purchases grey or finished textile fabrics or textile intermediaries and establishes contact with potential customers for the purpose of re-selling those goods. The merchant may or may not hold stock on his premises. (See also **textile agent** and **merchant converter.**)

textured pile
See **pile (carpet)**.

textured yarn
A continuous-filament yarn that has been processed to introduce durable crimps, coils, loops or other fine distortions along the length of the filaments.
Note 1: Most, but not all, texturing methods depend upon the thermoplastic properties of suitable manufactured fibres. The principal texturing procedures that are or have been used are:
(i) the yarn is highly twisted, thermally set and untwisted either as a process of three separate stages (now obsolescent) or as a continuous process (false-twist texturing). In an infrequently used alternative method (sometimes known as 'trapped-twist' texturing), two yarns are continuously folded together (see **fold**), thermally set, then separated by unfolding;
(ii) the yarn is fed through a nip into a **stuffer box** The yarn may be pre-heated or the stuffer-box may be heated (stuffer-box texturing);
(iii) the yarn is injected by a plasticising jet of hot fluid (usually hot air, sometimes steam) into a texturing tube or nozzle (hot-fluid jet texturing);
(iv) the yarn is plasticized by a passage through a jet of hot fluid and is impacted on to a cooling surface (impact texturing);
(v) the heated yarn is passed over a knife-edge (edge crimping);
(vi) the heated yarn is passed between a pair of gear wheels or through some similar device (gear crimping);
(vii) the yarn is knitted into a fabric that is thermally-set and then unravelled (knit-deknit, KDK texturing);
(viii) the yarn is over-fed through the turbulent air stream inside a jet assembly so that entangled loops are formed in the filaments; this method may also be applied to non- thermoplastic yarns (air-jet texturing; air-texturing);
(ix) the yarn is composed of **bicomponent fibres** of asymmetric cross-section and is subjected to a hot and/or wet process whereby differential shrinkage occurs.
Note 2: Procedures (i) and (v) in *Note 1* above give yarns of a generally high-stretch character. This is frequently reduced by thermally re-setting the yarn in a state where it is only partly relaxed from the fully extended condition, thus producing a **stabilized yarn** with the bulkiness little reduced but with a much reduced retractive power.
Note 3: Of the above procedures, only (i), (ii) and (viii) are currently of major commercial significance; (iv), (v) and (vi) are now rarely, if ever, used.

Tg
See **glass-rubber transition temperature.**

thermal setting
See **setting.**

thermally-bonded nonwoven fabric; thermobonded nonwoven fabric
Textile fabric composed of a **web** or **batt** of fibres containing heat-sensitive material, bonded by the application of heat, with or without pressure. The heat-sensitive materials may be in the form of fibres, bicomponent fibres or powders. (See also **calender bonding**, **point bonding** and **powder bonding.**)

thermofixation
See **baking** and **setting.**

thick end
See **mixed end.**

thickener
A substance used to increase the viscosity of a print paste or other fluid, in order to control its flow properties. Natural polymers (starch, alginates, etc.), chemical modifications thereof, synthetic polymers, emulsions, foams and clays can be used.

thin end
See **mixed end.**

thornproof tweed
A wool-type fabric produced from highly twisted yarns that are closely sett to give a firm, hard handle that is resistant to damage by thorns. Two-fold yarns consisting of differently coloured singles are commonly used.

thread
1. A textile yarn in general.
2. The result of twisting together in one or more operations two or more single or folded yarns.
3. A component of silk yarn. It is the product of winding together without twist a number of **baves** A three-thread silk yarn is the result of folding three such products together. (See also **silk, raw.**)
4. To pass sliver, yarn or fabric over, under, or through any element.

thread guide
See **yarn guide.**

thread interlacing (weaving)
The arrangement of the warp and weft over and under one another.
Note: The normal method of denoting an interlacing is to show the movement of a warp thread over and under weft threads thus:

$$\frac{3 \quad\quad 1}{\quad 2 \quad\quad 2}$$

which indicates warp passing over three, under two,over one, and under two weft threads. It is sometimes more convenient to describe a weave in terms of weft interlacings, in which case,

$$\frac{\quad 2 \quad\quad 2}{3 \quad\quad 1}$$

indicates the weft passing under three, over two, under one and over two warp threads.

thread trimmer (sewing machine)
A device fitted to a sewing machine which cuts the sewing thread(s) after completion of a run of stitches. A variety of different trimmers are available: for example, for lockstitch machines, where trimmers can cut either the upper or the lower thread or both, and for machines which produce a chain of stitches.

threads per unit length (woven fabric)
The number of weft yarns (picks) in a specified length of fabric.
Note 1: The traditional unit of length has been the inch but the value is now expressed as 'threads/cm' although the actual count may be made over 1cm , 2.5cm, 5cm, or 10cm, according

to the nature of the fabric.
Note 2: Counting may be carried out at the following stages of manufacture:
(i) In the loom: The location of the count should be agreed. It is usually taken between the fell of the fabric and the take-up roller, with the fabric under weaving tension.
(ii) Loomstate: The count is taken after the fabric has been removed from the loom and is relaxed from weaving tension, but before it is subjected to any further treatment that may modify its dimensions.
(iii) Finished: The count is taken when no further processing in the piece is contemplated.

threads per unit width (woven fabric)
The number of warp yarns (ends) in a specified width of fabric.
Note 1: The traditional unit of length has been the inch but the value is now expressed as 'threads/cm' although the actual count may be made over 1cm, 2.5cm, 5cm, or 10cm, according to the nature of the fabric.
Note 2: Counting may be carried out as for **threads per unit length (woven fabric)**.

three-for-one twisting
A system which inserts three turns of twist for each revolution of the twisting elements. This is achieved by combining the principle of **two-for-one twisting** with that of **uptwisting.** The feed package inside the two-for-one twister is driven to rotate at the same speed as the rotor but in the opposite direction.

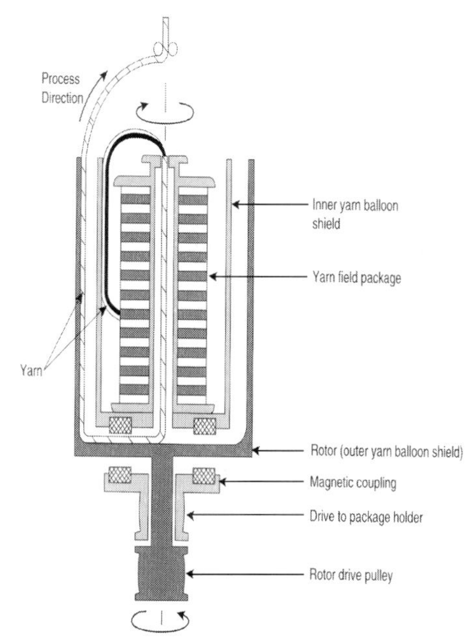

throat plate; needle plate (sewing machine)
That part of the **bed** of the sewing machine which has openings for the **needle** and for **feed dog** penetrations, and which provides localised support for the material. The openings vary in size and shape, depending on the sewing requirements.

throw
A term of Germanic and Anglo-Saxon origin meaning to twist or turn, used especially in the silk and manufactured fibre industries, to describe the twisting or folding of continuous-filament yarns.

throw *(continued)*
Note: The term 'throwster' was traditionally used to describe an individual or company specifically involved with these twisting processes but, in more recent times, the title has also been inherited by those who manufacture textured yarns by the false-twist method (see **textured yarn**, *Note 1* (i)).

throw (lace)
A traversing movement of a warp or patterning thread (caused by the shogging of the guide bar), which results in the laying of the thread across two or more **gaits** in the lace.

thrown silk
A yarn twisted from continuous-filament silk. Specific terms include **tram, organzine** and **crêpe.** (See also **throw.**)
Note: A yarn described as, for example, a six thread 20-22 denier thrown silk indicates a yarn consisting of 6 ends of 20-22 denier (22-24 decitex) continuous-filament raw silk twisted together.

thrum; through
A waste length of warp or fabric, or both, formed during the preparation of a loom for weaving or during sizing. A thrum may be formed:
(i) During the adjustment of a loom at the commencement of the weaving of the warp. When the loom is correctly adjusted, the portion of warp that contains picks inserted for testing the adjustment of the loom mechanisms is cut off.
(ii) During warp replenishment in a loom. The old warp is twisted or knotted to the new warp and, if drawn through the harness without weaving, a thrum may be formed from a portion of both warps. If the new warp is drawn through by weaving, the point in the woven fabric at which the twisted or knotted warp ends occur is called a through because the fabric is cut through to remove the thrum containing the imperfect fabric formed by the twisted or knotted warp ends.
(iii) During **looming** operations away from the loom. In the above cases, a thrum consists of portions of the old and new warp ends twisted or knotted together.
A thrum may also be:
(iv) A length of warp ends cut from a warp for the purpose of: (a) evaluating the percentage of applied size; (b) repairing end-breakages in the warp concerned.
(v) Any loose end(s) of warp.
(vi) A bundle of coarse yarns tied together by twine, for use in making a mop.

ticket number
See **sewing thread**, *Note 3*.

ticking
A general term applied to fabrics used for mattress covers, pillows, etc.

tie-and-dye; tie dye
A technique of dyeing for producing unique patterned effects by tying and/or knotting material before immersion in a dyebath, to restrict the penetration of dye locally.

tie-band
See *Note 2* under **leasing**.

tight end (defect)
A warp thread or part of a warp thread that is tighter than the adjacent ends which are tensioned normally.
Note: This may be due to weaving under greater tension or due to abnormal stretching of a yarn during some process prior to weaving. It may be caused by the presence of excess moisture, eg.,

during winding, and consequent contraction during finishing. (See also **fiddle string** and **shiner.**)

tight pick (defect)
The weft equivalent of **tight end (defect)**.

tight selvedge
See **selvedge, tight.**

tight spots
Faults in false-twisted textured yarns (see **textured yarn**, *Note 1* (i)) consisting of short isolated regions containing a local concentration of a few turns of twist in the original false-twist direction.
Note: The fault is caused by the failure of the yarn to untwist evenly during texturing.

tightness factor (weft-knitted fabrics)
See **cover factor (knitted fabrics)** and Table 2, p.400.

tights (knitted)
Close fitting knitted garments covering the lower trunk, legs and feet.

timing (weft knitting)

> **delayed needle timing**
> The setting of the point of knock-over of one set of needles on a two-bed knitting machine out of alignment with that of the other set so as to permit the formation of a tighter stitch.
>
> **sinker timing**
> The setting of the sinker movement relative to the point of needle knock-over on circular sinker-top weft knitting machines.
>
> **synchronized needle timing**
> The setting of the needles on a two bed machine so that the point of knock-over of one set is aligned with the point of knock-over of the other set.

tin, drying
See **drying cylinder.**

tinctorial value
See **colour value.**

tinctorial yield
See **colour yield.**

tinsel yarn
A textile yarn or thread, combined, coated, or covered with a shiny substance, often metallic (e.g., aluminium, occasionally gold or silver), to produce a glittering or sparkling effect. (See also **metallized yarn.**)

tinting
1. See **sighting**
2. The production of a pale colour in dyeing.

tip-sheared pile
See **pile (carpet)**.

tippy wool
Wool in which the tip portions of the fibres have been so damaged by weathering during growth as to have markedly different dyeing properties.

tissue
A term applied generally to woven fabrics. The term has been specifically applied to:
(i) A light-weight transparent or translucent woven fabric.
(ii) A general term for muslin or gauze fabrics used for window curtains and veils. The structure may be figured with feature threads or by the introduction of extra weft which will float loosely between the spot figures and afterwards be clipped or sheared off (see **clip spots**).
(iii) A figured, woven fabric in which the ground, typically **satin**, is constructed with one weft and additional patterning weft is **stitched** by a binding warp.

titre
A generic term for units of **linear density**.
Note: The corresponding term for units of specific length is count (see **count of yarn**).

tobacco cloth
See **cheese cloth.**

tog
A unit of thermal resistance used to quantify the thermal insulation of garments, bedding, etc.
Note: The SI unit of thermal resistance is the square-metre kelvin per watt ($m^2.K/W$) In order that the thermal resistance of clothing can be described in small integers, the practical unit, the tog, is one tenth of the fundamental unit. Thus:

$$1 \text{ tog} = 0.1 \text{ m}^2.\text{K/W}.$$

(See also **clo.**)

toile
1. The French word for 'fabric'.
2. Fabric of appropriate weight for the construction of prototype garments. Normally, unbleached cotton cloth is used.
3. In *haute couture*, the prototype garment cut out in toile (see 2), which may or may not be assembled.

tolerance (clothing)
1. The degree to which a garment measurement may vary from the specified value and still be regarded as acceptable.
2. See **ease.**

tom-tom
See **dolly** 2.

tone (colour)
The degree of luminosity of a **hue.**

top
1. Sliver that forms the starting material for the worsted and certain other drawing systems, usually obtained by the process of combing, and ideally characterized by the following properties: (i) the absence of fibres so short as to be uncontrolled in the preferred system of drawing; (ii) a substantially parallel formation of the fibres; (iii) a substantially homogeneous distribution throughout the sliver of fibres from each length-group present.
Note 1: Tops are usually produced by carding and combing, or by preparing and combing on worsted machinery, but recent years have seen the introduction of top-making by the cutting or

controlled breaking of continuous-filament tows of manufactured fibres, and the assembly of the resultant staple fibres into sliver in a single machine.
Note 2: The advent of manufactured fibres has meant the introduction of staple-fibre top into the flax, jute, spun silk, and other drawing systems.
2. The form or package in which sliver is delivered, e.g., ball top or bump top. (See also **oil-combed top** and **dry-combed top.**)

top and bottom feed
See **feed mechanisms (sewing)**.

top bars (lace machines)
steel bars working in the well of a Leavers machine and controlled by a top-bar jacquard.

top douping
See **leno weaving**, *Note 4*.

top dyeing
The dyeing of wool or other fibres as slubbing or top in package form.

top feed
See under **feed mechanisms (sewing)**.

top stitching
Exposed stitching normally sewn at a uniform distance from a seam to reinforce it or for decoration.

top-castle
The structure above a **cratch** which houses the pulleys that are part of the warp-tensioning system of some narrow fabric weaving machines.

Topham box
A device for twisting and winding a wet-spun continuous-filament yarn so as to produce a **cake**.
Note: Yarn is supplied through a reciprocating guide on the axis of a rotating hollow cylinder and is collected on the inner surface of the cylinder by centrifugal force. Thus, the cake is wound from its outside inwards.

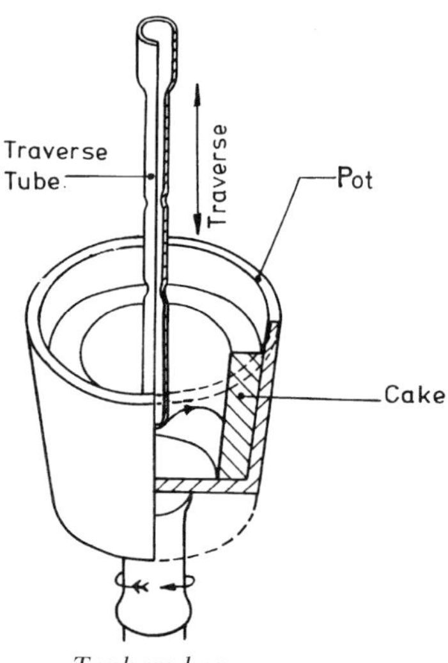

Topham box

topical finishing
The application of a liquor containing one or more chemical finishes to the surface of a textile substrate, typically textile fabric, using, for example, a lick roll, spray or foam application method.

topping
The application of further colorant not necessarily of the same **hue** or class to a dyed substrate in order to adjust the latter to the desired final colour.

torque direction
See **twist liveliness**, *Note 2*.

total pile mass; total pile weight
See **pile mass.**

touch and close fasteners
Fasteners consisting of two opposed tapes which repeatedly can be closed by the application of pressure and pulled apart manually. The fastening mechanism may be formed by minute hooks on the surface of one tape which interlock with minute loops on the face of the second tape, or a loop tape locking with mushroom-shaped protuberances on the face of the other. Typically, fasteners are made of polyamide, polyester or polypropylene.

tow (flax or hemp)
Any substantially clean fibre of less than scutched length.

tow (manufactured fibre production)
An essentially twist-free assemblage of a large number of substantially parallel filaments.

tow, machine
See **flax tow**.

tow, re-scutched
See **flax tow**.

tow, rug
See **flax tow**.

tow, straw
See **straw tow.**

tow-to-top conversion
See **converting.**

traceability
The ability to trace the history, application or location of an item or activity, or similar items or activities, by means of recorded identification.

tracking (carpet)
See **shading (carpet)**.

tram
A silk weft yarn comprising two or more threads folded together with 200 or 400 turns/m (5 or 10 turns/inch).

transfer bar (weft knitting)
A point bar used to transfer rib boarders or other knitted pieces on to the needles of a straight-bar machine (see **point (rib transfer)**) under **point (knitting)**).

transfer printing
Any process by which a design is transferred from one substrate, usually paper, to another. Several techniques have been used, for example, melt-transfer, film-release, and wet transfer, but vapour transfer (sublimation transfer) is the most important. Selected disperse dyes transfer in vapour form to thermoplastic fibres when the printed paper and fabric are brought into close contact in a calendar or press at 170°-220°C.
Note: Undesirable synthetic fibres in wool fabrics can be dyed by vapour transfer of disperse dye from a polyester fabric (Troyfill® process).

transferring (weft knitting)
See **barring-on (knitting)**.

transitional cotton
Cotton grown under the same conditions as **organic cotton**, but on land which has been farmed in this way for less than a defined transitional period (usually three years).

trap
1. An incident which may occur during weaving when the shuttle or some other object stops in the shed formed by the warp threads. A **warp-protector motion** should prevent normal beat-up but disfiguration of the fell of the cloth, or breakage of a number of the warp threads may occur.
2. A fault caused by 1 that is visible in the fabric.

trash (cotton)
The non-fibrous foreign matter present in bales of raw cotton other than abnormal items, such as stone, timber, pieces of old iron, etc. Fine trash is referred to as cotton dust. The particle sizes of trash and the various types of cotton dust have been defined as having the following diameters: trash over 500 microns (µm); dust 50 to 500 microns (µm); micro-dust 15 to 50 microns (µm); respirable dust under 15 microns (µm).
Note 1: Normal whole seeds, either ginned or unginned, are frequently excluded from this category but broken portions of them and also whole or broken undeveloped seeds are usually regarded as trash.
Note 2: The main component of trash is **chaff** and dirt in the form of soil or sand is another. Foreign (unwanted) fibres such as lengths of sisal, jute, hemp, and grass are sometimes regarded as trash but usually receive special reference when easily recognisable. (See also **motes.**)

trash content
The ratio of the mass of trash in a sample of raw cotton fibres to the total mass of the sample. The ratio is usually expressed as a percentage. (See also **trash (cotton)**.)

traveller
The metal or plastic component through which yarn passes on its way from the ballooning eye to the package surface in ring spinning or twisting. It is mounted on a ring and is dragged round by the yarn. (See also **ring spinning.**)

traverse
The distance between extreme positions of a reciprocating thread-guide in one cycle of its movement.
Note: The traverse may be a constant quantity, as in building a cheese or in winding on to a double-flanged bobbin; or it may be variable, as in building a ring bobbin or conical-ended

traverse *(continued)*
roving bobbin. In the latter cases the traverse during winding-on of the first complete layer of yarn or roving determines the **lift**, otherwise the traverse is smaller than the lift.

traverse point (warp knitting)
See **point (milanese machines)** under **point (knitting)**.

treadle loom
A weaving machine activated by a treadle connected to the main shaft.

tree cotton
Cotton of the species *Gossypium hirsutum Marie-Galante* grown as a perennial in Northern Brazil.

trevet; trevette; trivet
1. Small hand held instrument holding a sharp blade for cutting the pile of hand woven **velvet** in the loom.
2. An instrument for cutting the pile of velvets woven double.

triacetate (fibre) (generic name)
A manufactured fibre of cellulose ethanoate (cellulose acetate) wherein at least 92% of the hydroxyl groups of the original cellulose are ethanoylated (acetylated). (See also Classification Table, p.401.)

triaxial weaving
A system of weaving that interlaces two sets of warp ends and one set of picks in such a way that the three sets of threads form a multitude of equilateral triangles. The resulting fabric has excellent bursting, tearing, and abrasion-resistance.

Yarn path for triaxial weaving

trichomatic system
See **colour measurement**.

trick (knitting)
A slot for preserving the spacing of knitting elements.

trick (lace machines)
A device on a lace furnishing machine for preserving the spacing of the **jacks** and controlling their lateral movement. It consists of a series of thin metal stampings set to the gauge of the machine in an alloy base.

trick bar (lace machines)
A bar to which **trick leads** are attached over the lace-making width of a machine. It is shogged as part of the **foundation bar**.

trick lead (lace machines)
A number of **tricks** cast to the gauge of the machine in a lead-alloy base.

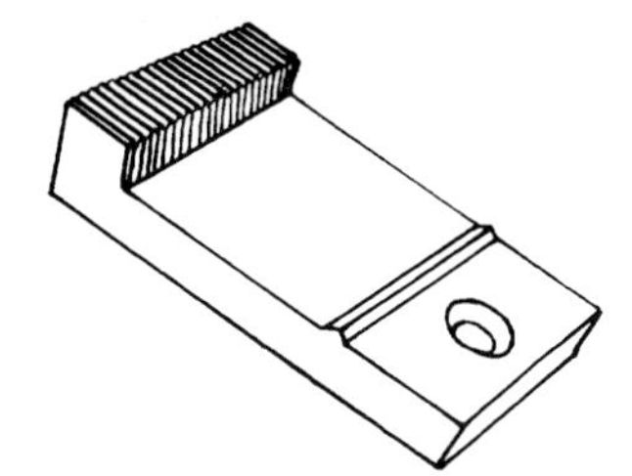

Trick lead (lace furnishing machine)

tricot warp-knitting machine
A warp-knitting machine generally using bearded or compound needles mounted vertically, or nearly so, in which the fabric is supported and controlled by **sinkers**. The fabric is removed from the knitting point at approximately 90° to the needles' movement (nearer the horizontal than the vertical). (See also **knitting machine.**)

tricot, warp-knitted
A warp-knitted fabric knitted with two full sets of warp threads, each set making a 1 and 1 lapping movement but in opposite directions. Additionally the term is now used generically to cover all types of warp-knitted fabric made on **tricot warp-knitting machines.**

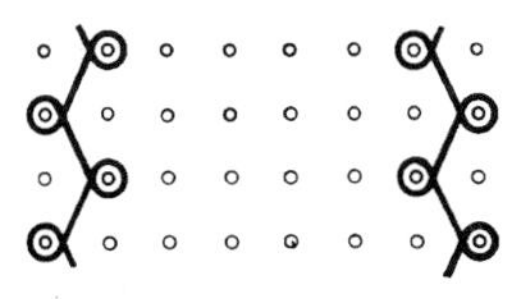

Warp-knitted tricot structure

tricotine
1. A weft-face woven fabric, with a cotton warp and worsted weft, displaying a fine flat twill line.
2. A plain-weave woven fabric, with fine horizontal ribs, silk warp and cotton weft.

trilam
A plain weave laminated fabric, made from polyester scrim coated with polyester film, for use

trilam *(continued)*
as sails. It is produced 1.37m wide from 1000dtex warp and 250dtex weft in varying thicknesses e.g. 1.1mm (116 g/m^2), 1.5mm (138 g/m^2) and 2.0mm (159 g/m^2).

trimming (knitting)
See *Note* under **boarding.**

trimming (making-up)
1. The attachment of various decorative effects, e.g., lace motifs, by sewing.
2. The removal of loose ends in the making-up of knitted goods.

trimming (narrow fabric)
Any **narrow fabric** used generally for the purpose of decorating or adorning furnishing fabrics or clothing, but often fulfilling a utilitarian function.

trimming loom
A machine that produces unattached wales of warp-knitted loops interlaced by weft threads that are carried across them by thread guides, and thus forms a fabric.

trimming machine
See **crochet-knitting machine.**

trimmings; haberdashery
Parts of a garment other than fabric and sewing thread, such as linings, shoulder pads, piping, ribbons, buttons and zips.

trimmings (lace)
See **edge (lace)**.

tristimulus values
The amounts of three defined primaries (usually blue, red and green) required to be mixed additively to match the colour of the object, under defined conditions.

trivet
See **trevet.**

trivinyl (fibre)
A manufactured fibre made from a synthetic terpolymer of cyanoethene (acrylonitrile), a chlorinated vinyl monomer and a third vinyl monomer, none of which represents as much as 50% of the total mass.
Note: This name was formerly an ISO **generic name** (See also Classification Table, p.401.)

tropical fabric
A light-weight suiting for use in hot climates.

trough
See **beck.**

troughed shed (weaving)
A warp shed in which the two sheets that form the shed are under different tensions in order to improve **cover.** This is frequently achieved by raising the back rest above the normal centre shed line, but alternative techniques may be used according to the weaving machine type: these include lowering the back rest, raising or lowering the height of the **heald frames** or raising or lowering the height of the cloth fell.

trouser binding
See **kick tape.**

trouser braid
A flat, usually black, braided **narrow fabric** with closely interlaced threads at the sides and more open interlacing in the centre, having a core thread on each side of a more loosely constructed centre. It is usually made of continuous-filament yarn and used on men's formal dress.

true cross
See **bias.**

true hemp
See **hemp, true.**

truth mark
An identification mark applied close to the ends of a **piece** of fabric by various means, such as weaving in a contrasting colour of yarn, using a marker pen, punching in a logo etc. The purpose of the mark is to show that the piece has remained intact between processes or between mill and customer.

tube
A hollow cylindrical or slightly tapered support, without a flange, on which yarn is spun or wound.
Note: A tube that extends throughout a yarn package and projects slightly at each end may be known as a 'through-tube'.

tube test
A form of combined **crimp retraction** and bulkiness test, used in the U.K. for the testing of **false-twist textured yarn**, in which the latent filament crimp of a cut hank of yarn is developed inside the confined space of a precision glass tube.

tubular welt (knitting)
See under **welt (knitting)**.

tuck loop (knitting)
A length (or lengths) of yarn received by a needle and not pulled through the loop of the previous course.

tuck ripple (weft knitting)
See **ripple (weft knitting)**.

tuck stitch (knitting)
A stitch consisting of a held loop and a tuck loop, both of which are intermeshed in the same course.

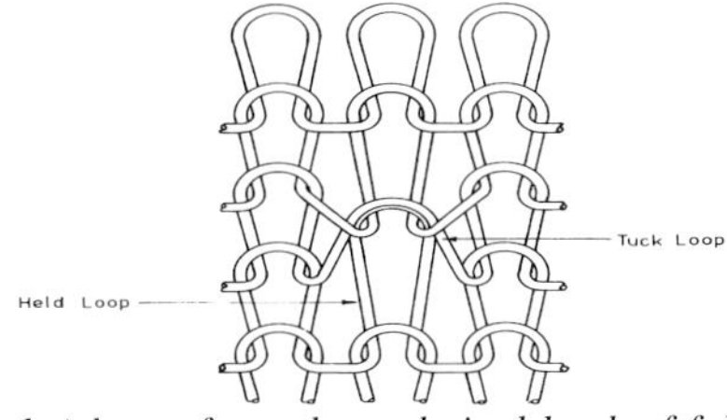

Tuck stitch (shown from the technical back of fabric)

tucking (defect)
A **tuck loop** or loops inadvertently produced in a knitted fabric.

tuft
An I, J, U, or W-shaped length of yarn, or a length of yarn in the form of a knot, of which the leg or legs form the pile of a carpet.

tuft bind test; tuft withdrawal test
A test for determining the force required to remove individual tufts of pile from textile floorcoverings or other pile fabrics.

tuft column; loop column
In carpets, tufts or loops running in a line essentially parallel to the direction of manufacture.

tuft length
The distance between the extremities of a basic unit tuft, after removal and straightening.

tuft row; loop row
In carpets, tufts or loops running in a line at right angles to the direction of manufacture.

tufted carpet
Carpet produced by inserting pile yarn, with needles, into a pre-made primary backing. Various secondary backings, e.g., hessian and foam, are then applied in a separate process (see **substrate (carpet)**).

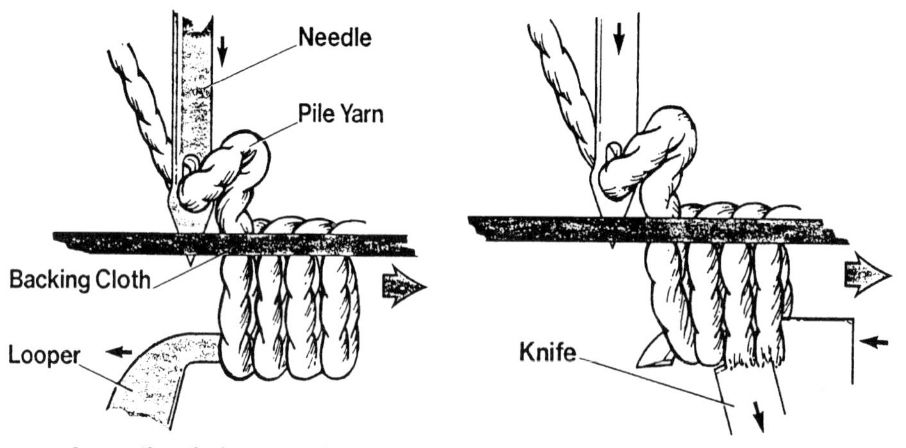

Loop pile tufted construction *Cut pile tufted construction*

tufting machine (carpet)
A machine for producing tufted carpets which may have one or more of the following basic components and/or patterning systems:
(i) Basic components

> **jerker bar; shaker bar**
> A yarn guide mounted on the needle bar assembly, for taking up the slack yarn on the needle upstroke of the tufting cycle.
>
> **knife**
> A flat sharpened blade which operates in conjunction with the looper to cut the formed loops to give a cut-pile product.

looper (hook)
A hook-shaped flat metal plate which, together with a needle, forms the loops during tufting processes.

needle (tufting)
An instrument for inserting pile yarn tufts into a pre-made substrate. As in a sewing machine needle, the eye and the point are at the same end. The needle is grooved on one side, to take the yarn, and chamfered on the other side to lead the looper to pass between the needle and the yarn.

needle bar
A rigid bar, running the width of the tufting machine and in which the tufting needles are mounted at fixed intervals. The needles may be in line (straight needle bar) or in two rows with alternate needles in separate rows (staggered needle bar).

pattern attachment
A device which controls the pile yarn feed so as to produce a high and low pile surface.

reed-plate
A flat comb-like plate with slots on one side spaced in accordance with the machine gauge.

(ii) Patterning systems

cross-over tufting
A patterning technique for multi-coloured carpets whereby different colours or textures of yarn are creeled and threaded through the needles in a pre-determined sequence to form one element of the design. The controlled lateral displacement of the needle bar/ bars forms the second element of the design. This technique is employed with a single or double staggered needle bar. The greater the traverse of the bar the larger the design.

eccentric disc
One of a series of grooved discs, one for each yarn end, mounted on a driven shaft. As the eccentrics rotate the tension of the yarn changes, and as a result differences in pile height are created.

overtufting
A patterning technique in which additional yarns are tufted into a pre-made carpet fabric.

scroll
A patterning attachment consisting of a number of yarn feed rollers which are capable of being driven at, usually, two or three different speeds to achieve loop or cut and loop pile carpet with two or three pile height.

slat
A patterning attachment consisting of two sets of intermeshing metal angular strips (also called slats), mounted on continuously moving roller chains. One set has a constant height, whilst the other set has a profile machined according to the required pile height in the design. A variant of this attachment has the profiled slat machined alternately high and low, and the pile yarns are directed to the high or low areas by electronically controlled guides.

sliding needle bar
A patterning device whereby the needle bar(s) are moved laterally during the tufting process in a precise manner. Movement of the needle bar(s) takes place only when the needles are lifted out of the primary backing cloth and individual stitches are controlled. The movement may be controlled by cam(s) or by computer with other mechanical means. This technique can be employed with a straight needle bar but more sophisticated patterns can be achieved when used in conjunction with a staggered needle bar. Even

tufting machine (carpet) *(continued)*

more complex designs are achieved when the technique is used in conjunction with two needle bars (double sliding needle bar) where each needle has a separate sliding needle bar operating mechanism.

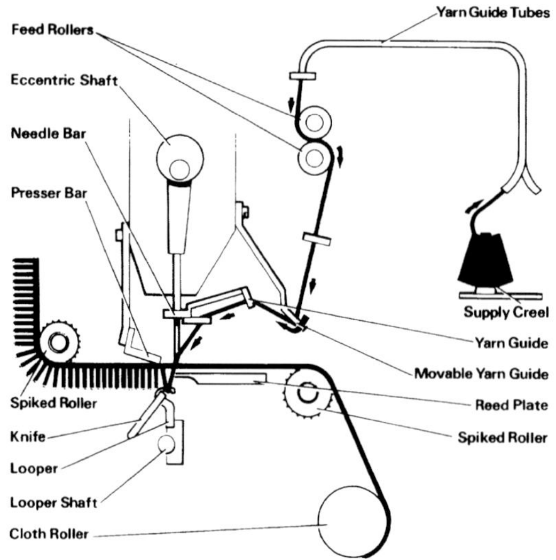

Tufting mechanism showing the main elements controlling the tufting actions

tug reed

A **gauze reed** that makes a reciprocating sideways movement to traverse the ends that pass through it.

tulle (lace)

A fine, soft, very light-weight, machine-made net with hexagonal mesh.

tulle, warp-knitted

A net with a hexagonal hole produced from two guide bars, the front bar knitting on one needle for a number of courses and then on the adjacent needle for a similar number of courses before returning to the original wale. The second bar lays-in so that any one thread moves with one front-bar thread. The lapping movements for a three-course tulle are as shown below:

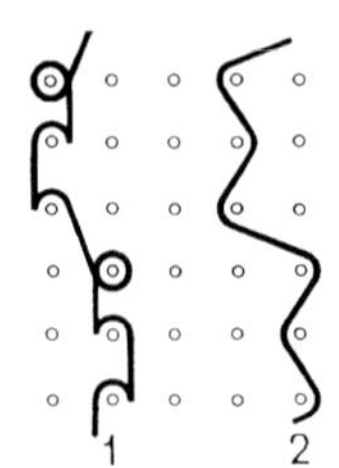

Warp-knitted tulle structure

tulle, woven

A very fine net fabric made in plain weave from silk yarns.

tumble felting

A method of felting hanks of wool yarn in either aqueous or other solvent media using rotary washing or dry cleaning machines.

tumbler
A frictionally driven, self-weighted, smooth wooden or metal roller that rests on the material supported by the carrier (see **carrier (spinning)**). It is used on some worsted drawing boxes and spinning frames to control the fibres during drafting.

turned welt (knitting)
See under **welt (knitting)**.

turning(s)
See **seam allowance.**

tussah silk
A coarse silk produced by a wild silkworm. There are three main types: *Antheraea mylitta* (largely Indian), *Antheraea pernyi* (largely Chinese), and *Antheraea yama-mai* (largely Japanese). It is brown in colour and is usually spun, since most cocoons cannot be reeled.
Note: The spelling 'tussah', although considered erroneous by etymologists, is in common usage in the textile industry for the name given to fibres and filaments.

tussore
A fabric woven from the coarse silk called **tussah.**

Twaddell
A scale used for the measurement of the relative density (RD) of liquids by hydrometry. The following formula expresses the relationship between relative density and degrees Twaddell (°Tw), for liquids more dense than water:

$$°Tw = 200\ (RD - 1).$$

The formula is only exact for comparisons at 15°C.

tweed
Originally a coarse, heavy-weight, rough-surfaced wool fabric for outerwear, woven in Southern Scotland. The term is now applied to fabrics made in a wide range of weights and qualities from woollen-spun yarns in a variety of **weave effects** and **colour-and-weave effects.**
Note: Descriptions of tweed not made substantially of wool need qualification.

twill
1. A weave that repeats on three or more ends and picks and produces diagonal lines on the face of fabric.
2. A fabric produced as above.
Note 1: The diagonal lines produced on the surface of a fabric by a twill weave are often referred to as the twill in such phrases as 'a prominent twill', 'a broken twill', 'unwanted twill'.
Note 2: Unwanted twill may arise as a defect in satin fabrics, the intensity of the unwanted twill depending on the fabric structure, the weave, and the number of ends per dent in the reed.

twill direction
The direction of a twill is generally described as a fabric is viewed looking along the warp. 'Twill right' then refers to the diagonal running upwards to the right (↗), and 'twill left' to the diagonal moving upwards to the left (↖). By analogy with twist direction in yarns, an alternative method is to describe 'twill right' as 'Z' and 'twill left' as 'S'.

twilled hopsack
See **hopsack**.

twillette
A twill fabric in which the weft predominates.

twine
1. See **bullion cord**.
2. Twisted **cordage** less than 4mm in diameter.

twine fringe
See **bullion fringe.**

twist
The condition of a yarn or similar structure when the component elements have a helical disposition such as results, for instance, from relative rotation of the yarn ends. A yarn may be twisted by down twisting (see **ring twisting**), **false-twisting**, **open-end spinning**, **self-twist spinning**, **two-for-one-twisting** or **uptwisting**. For all practical purposes twist is measured in turns per unit length, but for purely theoretical work its measurement in radians (the SI unit) often leads to much simpler expressions.

twist (yarn)
See **fold (yarn)**.

twist angle
The included angle between the path of a yarn element and the yarn axis.

twist blockage; twist congestion; twist hold-back; twist disturbance; twist accumulation
The inhibition of the free transmission of twist with or along a moving yarn or other linear textile structure, due to contact with a guide or other similar surface.

twist direction
Twist is described as S or Z according to which of these letters has its centre inclined in the same direction as the surface elements of a given twisted yarn, when the yarn is viewed vertically.

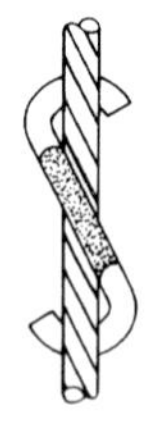
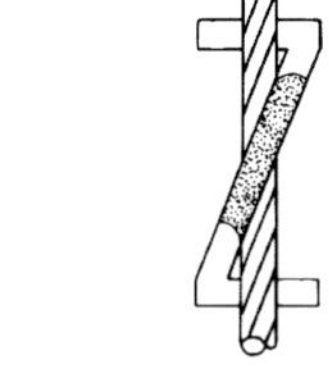

S-twist *Z-twist*

twist factor; twist multiplier; alpha metric (α_m)
A number (or value) given by the product of the twist level in a yarn (turns/unit length) and the square root of its count (direct system) or the inverse square root of its count (indirect system).
Note 1: The twist factor is directly related to the angle of twist helix the surface fibres have in a yarn for the same packing density. The higher the twist level, the greater is the helix angle and the larger the twist factor.
Note 2: The purpose of the twist factor is to calculate the level of twist to use in spinning, in order to maintain the same angle of twist helix and similar yarn characteristics when the yarn count is changed.
Note 3: Twist factor may also be applied to rovings and plied yarns.

twist level
The amount of twist per unit length of a yarn.

twist liveliness
The tendency of a yarn to twist or untwist spontaneously. (See also **snarly yarn.**)

Note 1: Examples of effects that may be caused by twist liveliness include snarling of yarns during processing and spirality in knitted fabrics.
Note 2: The direction of twist liveliness or torque, S or Z, is that of the twist change that takes place spontaneously when an end of yarn or hanging loop is allowed to rotate freely.

twist multiplier
See **twist factor.**

twist selvedge
See under **selvedge, woven.**

twist set; set twist; dead twist
Descriptive of the condition of a yarn in which an unbalanced twisting or untwisting torque has been dissipated or rendered latent by suitable treatment such as steaming.

twist yarn
See **worsted yarns, colour terms.**

twist-on-twist
Descriptive of a folded yarn in which the direction of the folding twist is the same as that of the single twist.

twisting-in
The operation of twisting ends of a new warp to the corresponding ends of an old warp to enable the supply to be maintained without re-threading.

twistless spinning
A system of yarn formation that relies on the use of a permanent or temporary adhesive to bond fibres together. (See also **spinning.**)
Note 1: Where temporary adhesive is used it is removed during fabric finishing, and the yarn (and fabric) strength is then obtained through lateral pressure produced by the interlacings in the fabric. A similar fabric construction can be achieved by using **wrap spun yarns** which have been produced with a soluble **binder**.
Note 2: Fibres may be bonded with an adhesive on to a filament core to form a yarn.

twistless yarn
A yarn without twist. (See also **continuous yarn felting**, **twistless spinning**, **zero-twist yarn**, **flat yarn** 1 and **wrap-spun yarn.**)

twitty
Descriptive of an irregular yarn in which local concentrations of twist have accentuated the irregular appearance.

twizzle
See **ballooning eye.**

two-by-two rib fabric, weft-knitted
See under **rib fabric, weft-knitted.**

two-for-one twisting
A system which inserts two turns of twist for each revolution of a twisting element. This is achieved by inserting one turn of twist between a stationary feed package and a rotating disc and a further turn between the latter and a balloon guide.
Note: If doubling is involved, either a single assembly-wound package or two separate single-

two-for-one twisting *(continued)*
wound packages may be used as the supply. When the single-wound packages are joined together, the system is called clip-cone two-for-one twisting.

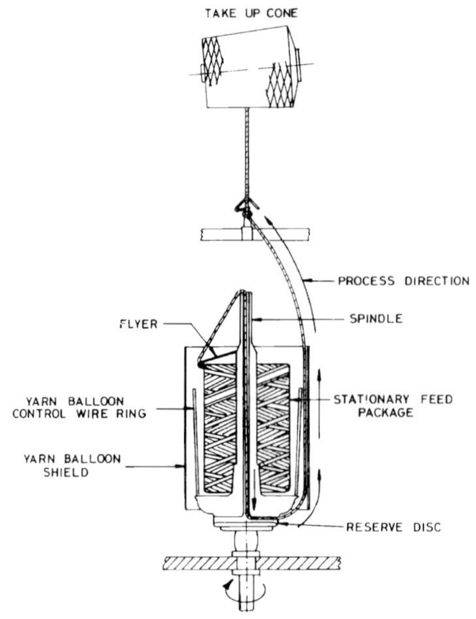

Two-for-one twisting

two-piece needle (machine knitting)
See under **needle (machine knitting)**.

two-stage twisting; stage twisting
A system of producing a yarn which consists of two stages: (i) inserting a low level of twist into a yarn or yarns by **ring twisting**; and (ii) taking the product of (i) and **uptwisting** to insert the desired amount of twist.

tying tape
See **bolduc.**

tying-in; tying back
See **knotting** 2.

tyre textiles

bead wrapping fabric
A rubber-coated cross-woven fabric which is wrapped around a rubberized bead coil.

belt
Two, or multiples of two, layers of tyre-cord fabric beneath the tread, lying at opposite angles close to the circumferential direction, with or without an additional layer with cord angle at 90° to the circumferential direction. Its purpose is to brace the carcase of a radial-ply tyre to stabilize and control its directional properties.

breaker fabric (crossply tyres)
One or more extra layers of **tyre-cord fabric** that lie between the crown of the carcase and the tread of a tyre. The breaker fabric may sometimes extend as far as the shoulder of the tyre.

casing, crossply
A casing having multiples of two plies extending from bead to bead, with alternate plies at bias angles opposite to the circumferential line, the bias angle increasing from sidewall to crown.

casing, radialply
A casing having one or more plies extending from bead to bead approximately 90° to the circumferential line.

chafer fabric
A fabric, coated with unvulcanized rubber, which is wrapped round the bead section of a tyre before vulcanization of the complete tyre, and whose purpose is to maintain an abrasion-resistant layer of rubber in contact with the wheel on which the tyre is mounted.
Note: Chafer fabrics originally were cross-woven cotton. For tubeless tyres they are usually resin-impregnated multi-filament mesh fabrics of rayon or nylon or alternatively nylon mono-filament mesh.

filler fabric; flipper fabric
A rubber-coated cross-woven fabric which is placed around the bead section assembly and serves to reinforce the join between apex and casing plies.
Note: In all-metallic radial-ply tyres this filler often consists of a ply of wire cords.

ply (tyres)
A layer of rubber-coated parallel cords.

tyre-cord fabric
A fabric that comprises the main carcase of a pneumatic tyre, and is constructed predominantly of a ply warp with a light weft to assist processing.

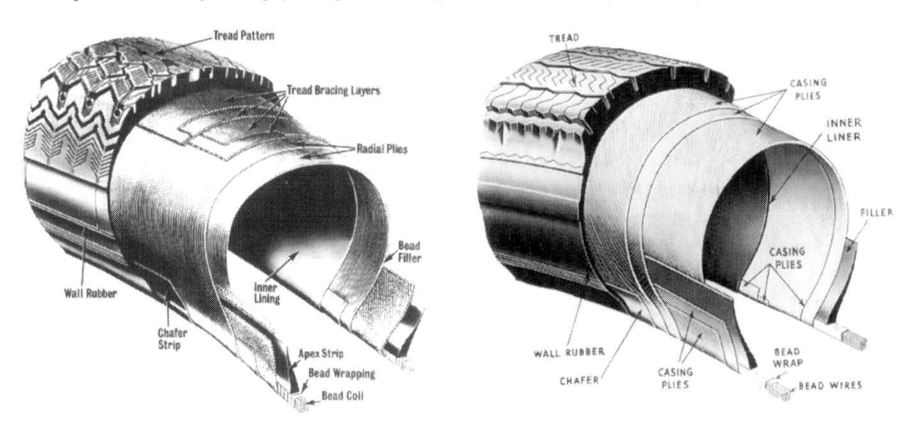

Radialply tyre *Crossply tyre*

tyre yarn
Yarn that is used in the manufacture of the textile carcase of rubber tyres. (See also **tyre textiles.**)

tyre-cord fabric
See under **tyre textiles.**

U%
A measure of the mass variability per unit length of yarn. It is statistically equivalent to the percentage mean deviation (PMD) and is often approximately equal to 0.8 CV%, providing the mass variations are normally distributed.

U-box
An upright U-shaped vessel for the continuous processing of **textiles**.
Note: The material enters one arm of the U, where it remains for a dwell period, and is then withdrawn through the other arm. (See also **J-box.**)

ultimate fibre
See **fibre ultimate.**

ultimate rupture
See **breaking point.**

ultra-fine fibre; ultra-fine microfibre
See **microfibre** 1, *Note 3*.

uncut pile
See **loop pile** under **pile (carpet)**.

underfelt (carpeting)
A felt, usually a needlefelt, used as an **underlay.**

underlap (warp knitting)
1. Lateral movements of the guide bar made on the side of the needle remote from the hook or beard; the amount of this movement is limited only by mechanical considerations.
Note: The terms: 'lap', 'shog', 'throw', 'rise', and 'fall' are also used to express general lateral motions of the guide bars without specific reference as to whether they are made in front of or behind the needles.
2. In the fabric, the connection between stitches in consecutive courses in a warp-knitted fabric.

underlay
A resilient layer of textile and/or other material placed between the textile floorcovering and the floor. Underlays include the following types:
(i) felt underlay: underlay made wholly of fibres entangled or matted together;
(ii) cellular underlay: underlay formed of a rubber or other polymeric foam, e.g., polyurethane, with or without being bonded to a fabric or plastic sheet;
(iii) rubber crumb underlay: underlay formed of crumb rubber with or without being bonded to a fabric or plastic sheet;
(iv) composite underlay: underlay composed of both fibrous and polymeric material, e.g., felt impregnated with rubber, foam laminated to felt.

undrawn tow
Extruded continuous-filament tow, the component macromolecules of which have a low degree of orientation.
Note: Undrawn tow represents an intermediate stage in the production of some synthetic fibres.

undrawn yarn
Strictly, any extruded continuous-filament yarn that has not been subjected to a drawing process, but this term is usually restricted to such yarns, the component macromolecules of which have a low degree of orientation.
Note 1: Undrawn yarn represents an intermediate stage in the production of some synthetic-fibre yarns. (See also **fully drawn yarn**, **draw-twist** and **draw-wind.**)
Note 2: Some undrawn filament yarns produced from liquid crystal polymers have a very high degree of orientation.

uniformity index
See under **fibre length.**

uniformity ratio
See under **fibre length.**

union dye
A dye, or a mixture of dyes, which will yield an apparently solid dyeing on the fibre mixture for which it is devised.

union fabric
A fabric made with warp of one kind of fibre and weft of another.
Note: Originally the term related to fabrics made from cotton warp and wool weft or from linen warp and cotton weft.

union yarn
A yarn made by twisting together yarns of different fibres.

unison feed
See **feed mechanisms (sewing)**.

unitary backing (carpet)
See under **substrate (carpet)**.

unripe cotton
See **maturity (cotton)**.

unweaving
The act of removing weft threads to correct weaving faults (picking-out) and the subsequent re-setting of the **fell** to the correct position (pulling-back) before the weaving machine is restarted.
Note: The terms pulling-back and picking-out may also be used to describe the whole operation.

upholstery cord; furniture cord
A cord consisting of two case cords and two gimp cords, which are twisted together.

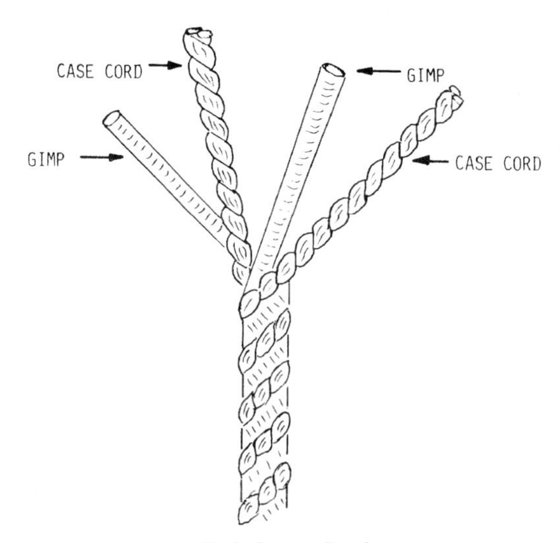

Upholstery Cord

upholstery web
See **chair web.**

upland cotton
A type of cotton (*Gossypium hirsutum*) which forms the bulk of the world's cotton crop. It varies in average staple length from about 22mm to about 32mm.

upper half mean length
See under **fibre length.**

uptwisting
A system of twisting one or more yarns by withdrawing them over-end from a rotating package.
Note 1: If more than one yarn is to be twisted, the yarns are first assembly-wound on to a single uptwister supply package.
Note 2: Uptwisting forms the second stage in **two-stage twisting**.
Note 3: Uptwisting is sometimes used to indicate the insertion of additional twist in a yarn, irrespective of method.

urea-bisulphite solubility
The solubility of wool in a neutral solution containing a specified amount of urea and sodium metabisulphite provides an index of the extent of change in its chemical properties brought about by certain agencies.

urena
A bast fibre similar to jute, from the plant *Urena lobata.*

usable width
That width of a fabric that can be utilized.

Utrecht velvet
See under **velvet, woven.**

V-bed knitting machine
See **flat knitting machine.**

valencia vesting
A fabric for waistcoats, woven with a cotton binder warp and a silk figuring warp with a worsted weft. It is imitative of Spitalfields figured silks and was popular in the early 19th century.

van Dyke braid
A **ric-rac braid** with the points at the edges forming an acute angle.

vandyke fabric
See **atlas fabric, single bar.**

variable cut device
An ancillary device on the crush-cutting type of converter (see **converting**) which cyclically varies the angle of approach of the tow to the cutting region in order to introduce a controlled variation of fibre length.

variable top feed
See **adjustable top feed** under **feed mechanisms (sewing)**.

vat
1. A vessel or tank.
2. A liquor containing a reduced (leuco) **vat dye.**
3. To dissolve a vat dye by the combined action of alkali and a reducing agent.

vat dye
A water-insoluble dye, usually containing keto groups, which is normally applied to the fibre from an alkaline aqueous solution of the reduced enol (leuco) form, which is subsequently oxidized in the fibre to the insoluble form.

veiling (lace)
Plain or ornamental nets, with relatively large meshes, used mainly for face veils or hat decoration.

velour (hats)
A rabbit-fur felt body with top surface carded in the wet state to produce a long soft pile.

velour, warp-knitted
A two or three-bar warp-knitted fabric in which a pile is produced higher than that normally associated with brushed or raised warp-knitted fabrics. Often the front bar movement is described by a 1-0/6-7 notation. The fabric is cropped after raising.

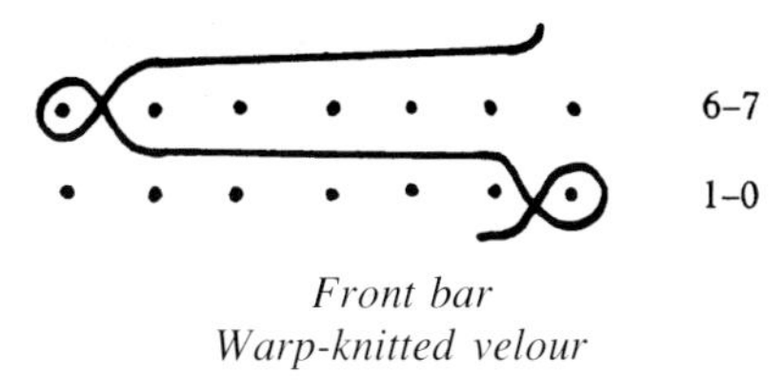

Front bar
Warp-knitted velour

velour, woven
1. A heavy pile fabric with the pile laid in a single direction.
2. A napped-surface woven fabric or felt in which the surface fibres are laid in a single direction to present a smooth appearance.
3. A terry fabric that has had the tops of the loops cut off in a process subsequent to weaving. It is also known as 'cropped terry pile' and 'sheared terry pile'.

velvet, woven
A cut warp-pile fabric, originally of silk, in which the cut ends of the fibres form the surface of the fabric. This effect is produced:
(i) from a pile warp lifted over wires and cut by a **trevet**;
(ii) from a pile warp lifted over wires which are withdrawn to cut the pile (see **wire (pile weaving)**);
(iii) by weaving two fabrics face to face with the pile ends interchanging from one fabric to the other: the pile ends are cut by a knife while still in the loom, giving separate pieces of velvet. (See also **plush.**)
Note: A wide range of named velvet structures exist. They have characteristic features created by the selected use of weaves, materials or finishing treatments.

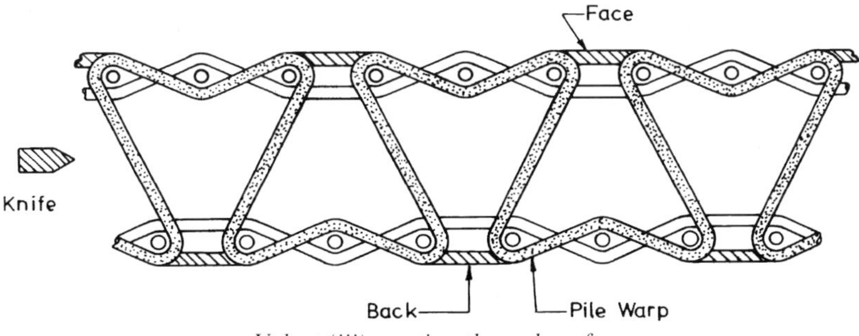

Velvet (iii): section through weft

velvet, woven *(continued)*

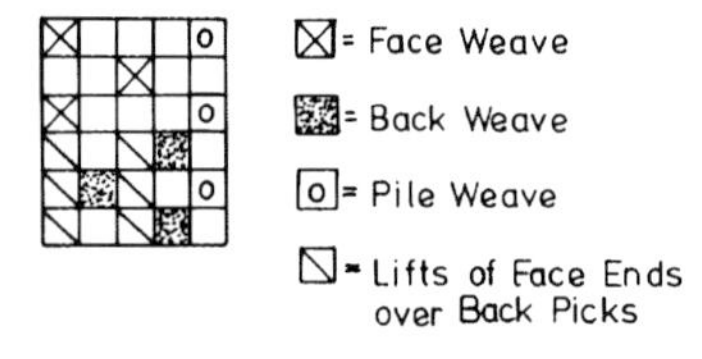

Velvet (iii): weave

panne velvet
A raised fabric in which the pile is laid in a single direction during finishing to give a very high lustre.

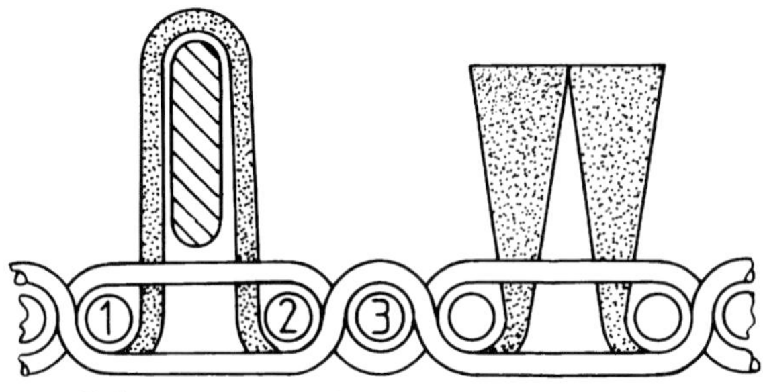

Velvet (ii) panne velvet: section through warp

G G P
G 1
W
G 2
G 3

G = Ground Thread
W = Wire in Section
P = Pile

= Ground End Raised
= Pile End Raised

Velvet (ii) panne velvet: weave

terry velvet
A fabric that is woven on the velvet principle (see **velvet, woven** (ii) above) but in which the pile is not cut.

Utrecht velvet
A furnishing velvet characterized by strong W-type anchorage (see diagram). The pile is usually, but not necessarily, of mohair and the ground commonly of cotton.

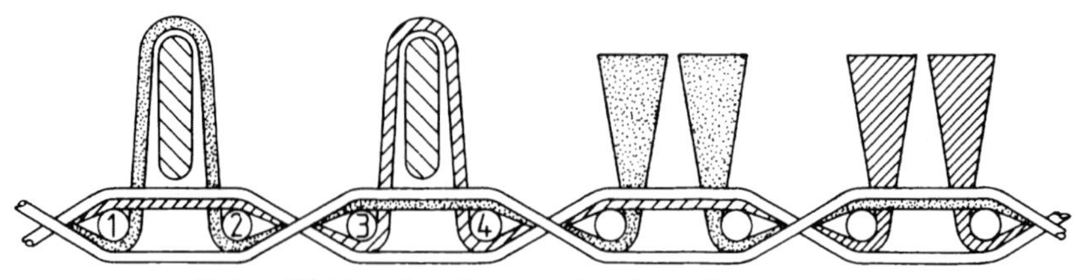

Velvet (ii) Utrecht velvet: section through warp

G P G P
G 1
W
G 2
G 3
W
G 4

G = Ground Thread
W = Wire Insertion
P = Pile Thread
= Ground End Raised
= Pile End Raised

Velvet (ii) Utrecht velvet: weave

velveteen

A cut weft-pile fabric in which the cut fibres form the surface of the fabric. The effect is produced by cutting the weft floats after weaving (see **fustian**).

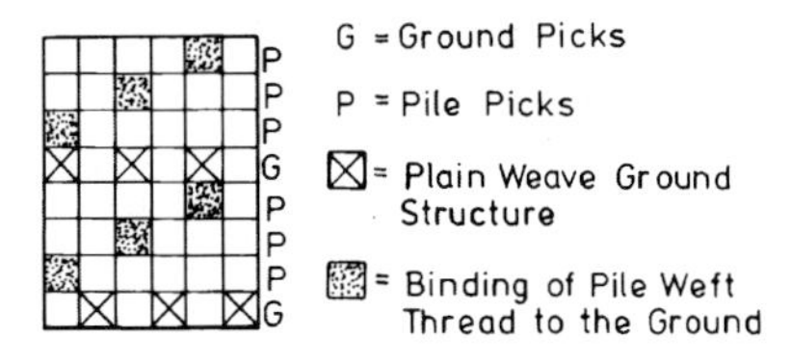

Velveteen weave

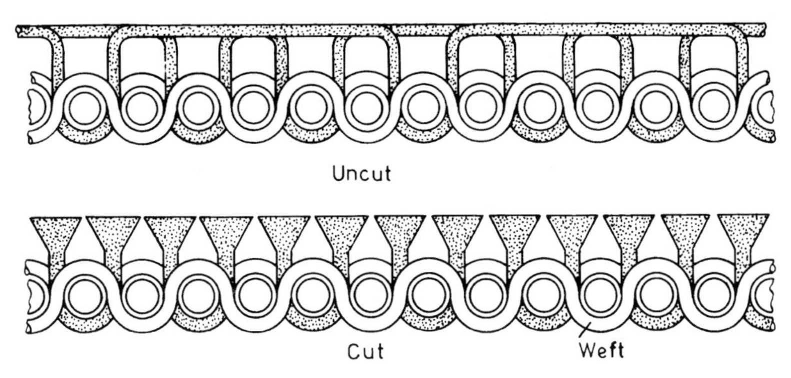

Velveteen: section through warp

Venetian fabric

1. (Cotton) An eight-end warp-faced satin, usually mercerized and schreinered, used as a lining fabric.
2. (Wool) A term applied to warp-faced fabrics, in five-end satin or modified satin weave (see diagram), from woollen or worsted warp and woollen weft, milled, lightly raised, and cropped to reveal a fine, steep twill.

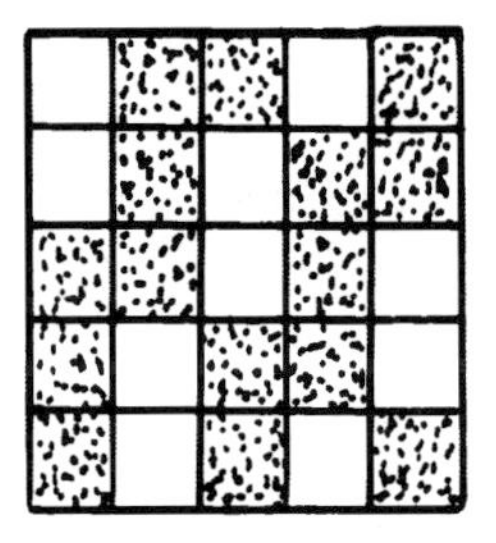

Venetian 2 weave

vicuna fibre (hair)

Fibre from the fleece of the vicuna (*Lama vicugna*).

Vigoureux printing

See **mélange printing.**

vinal (fibre) (generic name U.S.A.)

A manufactured fibre in which the fibre-forming substance is any long-chain synthetic polymer composed of at least 50% by weight of ethenol (vinyl alcohol) units and in which the total of the ethenol units and any one or more of various acetal units is at least 85% by weight of the fibre.

vinal (fibre) (generic name U.S.A.) *(continued)*
Note: The ISO **generic name** is **vinylal** (See also Classification Table, p.401.)

vinylal (fibre) (generic name)
A range of manufactured fibres composed of synthetic linear macromolecules of polyethenol (polyvinyl alcohol) of differing levels of acetalization. (See also **vinal** and Classification Table, p.401.)

vinyon (fibre) (generic name U.S.A.)
A manufactured fibre in which the fibre-forming substance is any long-chain synthetic polymer composed of at least 85% by weight of chloroethene (vinyl chloride) units.
Note 1: Vinyon is a **chlorofibre**.
Note 2: PVC (poly (vinyl chloride)) fibres are the most commonly met examples in this category. (See also Classification Table, p.401.)

virgin wool; new wool
Fibre from the fleece of a sheep or lamb that has not previously been spun into yarn or felted, nor previously been incorporated into a finished product.

virtual mass
See **added mass.**

viscose
The solution obtained by dissolving sodium cellulose xanthate in a dilute solution of sodium hydroxide.

viscose (fibre) (generic name)
A manufactured fibre of cellulose obtained by the viscose process (see **viscose**). (See also Classification Table, p.401.)

viscosity
The internal resistance to flow of a fluid. (See also **ball fall** and **fluidity**.)
Note 1: The dimensions of viscosity are $ML^{-1}T^{-1}$; the usual unit is the pascal second (Pa s=N s m^{-2}).
Note 2: The viscosity of a dilute solution of a polymer is commonly expressed in one of the following ways:
(i) viscosity ratio; relative viscosity: The ratio of the viscosity of a solution of defined polymer concentration to the viscosity of the solvent. This value is dimensionless.
(ii) specific viscosity: The viscosity ratio less unity. This value is dimensionless.
(iii) limiting viscosity number; intrinsic viscosity: The value obtained by extrapolating to zero concentration a set of values obtained by dividing the specific viscosity of a solution by the concentration of polymer in the solution for a series of solutions of different low concentrations. The dimensions are L^3M^{-1}, usually dl g^{-1}.
(iv) logarithmic viscosity number; inherent viscosity. The value obtained by dividing the natural logarithm of the viscosity ratio by the concentration for a polymer solution of defined low concentration. The dimensions are L^3M^{-1}, usually dl g^{-1}.

voile
A light-weight, approximately square-sett, open-textured, plain-weave fabric made from fine yarns of sufficient twist to produce a round, compact thread woven one-thread-per-dent unless the number of ends per unit length is so high as to render this impossible (see **voile yarn**). Typical construction: (i) cotton voile: 23x22; R15/2 x R15x2tex; 70 g/m^2; K=9.2+8.5; (ii) nylon voile: 40x36; 56x56dtex; 44 g/m^2; K=9.5+8.5.

voile yarn
Single or ply yarn of sufficiently high twist to produce a round, compact, resultant yarn.

wadding
A lofty sheet of fibres, which may be bonded, used for padding, stuffing, or packing.

wadding thread; stuffer thread; padding thread; filling thread
Additional warp or weft used in a fabric for the purposes of increasing its weight, bulk, firmness, or the prominence of the design. These threads are not visible on the fabric face.

waffle piqué
See **marcella.**

waistband
A narrow band of material around the waist of a garment.
Note: It may consist of cloth folded double and attached to the waist or it may be produced integrally with the garment. Depending on garment style, the waistband may be stabilised or elasticated.

wale (knitting)
A column of loops along the length of a fabric.

wale (lace)
The distance between the centres of two adjacent **pillars.**

wale density (knitted fabric)
The number of visible loops per unit length measured along a course.
Note 1: The traditional unit has been wales per inch but the value is now expressed as wales per cm.
Note 2: In certain constructions, the number of loops visible in one course may be different from that in another, and there may also be different numbers on the back and front of the fabric. Consequently, in such constructions, it is necessary to specify where the count is made.

wale spirality (weft knitting)
A distortion of a circular-knitted fabric in which the **wales** follow a spiral path around the axis of the tube. Spirality is caused by the use of yarn that is twist-lively, the direction and degree of spirality being determined by the direction and degree of **twist liveliness**. A comparable defect occurs in flat-knitted fabric. (See also **course spirality (weft knitting)**.)

wall (fibre)
The solid portion of the cotton fibre, divided into two parts: (i) primary wall: a thin skin on the surface of the fibre; (ii) secondary wall: the main part of the solid part of the fibre composed of layers of cellulose.

wall thickness, apparent
The apparent width of a fibre wall as seen when fibres are examined under a microscope.
Note: In the cotton fibre maturity test, the apparent wall thickness is assessed visually at the widest part of the fibres as a fraction of the maximum ribbon width.

warp; chain
1. Threads lengthways in a fabric as woven.
2. A number of threads in long lengths and approximately parallel, in various forms intended for weaving, knitting, doubling, sizing, dyeing, or lace-making.
3. To arrange threads in long lengths parallel to one another preparatory to further processing.
Note: In addition to **beaming** the following methods of warping are practised: ball warping, cross-ball warping, and chain warping. The primary stage of these methods of warping is the withdrawal of ends from a warping creel and their assembly in rope form, a form that may

warp; chain *(continued)*
conveniently be used for wet processing. For convenience of handling, this rope may be (i) wound into a ball (ball warping), (ii) machine-wound on to a wooden roller into a cross-ball cheese (crosss-ball or cheese warping), or (iii) shortened into a link chain (chain warping). A number of these ropes may be assembled into a complete warp on a beam in a dressing frame, or may be split and dressed and incorporated in warps made by other methods. (See also **section warping.**)

warp (lace machines)
Parallel threads wound in sheet form on to a warp beam to provide the main structural threads.

warp beam back frame
See **cratch.**

warp drawing
See **draw-warping.**

warp dressing
The operation of assembling on a beam yarns from a ball warp, beam warp, or chain warp immediately prior to weaving (see **dressing (warp preparation)**).

warp finings (lace)
A filling-in structure in Leavers lace obtained by two gait throws of warp yarn in opposite directions on alternate motions.
Note: An alternative Leavers-lace construction used only one warp, and the filling was obtained by two gait throws of the warp threads in opposite directions on alternate motions. This is no longer made.

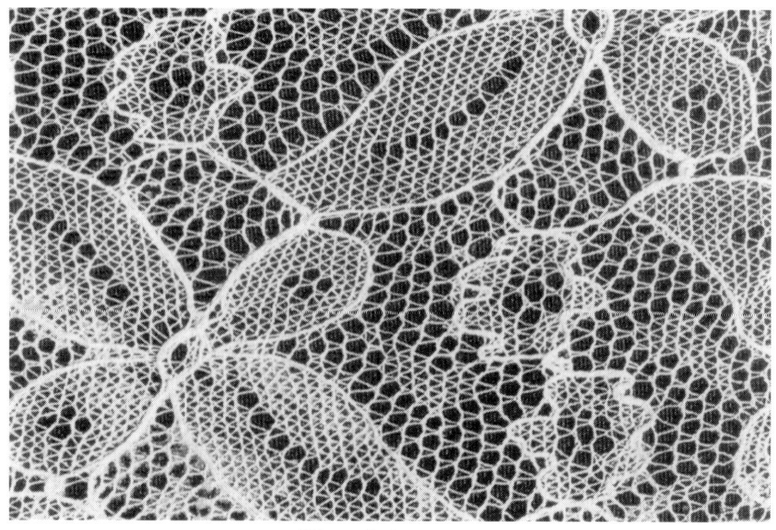

Warp finings

warp hairline
See **hairline.**

warp knitting
A method of making a fabric by normal knitting means in which the loops made from each warp thread are formed substantially along the length of the fabric. It is characterized by the fact that

each warp thread is fed more or less in line with the direction in which the fabric is produced.

warp lace machine
See under **lace machines.**

warp line
The line of a warp sheet on a weaving machine from the back rest to the fell.

warp rib
An effect obtained by weaving two or more picks as one and by using a warp **cover factor** approximately twice the weft cover factor, so that the warp is made to bend around the weft, which remains substantially straight. This leads to the formation of rounded warp-faced cords that run across the fabric. (See also **poplin** and **taffeta.**)

warp shed
See **shed.**

warp sheet; lash
The selected ends of pile warp yarns for a complete row of carpet pile prior to the insertion of the weft yarn.

warp snarl; weft snarl
A short length of warp or weft yarn that has twisted on itself owing to lively twist or insufficient tension. The snarling may occur during or prior to the weaving process. Weft snarls may also be referred to as weft curling.

warp stop motion
A device for stopping a machine in the event of a warp break. (See also **drop wire.**)

warp streak (defect)
In woven fabric, an elongated faulty area of fabric running parallel to the warp threads and containing warp yarn that differs in material, linear density, filament, twist, lustre, tension or crimp, size, colour, or shade from the adjacent normal warp.
Note: The term 'streak' implies that both edges of the faulty area are visible and that its length warp-way is short.

warp stretch
The amount of stretching sustained by warp yarn during sizing operations. It is usually expressed as a percentage of the original length of unsized warp.

warp stripe (defect)
In woven fabric, a stripe that runs parallel to the warp threads and contains warp yarn that differs in material, linear density, filament, twist, lustre, tension or crimp, size, colour or shade from the adjacent normal warp.
Note: The term 'stripe' implies that both edges of the faulty area are visible and that its length warp-way is appreciable.

warp-backed fabric
A woven fabric that contains two sets of warp threads and one set of weft threads. One warp and the weft together form the face, while the second warp is laid at the back of the fabric and is stitched into it at intervals so as to form one structure, without distorting the surface appearance.

warp-knitted fabric
Fabric composed of knitted loops in which the yarns forming the loops travel in a warp-wise

warp-knitted fabric *(continued)*
direction down the length of fabric generally parallel to the selvedge. Part of the yarn, between the loops which connect the wales together, is referred to as the underlap.
Note: The two sides of the fabric are referred to as the technical face (the side on which the knitted loops are prominent) and the technical back (the side on which the underlaps are prominent).

warp-protector motion
A mechanism provided to protect a warp from damage in the event of a shuttle being trapped in the shed (see **trap**). There are various types:
(i) mechanical:
(a) a fast-reed stop motion that arrests the movement of the **sley** prior to beat-up;
(b) a loose-reed stop motion (see **loose reed**);
(ii) electro-magnetic (see **electromagnetic warp protector**).

warper's beam
A **beam** on which yarn is wound in a warping machine.

warping
See **beam warping.**

warping creel
A creel for holding yarn packages, usually in tiers and from which an assembly of ends can be withdrawn for warp-making.

warping lease
See **lease, warping**.

warping mill
A machine used to make a warp. It has a large-diameter reel, which may be horizontal or vertical. On the horizontal mill, the warp is built up in sections (see **section warping**). On the vertical mill, a warp in the form of a rope is built up in stages and subsequently unwound as a **ball warp.**

wash and wear
See **easy-care.**

washer
A machine for removing impurities, excess dyes or chemicals from textiles by treatment in an aqueous medium. (See also **open width washer.**)

washing-off
Treatment of textile material in water or detergent solution to remove substances employed in previous processes.

waste (costing)
That proportion of the production of a machine or unit of production, normally expressed as a percentage of the whole, which cannot be sold as the intended product. There are two classes of waste:
(i) predictable waste which arises from the nature of the production process, e.g., **thrums** during the preparation for weaving, and **end-fents** arising in subsequent processes. Such waste occurs at every stage of manufacture and can be incorporated into the calculation of costs;
(ii) unpredictable waste which is due to material breakages and mechanical failures, the cost of which cannot accurately be forecast.

waste (cotton, wool and other staple fibres)
There are two classes of waste, known as 'hard' and 'soft', and their treatment differs according to the class. Hard waste is essentially that from spinning frames, reeling and winding machines and all other waste of a thready nature. Soft waste mainly comes from earlier processes where the fibres have little or no twist, and are neither felted nor compacted.

waste courses (weft knitting)
See **roving courses (weft knitting)**.

water mark
1. A **moiré fault** produced by the pressure of the surface of one layer of fabric on another.
2. An unwanted mark on a fabric caused by contamination with water prior to **tinting** or dyeing on a pad mangle or beam dyeing machine, which results in a reduction in dye uptake.
3. A pile orientation difference causing light reflection differences.

water penetration resistance
The ability of a fabric to withstand penetration by water in specified circumstances. Water may be under pressure (e.g., hydrostatic head test) or in the form of drops hitting the surface of the fabric.
Note: Different test methods may test different aspects of water penetration resistance and **water repellency**, and the test may not simulate the effects experienced in use.

water repellency
The relative degree of resistance of a fabric to surface wetting, water penetration, water absorption, or any combination of these properties. (See also **water-repellent.**)
Note: The term is used in relation to tests carried out with several very different pieces of apparatus and, therefore, the same parameters are not involved in every case (e.g., Bundesmann test, Wira shower test, Credit rain simulation tester, Spray rating test).

water-repellent
A state characterized by the non-spreading of a globule of water on a textile material.
Note: The term is not normally applied to a water-repellent finish impervious to air: this is generally referred to as **waterproof.**

watered
See **moiré fabric.**

watering
An operation used in grass bleaching that consists of spraying fabrics laid in a field with water.

waterproof
The ability of fabric to be fully resistant to penetration by water.
Note: The term is an absolute term and implies that the water penetration resistance of the fabric is equivalent to its hydraulic bursting strength.

wave braid
A **ric-rac braid** (often described as such), which has a curved 'S' shaped construction.

wave motion
A mechanical device used during the build-up of a warp on a mill or swift to ensure the inclusion of an extra length of any yarn that has a substantially lower extensibility than the yarn in the body of the warp. This extra length is needed to prevent tight ends or the yarn from breaking ('cracking') during finishing.
Note: A wave motion is used, for example, to insert an extra length of mercerized cotton striping

wave motion *(continued)*
yarn into a warp having a ground of wool yarn spun on the worsted system. The extra length is usually of the order of 6%.

wave shed; ripple shed
A **shed** in a **multiphase weaving machine** (see under **weaving machine**) where several sheds are formed in the direction of weft and move laterally. The sheds may be formed by healds driven individually or by a series of narrow heald frames at various stages of heald timing.

wave shed machine
See **weaving machine.**

waxing, warp
The application of wax to a warp sheet so as to improve its weaving performance. The wax, in emulsion or molten form, is usually applied by a trough and lick roller. The point of application may be between the swift and warp beam in **section warping**, or between the creel and beaming head in **beam warping** or immediately after drying in slasher sizing. Normally waxes must be readily removable in the subsequent finishing operation.
Note: Alternatively wax is sometimes applied to a warp by means of a wax rod placed in the nip between the warp sheet and the weaver's beam at the point where the yarn leaves the latter.

weathered piece
A piece of fabric that exhibits discoloration and soiling of exposed edges and folds caused by exposure during storage or in transit, or both.
Note: The term is used particularly with reference to worsted grey fabric. The defect may be difficult to remove and may result in irregular dyeing.

weathering
1. The action of atmospheric agencies or elements on substances exposed to them.
2. The discoloration, disintegration, etc., that results from this action.

weathering, artificial, test
Exposure to cyclic laboratory conditions involving changes in temperature, relative humidity and radiant energy, with or without direct water spray, in an attempt to produce changes in the material similar to those observed after long-term, continuous, outdoor exposure.
Note: The laboratory exposure conditions are usually intensified beyond those encountered in actual outdoor exposure in an attempt to achieve an accelerated effect. This term does not cover exposure to special conditions such as ozone, salt spray, industrial gases, etc.

weave
The pattern of interlacing of **warp** and **weft** in a woven **fabric.**

weave, plain; tabby
The simplest of all weave interlacings in which the odd warp threads operate over one and under one weft thread throughout the fabric with the even warp threads reversing this order to under one, over one, throughout.
Note 1: A plain weave does not necessarily result in a plain surface effect or design in the fabric, e.g., variation of the yarn linear densities warp to weft or throughout the warp and/or weft and variation of the thread spacing warp to weft can produce rib effects (see **taffeta**, **poult**, **faille** and **grosgrain**), while colour patterning of the warp and/or weft results in **colour-and-weave effects.**
Note 2: The area containing the two solidly filled squares in the square paper design indicates one **weave repeat**, i.e., 2 ends x 2 picks.

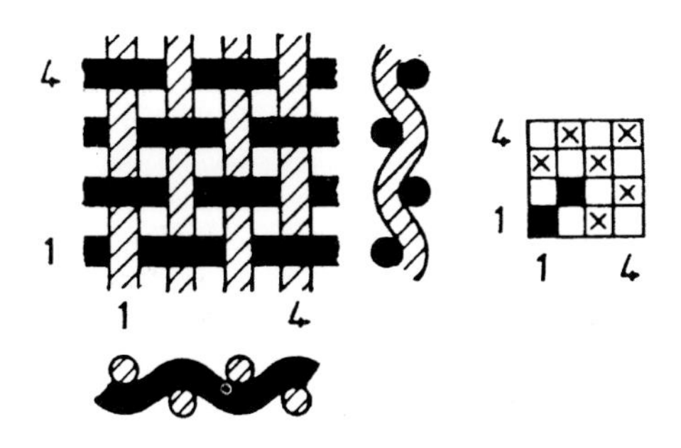

Plain weave

weave effect
The distinctive effect (e.g., **twill** or **honeycomb**) developed in a fabric by the **weave.**

weave number
See **weave repeat.**

weave repeat
The smallest number of ends and picks on which a weave interlacing pattern can be represented.

weaver's beam
A roller on which large flanges are usually fixed so that a warp may be wound on it in readiness for weaving.

weaver's bottom
Inflammation of the *ischial bursa* of handloom weavers who are seated at their work.

weaving
The action of producing fabric by the interlacing of warp and weft threads.

weaving bar
See **bar (woven fabric)**.

weaving lease
See **lease, weaving**.

weaving machine; loom
A machine used for producing fabric by **weaving.**

> **single phase weaving machine**
> A weaving machine in which the weft is laid across the full width of the warp sheet in a single phase of the working cycle. The main types of single phase weaving machines are:
> 1. Machines with **shuttles**:
> (i) **hand looms.**
> (ii) non-automatic weaving machines: power-operated machines on which the weft supply in the **shuttle** is changed by hand.
> (iii) automatic weaving machines; automatic looms: power-operated machines on which the shuttles or pirns are changed automatically.
> 2. Machines (power-operated) without shuttles; shuttleless weaving machines:
> (i) weaving machines with projectiles: machines in which the weft thread is gripped by jaws, fitted in a projectile, which is then propelled through the **shed.**
> (ii) weaving machines with rapier(s); rapier looms: machines in which the means of

weaving machine; loom *(continued)*

carrying the weft through the shed is fixed in the end of a rigid rod or in telescopic rods or in a flexible ribbon, this being positively driven. A rapier machine may have a single rapier to carry the weft across the full width or a single rapier operating bilaterally with a centrally located bilateral weft supply, or two rapiers operating from opposites sides of the machine.

(iii) jet weaving machines; jet looms: machines in which the weft thread is taken through the shed by a jet of air or liquid.

multiphase weaving machine

A machine in which several phases of the working cycle take place at any instant so that several picks are being inserted simultaneously. The main types of multiphase weaving machines are:

1. wave shed machines: weaving machines having different parts of the warp sheet in different phases of the weaving cycle at any instant. This type of shed makes it possible for a series of weft carriers or shuttles to move along in successive sheds in the same plane. When the weft carriers or shuttle carriers or shuttles travel a circular path through the **wave shed**, the machines are generally referred to as 'circular weaving machines' or 'circular looms'.

2. parallel shed machines: weaving machines having several sheds are formed simultaneously with each extending across the full width of the warp and with the shed moving in the warp direction.

weaving machine with projectiles

See **weaving machine.**

weaving machine with rapier(s)

See **weaving machine.**

web

1. The sheet of fibres delivered by a card (card web) or comber (comber web).
2. (Nonwoven) See **batt.**
3. A plain circular-knitted fabric.
4. A term applied to narrow woven fabrics, especially woven elastic.
5. A rarely used synonym for **warp** and for **fabric.**
6. A length of lace of full machine width in the **grey** state.

webbing

1. A woven **narrow fabric**, the prime function of which is load bearing, generally with multiple plies.
2. See **cobwebbing.**

weed control matting

A fabric, laid directly on to the soil, which prevents weed growth by excluding sunlight. Slits in the fabric are used to allow seedlings or existing plants to grow. The fabric is sufficiently permeable to allow water to penetrate to the soil and to allow the soil to breathe, whilst limiting evaporation of water from the soil.

Note: Some types also act as a mulch.

weft; woof; shute; shoot; filling

1. Threads widthways in a fabric as woven.
2. Yarn intended for use as in 1.

weft accumulator

A **yarn accumulator** used on shuttle-less weaving machines.

weft bar
See under **bar (woven fabric)**.

weft bobbin
See **pirn.**

weft carrier
A yarn carrier providing a supply of weft and driven positively through a shed.

weft crackiness; shires
A defect in woven fabrics in which fine weftway cracks or ribs (cracky weft) give the appearance of 'lines' distributed randomly across the whole part of the fabric width. They are usually associated with a slightly uneven pick spacing and are caused by varying friction between the warp and weft resulting in an uneven beat-up.

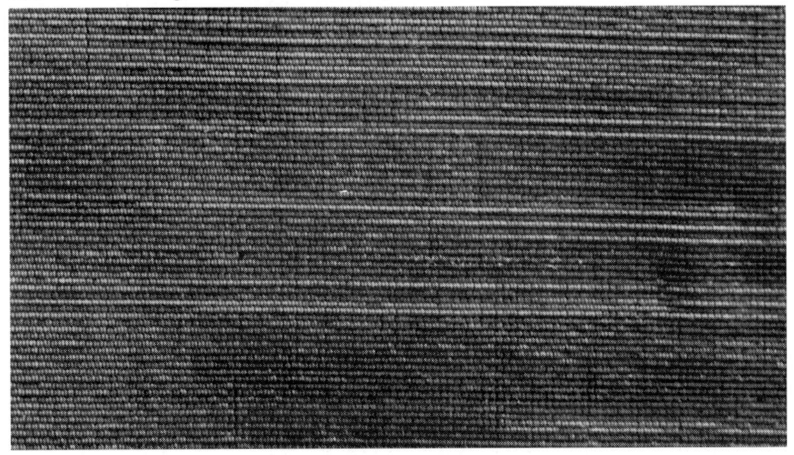

Weft crackiness

weft curling
See **warp snarl.**

weft detector
An electronic device for: (i) monitoring the presence of weft during weaving, normally on non-shuttle weaving machines; (ii) monitoring the time of arrival of the weft on relay air-jet weaving machines; (iii) indicating that the weft on a pirn in a shuttle is becoming exhausted.
Note: Mechanical devices which achieve the same objectives are **weft forks** and **feeler motions.**

weft fork
A mechanical device for monitoring the presence of weft during weaving, normally on shuttle looms. (See also **weft detector.**)
Note: If this device operates at one side of the warp, between the edge of the warp and the boxes, it is referred to as a 'side weft fork', but if it operates anywhere between the two edges of the warp, it is referred to as a 'centre weft fork'.

weft hairline
See **hairline.**

weft insertion (warp knitting)
1. Descriptive of a machine in which weft threads are introduced between the back of the needles and the warp threads, across the complete width of the fabric.

weft insertion (warp knitting) *(continued)*
2. Descriptive of a fabric that contains weft threads across the complete fabric width, each being positioned between the knitted loops and the underlaps of the fabric.

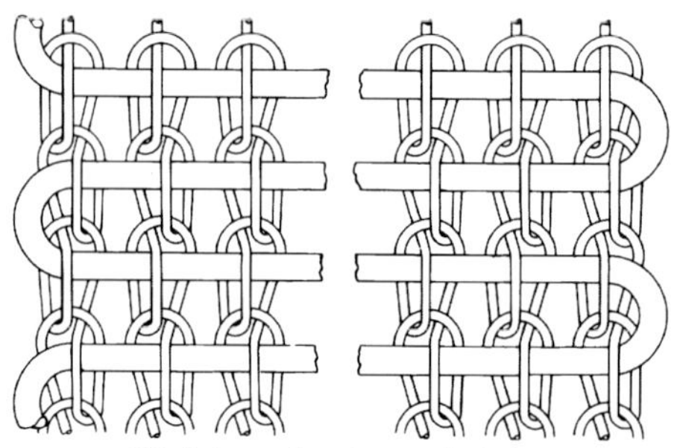

Weft insertion (warp knitting)

weft knitting
A method of making a fabric by normal knitting means in which the loops made by each weft thread are formed substantially across the width of the fabric. It is characterized by the fact that each weft thread is fed more or less at right angles to the direction in which the fabric is produced.

weft loop (defect)
A short length of weft yarn that is looped on the surface of the fabric or doubled back on itself in the fabric without snarling. The defect is associated with weft yarn that is not lively and may be caused by inadequate tension control in the shuttle, by shuttle bounce, by the reed being unsuitable for the fabric being woven, or by faulty setting of the **weft fork** etc. In the last case, it may be referred to as a 'centre loop' defect.

weft rib
An effect produced by weaving two or more ends as one and by using a weft **cover factor** approximately twice the warp cover factor, so that the weft is made to bend round the warp, which remains substantially straight. This leads to the formation of rounded weft-faced cords running down the fabric.

weft snarl
See **warp snarl.**

weft stop motion
A device for stopping a **weaving machine** in the event of a weft break.

weft straightener
A device for correcting the misalignment of fabric weft, especially if the latter is bowed and/or skewed (see **bow (weaving)** and **skew**).

weft streak (defect)
An elongated faulty area of fabric running parallel to the weft threads and containing weft yarn which differs in material, linear density, filament, twist, lustre, tension or crimp, size, colour, or shade from the adjacent normal weft.
Note: The term 'streak' implies that both edges of the faulty area are visible and that its length weft-way is less than the fabric width.

weft-backed fabric
A woven fabric that contains one set of warp threads and two sets of weft threads. The warp and one weft together form the face, while the second weft is laid at the back of the fabric and is

stitched into it at intervals so as to form a single structure, without distorting the surface appearance.

weft-knitted double jersey
See **double jersey, weft-knitted**.

weftless tape
See **bolduc.**

weighting
The addition of metallic salts to silks to increase the mass and impart a firmer handle.

well (lace machines)
The space, running the whole width of the machine, between the front and back combs. In this space, the positioning of warp and pattern threads is effected by the **guide bars.**
Note: The width of the well in the Leavers machine is determined by the maximum number of **steel bars** to be worked.

welt (clothing)
A piece of material, normally folded to form a band, applied to finish the edge of a pocket or the hem of a garment.

welt (knitting)
A secure edge of a knitted fabric or garment made during, or subsequent to, the knitting process. Welts made during the knitting process are usually at the starting end of the fabric and are formed parallel to the course. Seamed welts, which are made after the knitting process, may occur in any position in the fabric.

> **inturned welt**
> A **welt** consisting of a double fold of plain fabric made on a circular stocking machine. Sinker loops from one of the first few courses are retained while the welt fabric is knitted and are later intermeshed with alternate needle loops of a subsequent course.
>
> **reverse welt**
> A **roll welt** in which the plain courses are intermeshed towards the reverse side of the fabric. This welt is used particularly for stockings with turnover tops.
>
> **roll welt; English welt**
> A **welt** made on a rib basis, in which all the courses of loops except the first and last are intermeshed in the same direction towards the face side of the fabric. In making such a weft on 1x1 rib, the first and last courses are knitted on both sets of needles and the intermediate courses are knitted on only one set of needles.

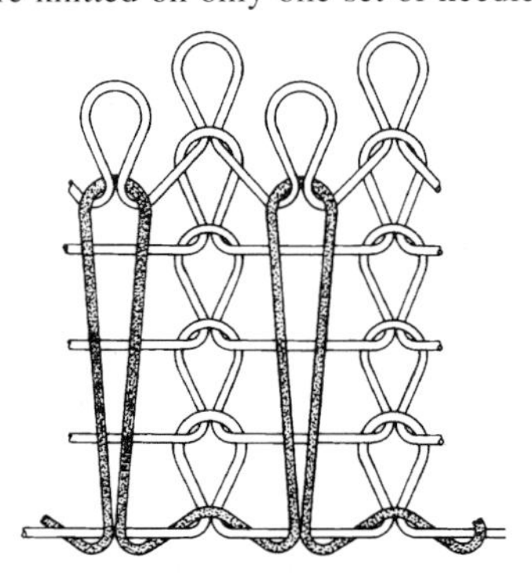

Roll welt

welt (knitting) *(continued)*

tubular welt; French welt

A **welt** made on a rib basis, in which the number of courses with loops intermeshed in one direction is equal to the number of courses with loops intermeshed in the other direction. In making such a welt on a 1x1 rib, the first and last courses are knitted on both sets of needles and the intermediate courses consist of an equal number of plain courses on each set of needles.

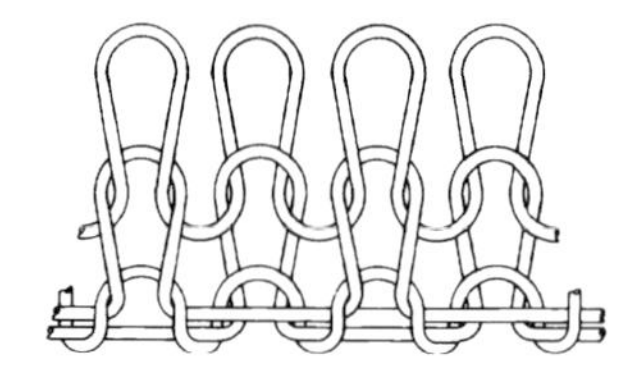

Tubular welt

turned welt

A **welt** that consists of a double fold of plain fabric and is made usually on a straight-bar knitting machine. All or alternate sinker loops of the first course are retained while the weft fabric is knitted and later intermeshed with the needle loops of a subsequent course.

welt (woven)

A fabric showing rounded cords in the weft direction, with pronounced sunken lines between them that are produced by the nature of the weave. The weave on the face of the cords is plain. There are warp floats the width of the cords on the back. Wadding picks are used to accentuate the prominence of the cords.

Note: For many years, the term **piqué** has been applied to a much less expensive white fabric made in a light-weight **Bedford-cord** weave.

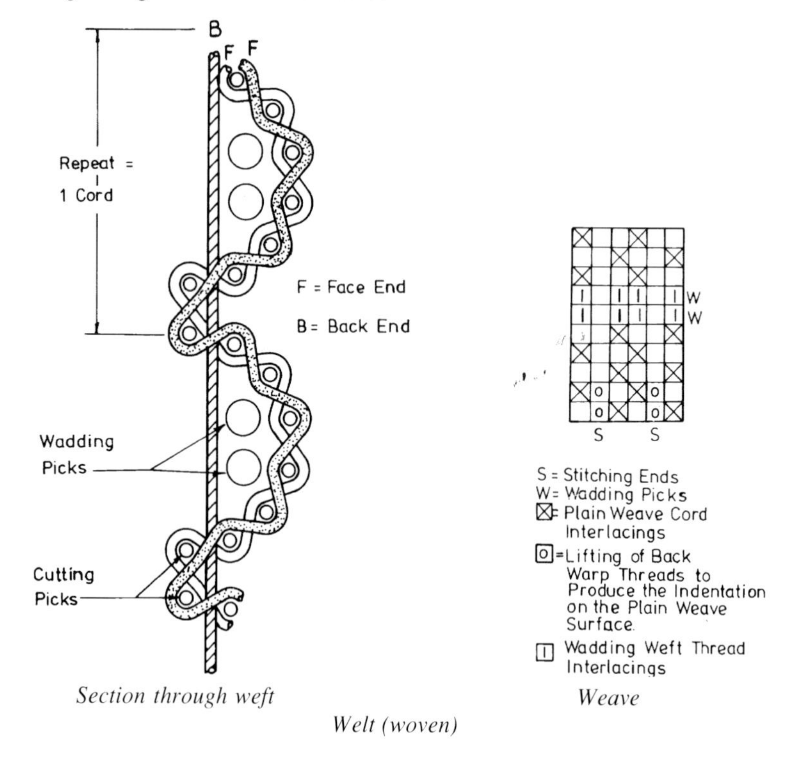

Section through weft

Weave

Welt (woven)

welt ripple (knitting)
See **ripple (weft knitting)**.

welting (seaming)
See **welt (knitting)**.

welting cord
A cord consisting of two or more gimps twisted together. The resultant cord is hard and wiry.

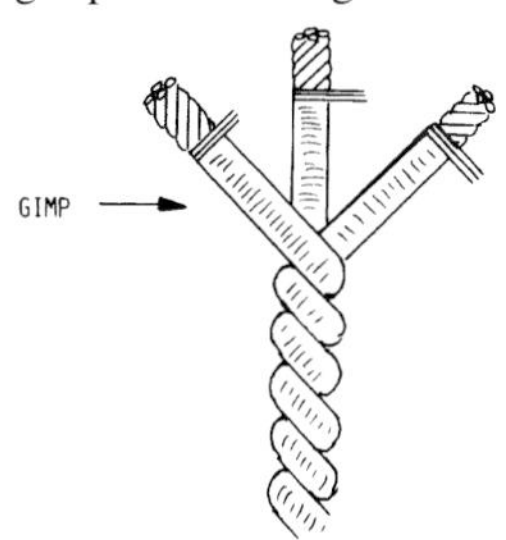

Welting Cord

wet cleaning
An aqueous process for cleaning articles made from materials liable to be damaged by dry cleaning solvents.

wet doubling (yarns)
The combination by twisting together of two or more **single yarns** which have been wetted out by immersion in water or water with suitable additives or alternatively damped by contact with a transfer medium for similar fluids, before the folding twist is inserted to form a plied yarn (see **folded yarn**).

wet fixation
A finishing process for cellulosic-fibre fabrics that improves the wet crease recovery and smooth drying properties, but not the dry crease recovery.

wet laying
The process of forming a fibre sheet by papermaking techniques, for **nonwoven fabric** production. (See also **wet-laid nonwoven fabric.**)

wet pick-up; pick-up
The weight of liquor taken up by a given weight of fabric after impregnation, spraying, or coating. The percentage pick-up (wet pick-up) is given by: $100(w_2-w_1)/w_1$
where w_1 is the weight of material before impregnation and w_2 is the weight of wet material after impregnation.
Note: **low wet pick-up** has a special application in respect of easy-care finishing (see **critical application value**).

wet spinning (manufactured fibre production)
The conversion of a dissolved polymer into filaments by extrusion into a coagulating liquid.
Note: The extrusion may be directly into the coagulating liquid or through a small air-gap. In the latter case it may be known as 'dry-jet wet spinning' or 'air-gap wet spinning'.

wet-back
The return flow of liquid through the **coverstock** from an absorbent pad used in hygiene products.

wet-laid nonwoven fabric
A fabric made from a fibre sheet formed by **wet laying**, normally followed by adhesive bonding. The material may contain a high percentage of non-textile fibres, e.g., wood pulp.

wet-on-wet
A **padding** process in which the material to be padded is wet, e.g., from a previous process stage.

wet-spun
1. Descriptive of a yarn of flax, hemp, or similar fibre spun from roving that has been thoroughly wetted out in hot water immediately prior to the drafting operation.
2. Descriptive of manufactured filaments produced by **wet-spinning.**

wether
A castrated male sheep.

wevenit
See **double piqué, weft-knitted.**

wharve; wharl; whorle
The pulley or boss on a spindle or false-twisting unit that is driven by tape, cord (banding), belt, or rope.

wheel (lace machines)
A term used in the West of England for a brass bobbin (see **bobbin, brass**).

wheel feed
See **feed mechanisms (sewing)**.

whip roller
See **back rail.**

whipcord
A term applied to fabrics of a wide range of qualities and commonly made of cotton or worsted. The characteristic feature is a more or less bold upright warp twill (often a 63° steep twill), which is accentuated by a suitable weave structure, more ends per unit length than picks, and a clear finish to an extent which causes the twill or warp threads to form a cord-like effect.

Whipcord (magnification x 2)

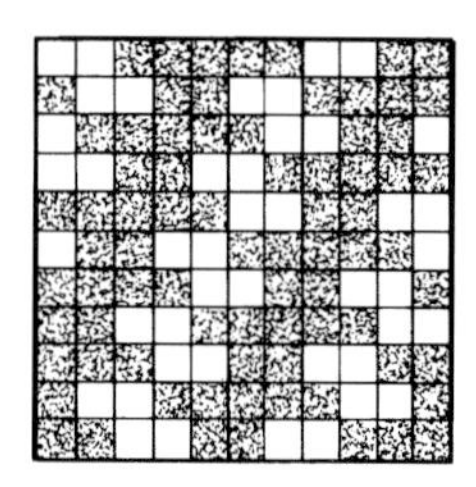

Whipcord weave

whorle
See **wharve.**

wick
A woven or braided fabric, or a yarn or a group of yarns, having outstanding capillary properties.

wicking
The passage of liquids along or through a textile material or the textile element of a coated fabric or along interstices formed by the textile element and the coating polymer of the coated fabric.

width (lace machines)
The maximum lace-making width of Leavers, Leavers-type, furnishing, bobbinet and raschel machines.

width of crotch; width of crutch
See under **crotch.**

width, fabric
Unless otherwise specified, the distance from edge to edge of a fabric when laid flat on a table without tension. In the case of commercial dispute the measurement should be made after the fabric has been conditioned in a **standard atmosphere for testing.** When buying and selling fabric it is normal to specify the basis on which the width is to be assessed, e.g., overall, within limits, or **usable width.**

Wigan
A plain-weave, cotton fabric of low-to-medium quality traditionally about 135 g/m^2. Typical construction: 17x23; 25x31tex; 127 g/m^2; K=8.5+12.8.

Wigan finish
A firm, starched, plain-calendered finish without lustre, applied to light-weight sheetings and print fabrics.

wild silk
See **silk, wild.**

wildness
A ruffled appearance of the surface fibres in **slivers**, **slubbings**, **rovings**, and yarns.
Note: Wildness may be due to the processing of these products under dry atmospheric conditions, which causes increased inter-fibre friction and static-electricity troubles. The static charges cause mutual fibre repulsion and prevent fibres from taking up normal orderly positions in the respective products.

willey
A machine consisting of bladed or pinned rollers for opening, cleaning and mixing staple fibre material as a preliminary to **scouring** (for greasy wool) or **carding** (for most animal and synthetic fibres).
Note: Various machines of different designs exist to meet specific material requirements. These include battering willey, cockspur willey, double-cylinder willey, dust willey, fearnought, single-cylinder willey, teazer, tenterhook willow, and wool willey.

willey; willow; teaze
To open and disentangle fibres prior to **scouring** and/or **carding.**

Wilton, carved
A style of Wilton carpet where a textured design is achieved by omitting tufts. Designs are

Wilton, carved *(continued)*
usually single or two colour and the tufts are omitted by the jacquard selecting a **frame** empty of yarn.

Wilton, figured
A carpet, usually woven on a jacquard loom, that bears a design obtained by the use of two to five **frames**, each of a different colour. Additional colours may be obtained by substitution (planting) of colours in any frame.

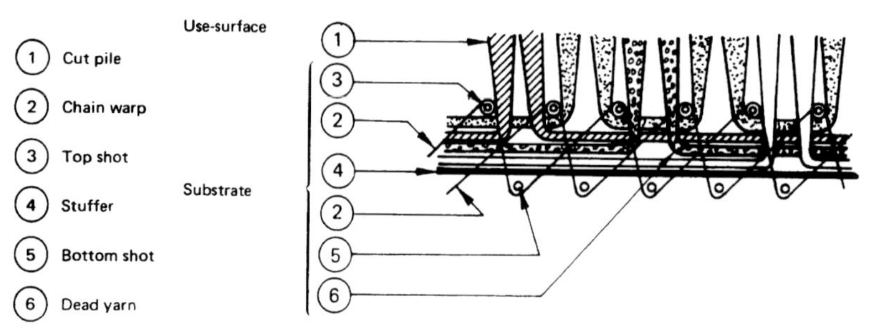

Figured Wilton weave carpet, two-shot (warpways section)
The above diagram is derived from figure 3 of BS 5557

Wilton, plain
A **Wilton carpet** that bears no surface design. It is normally of a single colour, and if it has a cut pile it may be described as 'plain velvet'. In a variation, hard-twist yarn is used.

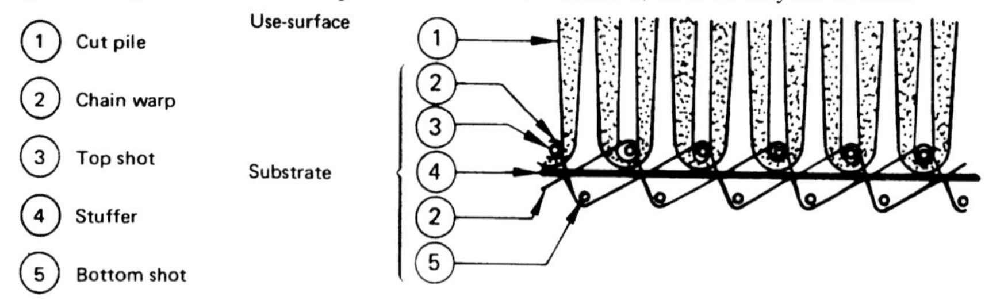

Cut-pile, plain-woven carpet, two-shot (warpways section)
The above diagram is derived from figure 1 of BS 5557)

Wilton, plain, hard-twist
See **curled pile** under **pile (carpet)**.

Wilton carpet
A woven carpet in which the pile threads run continuously into the carpet and are treated as an integral part of the weaving process, being raised above the surface of the backing to form a pile by means of wires or hooks (warp wires), or by being stretched between two backings (face-to-face weaving). During the weaving process, the pile may be left as a loop, or cut by a bladed wire, or, in the case of face-to-face weaving, formed by the separation of the two fabrics.

wince
See **winch.**

wincey
A light-weight fabric of the flannel type, finer in texture than baby flannel. Originally it was made with a cotton warp and a wool weft but it may now be made from mixture yarns containing

wool. The mass per unit area was traditionally about 136 g/m^2.

winceyette
A light-weight fabric, originally and usually of cotton, raised on one or both sides, the weave usually being plain or twill. It is used chiefly for children's and women's nightdresses.
Note: It may be woven grey then bleached, piece-dyed, or printed. Yarn-dyed fabrics are also made. It is similar to **flannelette** but lighter in weight, the mass per unit area being traditionally 136-170 g/m^2.

winch; wince
A horizontal rotor, commonly circular or elliptical in cross-section, which draws fabric over it, in rope form or in open width, during processing. (See also **winch dyeing machine.**)

winch dyeing machine
A dyeing machine consisting of a dye vessel fitted with a driven **winch** (usually above the liquor level) which rotates and draws a length of fabric, normally joined end to end, through the liquor.

wind ratio
The number of wraps wound on a take-up package while the traverse completes a full stroke in one direction.

winder
A machine used for transferring yarn from one package to another.

winding fallers
Fallers that carry, stretched taut between them, a wire (the winding faller wire) that serves to guide the yarn on to the chase of a **cop.** (See also **mule spinning** diagram.)

winding-on angle
See **angle of lead.**

window-blind Holland
See **holland** 2.

Wira shower test
See **water repellency.**

wire (pile weaving)
A metal strip or rod, that is inserted during weaving between raised pile warp threads and the foundation of a fabric to form loops of pile above the foundation. It is either 'bladed' (a small

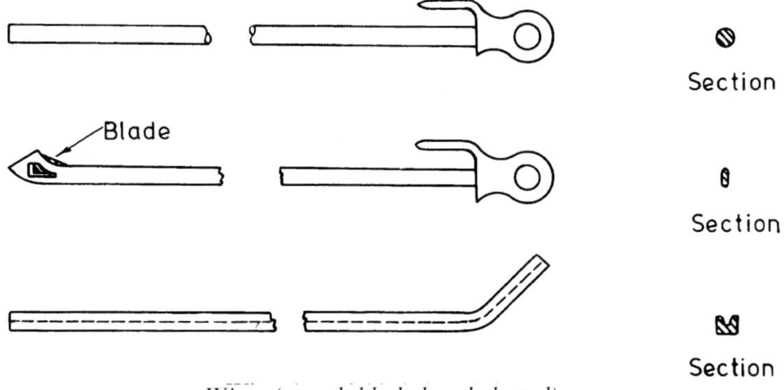

Wires (round, bladed and slotted)

wire (pile weaving) *(continued)*
blade on the upper edge of the strip at one end) so that when it is withdrawn the loops are severed to form a cut pile, or 'unbladed', when a loop pile is left on withdrawal. It can be round (for the production of 'cord' carpets) or flat (for Brussels and Wilton carpets). 'Profile' wires are flat wires having an upper edge of irregular outline, and are used to produce loops of different heights across the width of the carpet. When the wire is withdrawn, loops of the preceding row corresponding to low parts of the wire are robbed of yarn and variations in pile height are thereby produced. 'Slotted' wires are used in the hand-weaving of velvet and for the production of high-pile rugs. The wires are grooved along the upper surface, and a trevet or knife, guided by the slot, is drawn by the weaver across the loops to form a cut pile.

wire heald
See under **heald.**

wire profile
A **wire** used to form loops in Wilton weaving. It is characterized by a profile that varies in height along its length and is used for the production of pile loops of different height across the width of the carpet.

woof
See **weft.**

wool
1. The fibrous covering of a sheep (*Ovis aries*) (see *Note* under **hair**).
2. Appertaining to wool generally.

wool classing; classing
A process by which whole fleeces are separated into different classes before being baled and sold.

wool sorting
A process by which fleece or skin wool is divided up into various qualities.

wool waste
See **waste.**

woollen
Descriptive of yarns, or fabrics or garments made from yarns, which have been produced on the condenser system, wholly from wool fibres, new or otherwise.
Note 1: As an adjective appertaining to wool generally, the term 'wool' and not 'woollen' is recommended.
Note 2: The trade term 'woollen-spun' is applied to staple yarn produced by carding, condensing, and spinning on machinery originally designed for the processing of wool into yarn: it is descriptive of processing technique and not fibre content. Since the yarn may not contain any wool fibre, the alternative **condenser spun** is preferable.

woollen (condenser) carding
See under **carding.**

woollen, blended
Descriptive of yarns spun on the condenser system and having wool as the main component, or descriptive of fabrics or garments made from such yarns.

woollen-spun
See **woollen**, *Note 2.*

woollen-type fabric
A fabric manufactured wholly from woollen-type (**woollen-spun**) yarns, or from cotton warp and woollen-type (woollen-spun) weft, and which may or may not contain decoration threads of other fibres.

workers
Wire-covered rollers that form part of a roller and clearer card and are used to perform the **carding** action jointly with the **swift**. (See also **strippers.**)

workwear
Apparel designed either to provide a measure of general or specific protection in use, or to define a corporate identity (generally referred to as 'corporate clothing').
Note: Apparel designed for protection against specific hazards is more usually described as 'protective clothing'.

worsted
Descriptive of yarns spun wholly from combed wool in which the fibres are reasonably parallel, and fabrics or garments made from such yarns. In most countries fabrics with a small proportion of non-wool decorative threads can be described as worsted. (See also **worsted-spun.**)

worsted, blended
Descriptive of yarns in which the fibres are reasonably parallel and have combed wool as the main component, or descriptive of fabrics or garments made from such yarns.

worsted carding
See under **carding.**

worsted fabric
A fabric manufactured wholly from worsted yarns, except that decoration threads of other fibres may be present.

worsted yarns, colour terms
In all the definitions given below, the expression 'colour' includes black and white.

> **double marl**
> A yarn which has been produced from a roving of one end of two colours (as in **single marl**) twisted with one end of another two colours that have been roved together (see **roving**).
>
> **half-marl**
> A yarn which has been produced from a roving of one end of mixture shade or solid colour, twisted with one end of two colours (as in **single marl**).
>
> **mélange**
> A yarn spun from tops that have been mélange-printed (see **mélange printing**).
>
> **mixture; ingrain**
> A yarn made from fibres of two or more colours blended together.
>
> **single marl**
> A single yarn which has been produced from a roving of two different colours.
>
> **single mottle**
> A single yarn made as a **single marl** in respect of colour combination, but with the marl effect being obtained by spinning from two half-weight rovings of different colours into the single end. The effect is a sharper contrast of colour than in a single marl.

worsted yarns, colour terms *(continued)*

solid colour
A yarn made from fibres of a single colour.

twist
A yarn comprising two single ends of different colours twisted together, the single ends being either solid colours or mixtures.
Note: In woollen spinning a twist yarn is known as a **marl yarn**

worsted-spun
A term applied to yarn spun from staple fibre processed on worsted-spinning machinery by carding or preparing, combing, and drafting; or by converting a continuous-filament tow and drafting; or from a combination of slivers or rovings from both systems.
Note: This definition is descriptive of processing technique and not fibre content.

worsted-type fabric
A fabric manufactured wholly from worsted-type (**worsted-spun**) yarns, or from cotton warp and worsted-type (worsted-spun) weft, and which may or may not contain decoration threads of other fibres.

wound dressing materials; surgical dressing
Materials, including textiles, whose functions are to provide protection to a wound against infection, absorb blood and exudate, and promote healing. Common wound dressing materials are composite structures which consist of a wound contact layer, an absorbent layer, and a flexible base material.

wound packages (yarns)
Yarn wound on formers which facilitate convenient handling.
Note: In some cases the former may be withdrawn before further processing.

woven flat ruche
See under **ruche**.

woven label
A length of **narrow fabric** woven on a jacquard loom, incorporating names, logos and/or text to identify the article to which the label is attached, and to give instructions and other details.
Note: Woven labels can be produced on narrow fabric looms or slit from a fabric.

wrap spinning
See **hollow spindle spinning** and **spinning.**

wrap stripe
See **embroidery-plated fabric, knitted** under **plated fabric, weft-knitted.**

wrap yarn
1. A fibrous yarn covered with other yarn(s) to bind projecting fibre ends to the main body.
Note: It is commonly used for interlinings to prevent fibre ends from penetrating the outer fabric.
2. Any yarn used in **embroidery plating** or wrap striping.

wrap-spun yarn
A yarn with a twistless core wrapped with a binder and normally produced by **hollow spindle spinning.**

wrapper fibres; belly bands
Fibres which wrap around the main body of a staple fibre yarn during yarn formation in the

production of **open-end** and **fasciated yarns.**

wrinkle
An unwanted short and irregular crease in a fabric.
Note: The distinction between a wrinkle and a crease is often not clear but creases are generally sharper and longer than wrinkles.

wrinkle resistance
See **crease-resistance.**

wrong denting; wrong sleying
The drawing of one or more ends through a reed in an incorrect order.

wrong draft (weaving)
The drawing-in of one or more ends through the healds or harness in an incorrect order.

wrong end
See **mixed end.**

wrong lift
An incorrect interlacing of warp and weft threads caused by a heald shaft or jacquard harness cord being in the incorrect position as a result of a fault in the pattern chain (pattern stitching) or faulty mechanical action (machine stitching).

wrong picking
See **mispick.**

wrong sleying
See **wrong denting.**

yak fibre (hair)
Fibre from the yak (*Bos (Poëphagus) grunniens*).

yarn
A product of substantial length and relatively small cross-section consisting of **fibres** and/or **filament(s)** with or without twist.
Note: Assemblies of fibres or filaments are usually given other names during the stages that lead to the production of yarn, e.g., **tow**, **slubbing**, **sliver**, or **roving.** Except in the case of continuous-filament or tape yarns, any tensile strength possessed by assemblies at these stages is generally the minimum that can hold them together during processing.

yarn accumulator
A device which stores a controlled amount of yarn between a supply package and a fabric producing machine. It is used to minimise the effect of tension variation in the yarn as it is unwound from the supply package in order to improve tension control during fabric formation. (See also **weft accumulator.**)

yarn carrier (flat or straight-bar knitting machines)
The final element that guides yarn to the knitting instruments.

yarn clearer
A device designed to cut or break yarn when it detects changes in mass or diameter which exceed set limits. Yarn clearers are normally part of, or an addition to, a winding machine.

yarn count
See **count of yarn.**

yarn evenness
Variation in mass per unit length of yarn (along the length of the yarn). (See also **U%**.)

yarn guide; thread guide; feeder (knitting)
An element which guides the yarn.

yarn linear density
See **count of yarn.**

yarn number
See **count of yarn.**

yarn setting
See **setting.**

yarn sizing, single-end
The application of size to textile yarns in single-end form during winding from one yarn package to another.
Note 1: The words 'single-end' here embrace any textile yarns, whether single or plied.
Note 2: Methods of application include:
(i) bobbin-to-bobbin sizing: Double-flanged bobbins arranged in a battery and fitted with yarn-traversing motions are rotated to pull yarn from a similar battery of unrolling bobbins. In transit, each end of yarn passes in contact either over or under a roller immersed in size solution in a suitable trough: hot air is circulated over the yarn and take-up bobbins to help in drying the applied size.
(ii) Yarn from a supply cheese or bobbin (e.g., an uptwister take-up package) is wound on to a sized-yarn package by a winding spindle of a precision winding machine. In transit, the yarn makes contact with a size roller partially immersed in size solution.
Note 3: The term 'gumming' applies to the process in *Note 2* (ii) and is derived from the gummy nature of some size solutions, and from the adhesive properties of size. Size, however, is not synonymous with gum and for this reason many mill technologists consider that the term 'gumming' should be discouraged.
Note 4: Size is applied to some rayon yarns in cake form but this method is covered by the term cake sizing.

yarn slippage
The unwanted movement of yarns in a woven fabric due to the warp and weft yarns slipping over each other.
Note 1: Yarn slippage may occur during garment manufacture, or in garment components during wear, as a result of excessive stress or abrasion.
Note 2: Yarns in a fabric have a greater or lesser tendency to slip depending on factors such as yarn type, weave, ends and picks per unit length, fabric weight, and the type of finishing process.

yellowing
1. See **oxidized oil staining.**
2. The yellow discoloration that may develop on textile materials during processing, use, or storage.

yield point
The point on the stress-strain curve corresponding to the **elastic limit.**

yoke
A small, upper component of a garment attached to the main body of the garment from which the rest of the garment is suspended. It forms a style feature in blouses, dresses, coats, trousers and skirts.

Yorkshire dressing
See under **dressing (warp preparation)**.

Z-twist
See **twist direction.**

zephyr
A fine fabric of plain weave used for dresses, blouses, and shirtings and made in various qualities. A typical zephyr has coloured stripes on a white ground and exhibits a cord effect made by the introduction of coarse threads at intervals.

zero-twist yarn
1. A continuous-filament single yarn in which there is no twist.
2. A multi-fold yarn in which there is no folding twist.
Note 1: Some fibrous yarns are described as twistless, since the fibres may be held together by adhesive temporarily e.g., until incorporated in fabrics. Varieties of core-spun yarn and scaffolding yarn have appeared with this description after solvent-removal of one component.
Note 2: Zero-twist continuous-filament yarns usually become partially twisted by over-end withdrawal, e.g., from a pirn in a loom shuttle. (See also **twistless yarn.**)

zibeline
A heavy coating fabric with a long shaggy nap laid in a single direction. It is usually made of woollen yarns in strong colours.

zig-zag machine
See **swing needle machine.**

zip; zip fastener; zipper
A fastening device operating by means of two parallel rows of metal or plastic teeth on either side of a closure that are interlocked by a sliding tab. The teeth are carried in two **zip fastener tapes** that run the length of the zip and which are usually stitched into apparel and other textile products.

zip fastener tape; slide fastener tape; zipper tape; zip-tape
A **narrow fabric** which has a high lateral strength, and may be woven or knitted, being so constructed as to provide at one selvedge, for the suitable embodiment of the fastener elements (see **zip**).

zoll
See **gauge (knitting)**.

SYSTEMS FOR YARN NUMBER OR COUNT

1. Direct System
(Number of mass units per defined length unit)

System	Symbol	Abbreviation for unit	Standard mass unit	Standard length unit	Tex equivalent
SI system (and commonly used decimal multiples and fractions)					
Tex	Tt	tex	gram	1 kilometre	1
(decitex)		dtex	gram	10 kilometres	0.1
(millitex)		mtex	gram	1000 kilometres	0.001
(kilotex)		ktex	gram	1 metre	1000
Other direct systems					
Denier	Td	d	gram	9 kilometres	0.1111
Linen (dry spun), Hemp, Jute	Tj	-	pound	14,400 yards (1 spyndle)	34.45
Woollen (Aberdeen)	Ta	-	pound	14,400 yards	34.45
Woollen (American grain)	-	-	grain	20 yards	3.543

Note: In the *Tex system*, the linear density of yarns spun from natural or manufactured staple fibres is generally described in *tex* units. For continuous-filament yarns and for individual fibres, *dtex* units are commonly used. The linear density of slivers and rovings and of continuous-filament tows is commonly described in *ktex* units. Heavy industrial monofilaments are usually described in terms of their diameter.

SYSTEMS FOR YARN NUMBER OR COUNT

2. Indirect System or Specific Length (Number of length units per unit mass)

System	Symbol	Standard length unit	Standard mass unit	Tex equivalent
Asbestos (American)	Na_A	100 yards (1 cut)	1 pound	4961
Asbestos (English)	Ne_A	50 yards	1 pound	9921
Cotton (Bump yarn)	N_B	1 yard	1 ounce	31003
Cotton (English)	Ne_C	840 yards (1 hank)	1 pound	590.5
Glass (US and UK)	N_G	100 yards	1 pound	4961
Linen (Wet or dry spun)	Ne_L	300 yards (1 lea)	1 pound	1654
Metric	Nm	1 kilometre	1 kilogram	1000
Spun silk	N_S	840 yards	1 pound	590.5
Typp	Nt	1000 yards	1 pound	496.1
Woollen (Alloa)	Nal	11520 yards (1 spyndle)	24 pounds	1033
Woollen (American cut)	Nac	300 yards	1 pound	1654
Woollen (American run)	Nar	100 yards	1 ounce	310
Woollen (Dewsbury)	Nd	1 yard	1 ounce	31003
Woollen (Galashiels)	Ng	300 yards (1 cut)	24 ounces	2480
Woollen (Hawick)	Nh	300 yards (1 cut)	26 ounces	2687
Woollen (Irish)	Ni_W	1 yard	0.25 ounces	7751
Woollen (West of England)	Nwe	320 yards (1 snap)	1 pound	1550
Woollen (Yorkshire)	Ny	256 yards (1 skein)	1 pound	1938
Worsted	Ne_W	560 yards (1 hank)	1 pound	885.8

Equivalents Used Occasionally

Cotton (English)	Ne_C	120 yards (1 lea)	1000 grains	590.5
Woollen (Yorkshire)	N_Y	1 yard	1 dram	1938
Worsted	Ne_W	80 yards (1 lea)	1000 grains	885.8

SYSTEMS FOR YARN NUMBER OR COUNT

Conversions

Linear Density Systems

To convert to tex from a value in another linear density system, multiply by the appropriate tex equivalent. To convert from tex to a value in another linear density system, divide by the appropriate tex equivalent.

Specific Length or Count System

To convert to tex from a value in a specific length or count system, divide the appropriate tex equivalent by the count. To convert from tex to a value in a specific length or count system, divide the appropriate tex equivalent by the value in tex.

All other conversions can be effected by first converting to tex.

Precise conversions can be based on the following exact equivalents, where the mass units are in the avoirdupois system:

1 yard	(yd)	=	0.9144	m
1 pound	(lb)	=	0.45359237	kg
1 ounce	(oz)	=	$^1/_{16}$	lb
1 dram	(dr)	=	$^1/_{16}$	oz
1 grain	(gr)	=	$^1/_{7000}$	lb

Folded and Cabled Yarns

In the Tex System the resultant yarn number is given with the prefix R. Thus R40/2tex is 20tex x 2 (ignoring folding twist contraction).

In Indirect Systems of Count, folded yarn is described numerically as (e.g. Two-fold Twenties) 2/20 or 20/2. Cabled yarn is numbered (e.g. Three-fold, Two-fold Sixties) 3/2/60. In Direct Systems the resultant count is usually given. Thus Three-fold 300 Denier is designated 900/3.
Note 1: The gauge or size of some monofilament yarns is designated by diameter.
Note 2: Yarns made by cutting film are frequently designated by width.

SI UNITS AND CONVERSION FACTORS

Table 1. SI Units and Conversion Factors for Mill and Commercial Transactions

Quantity	SI Units and their appropriate decimal multiples	Unit symbol	To convert to SI units multiply value in unit given by factor below	
Length	millimetre	mm	inch	25.4
	centimetre	cm	inch	2.54
	metre	m	yard	0.9144
Width	millimetre	mm	inch	25.4
	centimetre	cm	inch	2.54
	metre	m	yard	0.9144
Area	square metre	m^2	yard^2	0.8361
Volume	litre	L	pint	0.5682
			gallon (UK)	4.546
Mass	kilogram	kg	pound	0.4536
	tonne	t	ton	0.9842
Thickness	millimetre	mm	inch	25.4
Linear density	tex*	tex	—	—
	millitex	mtex	—	—
	decitex	dtex	—	—
	kilotex	ktex	—	—
Threads in fabric:				
length	number per centimetre**	picks/cm	picks/inch	0.3937
width	number per centimetre**	ends/cm	ends/inch	0.3937
Warp threads in loom	number per centimetre	ends/cm	ends/inch	0.3937
Stitch length	millimetre	mm	inch	25.4
Courses per unit length	number per centimetre	courses/cm	courses/inch	0.3937
Wales per unit length	number per centimetre	wales/cm	wales/inch	0.3937
Mass per unit area	gram per square metre	g/m^2	ounces/yard^2	33.91
Twist level	turns per metre***	turns/m	turns/inch	39.37

* The Tex system is fully described in BS 947. It is based on the principle that linear density in tex expresses the mass in grams of one kilometre of yarn. Hence millitex, decitex and kilotex express the mass in mg, dg and kg of one km. BS 947 gives conversion factors for all the recognised linear density and count systems. The rounding procedure in Appendix A of BS 947 is superseded by BS 4985, which is based on ISO 2947, and gives rounded Tex system equivalents for the six main systems.

** For particularly coarse fabrics, the unit 'threads/10 cm' may be used if there is no possibility of confusion.

*** In some sectors, twist level is expressed as turns/cm.

The information in the table above has been derived from PD 6469: 1973 *Recommendations for programming metrication in the textile industry*, which has now been withdrawn by BSI.

SI UNITS AND CONVERSION FACTORS

Table 2. SI Units and Conversion Factors for Laboratory Use

Quantity	SI units and their appropriate decimal multiples	Unit symbol	To convert to SI units multiply value in unit given by factor below	
Diameter	micrometre millimetre centimetre	μm mm cm	1/1000 inch inch inch	25.4 25.4 2.54
Cover factor (woven fabrics)	threads per centimetre x $\sqrt{tex}$ x 10^{-1}	$\frac{(threads / cm)\sqrt{tex}}{10}$	$\frac{threads / inch}{\sqrt{cotton\ count}}$	0.957
Cover factor (weft-knitted fabrics)	$\sqrt{tex}$ divided by stitch length in mm	$\frac{\sqrt{tex}}{stitch\ length(mm)}$	$\frac{1}{stitch\ length\ (inch)}$ x $\frac{1}{\sqrt{worsted\ count}}$	1.172
Twist factor (or multiplier)	turns per metre x $\sqrt{tex}$ x 10^{-2}	$\frac{(turns / m)\sqrt{tex}}{100}$	$\frac{turns / inch}{\sqrt{cotton\ count}}$	9.57
Breaking force	millinewton newton decanewton	mN N daN	gf lbf kgf	9.81 4.45 0.98
Tearing strength	newton	N	lbf	4.45
Tenacity	millinewton per tex	mN/tex	gf/den	88.3
Bursting pressure	kilonewtons per square metre	kN/m^2(kPa)	$lbf/inch^2$	6.89

The information in the table above has been derived from PD 6469: 1973 *Recommendations for Programming metrification in the textile industry* which has now been withdrawn by BSI.

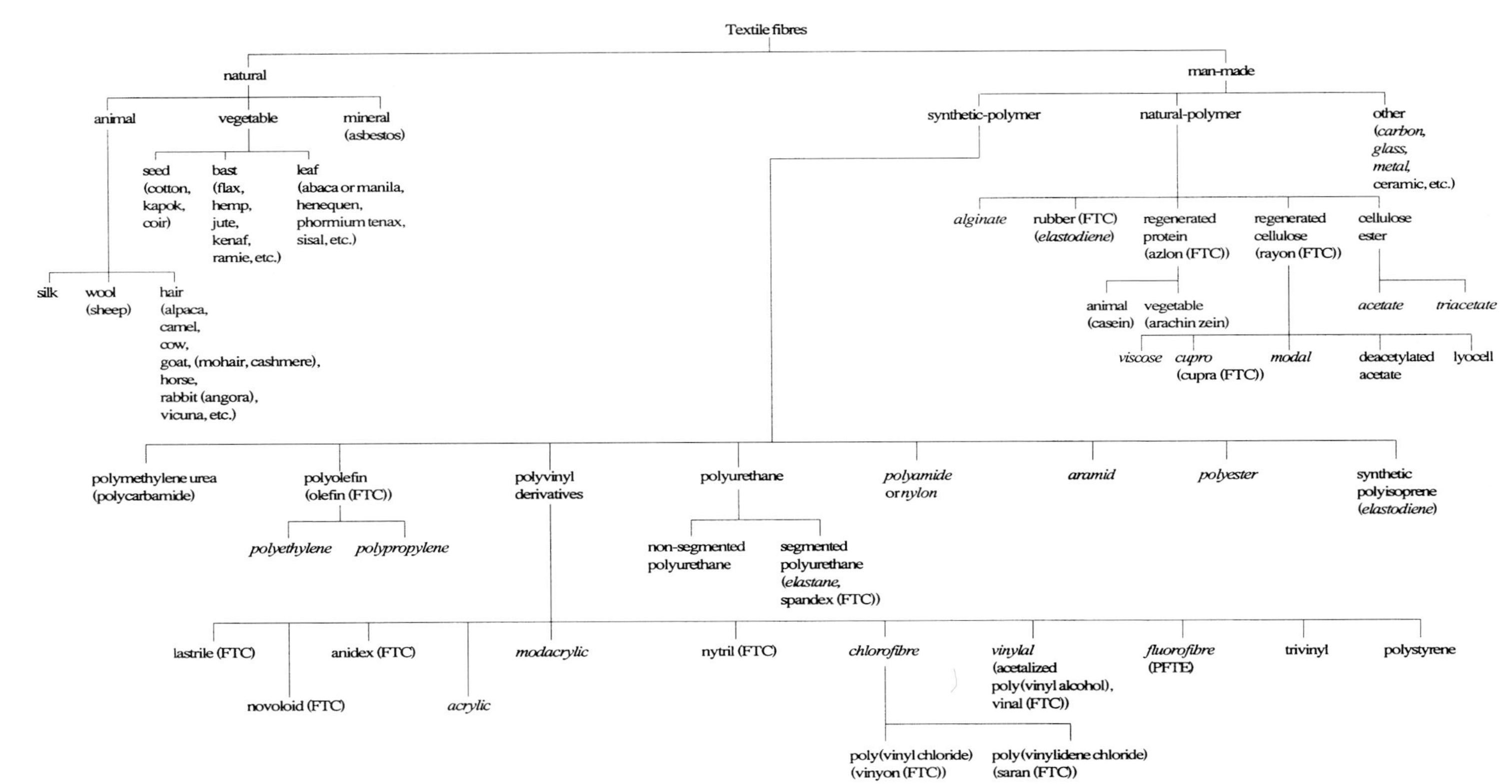

Terms in italics are the generic names recommended by the International Organization for Standardization in ISO 2076: 1989(E). Terms approved by the U.S. Federal Trade Commission that do not conform with the ISO list are identified by (FTC) after the term. (See also **generic name**.)